Techniques for Measurement and Removal of Dioxins and Furans

Techniques for Measurement and Removal of Dioxins and Furans

Vaishali V. Shahare
Assistant Professor
Rajdhani College
University of Delhi
Delhi

CRC Press is an imprint of the
Taylor & Francis Group, an **informa** business
A SCIENCE PUBLISHERS BOOK

Cover illustration provided by the author of the book, Dr. Vaishali Shahare.

CRC Press
Taylor & Francis Group
6000 Broken Sound Parkway NW, Suite 300
Boca Raton, FL 33487-2742

First issued in paperback 2021

Version Date: 20170831

ISBN-13: 978-0-367-78165-1 (pbk)
ISBN-13: 978-1-4987-7149-8 (hbk)

Library of Congress Cataloging-in-Publication Data

Names: Shahare, Vaishali V., author.
Title: Techniques for measurement and removal of dioxins and furans / Vaishali V. Shahare, assistant professor, Rajdhani College, University of Delhi, Delhi.
Description: Boca Raton, FL : CRC Press, 2017. | "A science publishers book."
Identifiers: LCCN 2017030553 | ISBN 9781498771498 (hardback)
Subjects: LCSH: Dioxins--Environmental aspects. | Dioxins--Measurement. | Furans--Environmental aspects. | Furans--Measurement. | Hazardous substances--Risk assessment. | Chemicals--Safety measures.
Classification: LCC TD196.D56 S47 2017 | DDC 628.5/2--dc23
LC record available at https://lccn.loc.gov/2017030553

Visit the Taylor & Francis Web site at
http://www.taylorandfrancis.com

and the CRC Press Web site at
http://www.crcpress.com

Preface

The book is a comprehensive volume on the group of chemical substances the Polychlorinated dibenzo-p-dioxins (PCDDs) and Polychlorinated dibenzofurans (PCDFs). Due to the large number of congeners and the extreme toxicity of a few of these congeners highly specific, selective and sensitive analytical methods are needed for their identification and measurement. The requirements for generating good analytical data on dioxins and furans are: representative sampling, high selectivity, high specificity, high sensitivity, safe and reliable quantification, and good reproducibility and confirmation. The number of the fully chlorinated dioxins and furans congeners is 75 and 135 respectively, and same number of the fully brominated analogues also exist. Moreover, the number of the mixed bromo-chloro compounds is 4600. However this book focuses only upon dioxins and furans. The compounds included in this volume have given widespread cause for concern in relation to their environmental persistence and high toxicity, and their potential for adverse effects on humans wildlife and environment. These have been banned by United Nation Environment Programme (UNEP) in May 1995. The book talks about various countries which are a party to Stockholm Convention and taken a leading role to reduce and/or eliminate POPs and their releases on a regional and global basis. This book address the sources and formation of dioxins and furans; global contamination and movement of dioxins and furans in the environment; health effects of dioxins and furans; analytical method and development of removal techniques for dioxins and furans. Aspects of analysis and quality assurance/control are discussed in brief. This volume highlights the expertise of some of the most distinguished workers in this field to review the knowledge of environmental concentrations and pathways, human toxicity and ecotoxicology, emission and control methods for dioxins and furans.

Despite the fact that dioxins and furans are not manufactured except for laboratory use, they remain relatively abundant in the environment because of their intentional or unintentional production. A further consequence of their lipid solubility is the tendency to concentrate within food chains and hence pose the greatest level of risk to human health and the environment. Historically, one of the most important source categories for 'dioxin' emissions has been the combustion of waste in incinerators. The consequences of the exposures have been estimated in detail. The book draws attention to the concentrations of dioxins and furans in various matrices signifying that they are still circulating within the environment. In the last chapter, the available operating procedures and control technologies for minimizing such emissions are described which gives detailed insight into the processes involved and their relative efficiencies.

Since the dioxins and furans problem is evident, stringent controls have been applied in developed countries to their emissions and much has been discussed about the best possible techniques for controlling PCDD and PCDF formation and emission. Reduction of dioxin emissions from industrial plants has become necessary as the impact of dioxin emissions from certain industrial plants is far higher than from hazardous waste incineration plants. Due to the ubiquitous occurrence of PCDDs and PCDFs and their transport across borders permanent solutions can be achieved only by measures with far-reaching impacts.

This volume gives a unique and valuable compilation of information on an extremely important group of environmental pollutants.

Acknowledgement

I wish to dedicate this book to my beloved husband **Dr. Virendra B. Shahare**, for his wholehearted encouragement, support, patient bearing and moral support. He has been my inspiration throughout my career and writing this book. I wish to thank my wonderful children, my daughter **Anshika V. Shahare** and my son **Atharva V. Shahare** for always making me smile and for understanding me on those weekends when I was writing this book instead of playing with them. I also express my special thanks to my mother-in-law for her warm support.

Contents

List of Figures

List of Tables

List of Abbreviations

2,3,7,8-TCDD	2,3,7,8-Tetrachlorodibenzo-p-dioxin
2,3,7,8-TCDF	2,3,7,8-Tetrachlorodibenzofuran
2,4,5-TCP	2,4,5-Trichlorophenol
μg	Microgram
μL	Microliter
ADP	Air dried pulp
EI	Electron impact
GAC	Granular activated carbon
GC-ECD	Gas chromatograph-electron capture detector
GC-MS	Gas Chromatograph-Mass Spectrometer
HpCDD	Heptachlorinated dibenzo-p-dioxin
HxCDD	Hexachlorinated dibenzo-p-dioxin
IARC	International Agency for Research on Cancer
I-TEQ	International Toxic equivalent
m/z	Mass to charge ratio
NA	Not applicable
ND	Not detected
ng	Nanogram
N-TEQ	Nordic Toxic equivalent
OCDD	Octachlorodibenzo-p-dioxin
OCDF	Octachlorodibenzofuran
PCBs	Polychlorinated biphenyls
PCDDs	Polychlorinated dibenzo-p-dioxins
PCDFs	Polychlorinated dibenzofurans
PeCDD	Pentachlorinated dibenzo-p-dioxin
PFTBA	perflourotributylamine
pg	Picogram
pg/g d.w.	Picogram per gram of dry weight
pg/kg bw/day	Picogram per kg body weight per day

POPs	Persistent organic pollutants
PPM	Pulp and paper mill
ppq	Parts per quadrillion
ppt	Parts per trillion
PVC	Polyvinylchloride
S/N	Signal to noise ratio
TEF	Toxic equivalency factor
TEQ	Toxic equivalent
TOF	Time of flight
UNEP	United Nations Environmental Programme
UNIDO	United Nations Industrial Development Organisation
USEPA	Unites States Environmental Protection Agency
UV	Ultraviolet

CHAPTER 1

Introduction

1.1 ANALYTICAL TECHNIQUES

An analytical technique is a method that is used to determine the concentration of a chemical compound or chemical element. There are two major types of analysis, qualitative analysis, which seeks to establish the existence of a given element or compound and quantitative analysis, which seeks to establish the amount of a given element or compound. There are bewildering arrays of analytical methods available to separate, detect, and measure chemical compounds. The oldest methods required painstaking separation of substances in order to measure the weight or volume of an organic pollutant. There are a wide variety of techniques used for analysis, from simple weighing (gravimetric) to titrations (titrimetric) to very advanced techniques using highly specialized instrumentation. The most common techniques used in analytical chemistry are the titrimetry, gravimetric, electrometric, spectrometry, and chromatography. Newer methods include gas chromatography-mass spectrometry (GC-MS), where volatilisation of a sample occurs in the first step, and measuring of the concentration occurs in the second. The first step may also involve a separation technique, such as chromatography, and the second, a detection/measuring device.

Most of the organic pollutants are present at very low levels and require highly sophisticated instruments for their analysis viz. Gas-liquid chromatography (GLC). GLC is based on partition equilibrium of analyte between a liquid stationary phase and a mobile gas. It is useful for a wide range of non-polar analytes, but poor for thermally labile molecules. In an analytical method, identification tests are intended to ensure the identity of an analyte in a sample. The objective of the analytical procedure should be clearly understood since this governs the validation characteristics which need to be evaluated. The typical validation characteristics which should be considered are accuracy, precision (repeatability and intermediate precision), specificity (specificity is the ability to assess unequivocally the analyte in the presence of components which may be expected to be present, typically these may include impurities), detection limit, quantitation limit, linearity, and range. The analytical procedures refer to the way of performing the analysis. It describes in detail the steps necessary to perform each analytical test. This may include the reference standard, reagents preparation, use of apparatus, generation of the calibration curve, use of the formulae for calculation, etc. Analytical methods rely on scrupulous attention to cleanliness, sample preparation, precision and accuracy.

A standard method for analysis of analytes concentration involves the creation of a calibration curve. Series of standards across the range of concentrations that are of interest are created. Care is taken that these concentrations are in the detection range of the technique (instrumentation) which is being used. These standards will have a precisely known concentration of the element or compound under study. Running each of these standards several times using the chosen technique will produce a series of readings, each set indicative of one of the known concentrations. By plotting these points (reading *vs* concentration) on a graph, it is possible to plot a line of concentration *vs* reading across the detection range of that technique. Thus, when the sample is run and a reading obtained, the experimenter can simply refer to the graph to read off the concentration. If the concentration of element or compound in the sample is too high for the detection range of the technique, it can simply be diluted in a pure solvent. If the amount in the sample is below an instrument's range of measurement, the method of addition can be used. In this method, a known quantity of the element or compound under study is added, and the difference between the concentration added, and the concentration observed is the amount actually available in the sample.

1.2 POLYCHLORINATED AROMATIC HYDROCARBONS-DIOXINS AND FURANS

Persistent chemicals, which accumulate in the biomass, are among the most important types of chemical pollutants. Aromatic halogenated organics are among the most widespread and persistent chemicals present in the ecosystem (Faust 1964; Menn 1960; Stringer 1960). Chemicals, which persist in the environment, are typically those, which have been introduced directly, for example, Polychlorinated biphenyls (PCB) or indirectly and unintentionally formed, for example, Polychlorinated aromatic hydrocarbons-dioxins and Furans (PCDDs and PCDFs) and are resistant to environmental breakdown. These compounds make their entry into the water resources via wastewater discharges and consequently contaminate water sources. Such organic compounds have been detected in soil matrices, paper and pulp mill effluents, and sludge and flyash of the incinerators. Toxicological studies have linked some of these compounds to adverse human health effects. These compounds are found virtually everywhere on earth, with the main transport mechanism being atmospheric dispersion and deposition (Mukerjee 1998).

PCDDs and PCDFs are formed in a multitude of industrial and non-industrial processes in which sources of chlorine, organic matter and sufficient energy are incorporated (Fiedler 1998; Tuppurainen et al. 1998; Rappe 1993). Minor formation takes place in forest fires, composting and other natural processes (Prange et al. 2003; Ferrario et al. 2000; USEPA 1998). Although the rise of industrial production in the 1930s and 1940s led to elevated levels of PCDDs and PCDFs in the environment, scientists have detected some Dioxins and Furans in the environment that predate industrial society. EPA and the United States Department of Agriculture documented the discovery of traces of Dioxins in certain kinds of clay deposits (USEPA 1998). Scientists have also noted the occurrence of low levels of Dioxins and Furans in peat bogs in Canada that appear to be from natural sources (Bonn 1998; Silk et al. 1997). Wood fires are one of the possible sources of Dioxins in nature. Estimates vary from a few grams a year to several kg. PCDD is found in the bark of thousands of years old redwoods and in 6,000 years old human bodies. In a recent PCDDs scare in chicken meat in US, the origin of the PCDDs were found in millions of years old clay layers, not influenced by any man-made sources (Ferrario et al. 2004).

However, most of the PCDDs and PCDFs are of anthropogenic origin: humans bear responsibility for their formation through combustion processes and the manufacture of chlorinated chemicals (Dyke and Amendola 2007; Breivik et al. 2004; Green et al. 2001; Alcock and Jones 1996). From these sources, PCDDs and PCDFs are released into air, land and water. Despite the poor mobility of PCDDs and PCDFs, their long half-live does facilitate transportation and redistribution between environmental compartments (Sinkkonen and Paasivirta 2000; Cole et al. 1999). In certain human populations, the exposure to PCDDs and PCDFs is close to the concentrations that potentially can cause adverse health effects (Nau 2006; Pelclova et al. 2006; Büchert et al. 2001; WHO 2000; Alaluusua et al. 1999). Therefore, it is essential to identify the main sources of PCDDs and PCDFs, including direct emissions and emissions from reservoir sources such as contaminated soils and sediments (Commission of the European Communities 2001).

Recent PCDD/F emission inventories have utilized an integrated approach to estimate emissions to the air, land and soil. Direct releases to water have only been qualitatively estimated (QuaB et al. 2000; Wenborn et al. 1999). The inventories suggest that emissions of PCDD/Fs to land, including landfills, may be as high as or even higher than emissions into the air (Quass et al. 2004; Hansen and Hansen 2003; Haynes and Marnane 2000; Wenborn et al. 1999; Dyke et al. 1997). The greatest releases are attributed to the production and the use of organochlorine pesticides (Wenborn et al. 1999). Pesticides and other chlorinated chemicals have an especially high potential of penetrating into water systems. There are many chlorinated chemicals that may contain PCDD/F impurities, which makes it extremely difficult to estimate the historical and current emissions.

Polychlorinated aromatic hydrocarbons consist of two series of chemicals viz. polychlorinated dibenzo-p-dioxins (PCDDs) and polychlorinated dibenzofurans (PCDFs) (Fig. 1.1) which have similar chemical and physical properties; they are ubiquitous in the environment. PCDDs and PCDFs commonly known as Dioxins and Furans are chemically classified as halogenated aromatic hydrocarbons. They are not produced intentionally or deliberately, but are formed as a by-product of chemical processes. There are 75 congeners of the Dioxin family and 135 congeners of the Furan family. Of the 210 congeners, only 17 congeners are toxic which have chlorine at their 2,3,7 and 8 position. All have varying degrees of toxicity in comparison to the most toxic 2,3,7,8-tetrachlorodibenzo-p-dioxin (2,3,7,8-TCDD). The most widely studied compound is 2,3,7,8-TCDD which is often referred to simply as Dioxin and is the reference for a number of compounds which are similar structurally and have Dioxin-like toxicity. In general, these compounds have low water solubility, low vapour pressure, are very stable and tend to bioaccumulate. There are also a number of Dioxin-like PCBs, polybrominated biphenyls and mixed chlorinated and brominated congeners with Dioxin-like properties. Compounds in these families have different properties, depending on the number and position of halogen atoms in the molecule. Halogenated aromatic hydrocarbons are ubiquitous environmental contaminants that have been characterized as potent toxicants and carcinogens (IARC

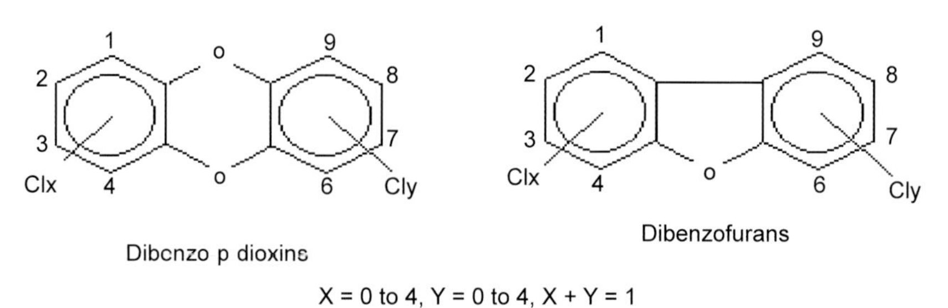

Fig. 1.1: Structures of dibenzo-p-dioxin and dibenzofuran.

2000). These compounds elicit diverse biological effects in laboratory animals such as promotion of carcinogenesis, immune suppression and impairment of reproduction. They are probable human carcinogen. TCDD is possibly one of the most potent toxins ever evaluated by the United States Environmental Protection Agency (USEPA). Polychlorinated aromatic hydrocarbons score negatively in most genotoxicity assays, including the Ames (Salmonella) assay. Although their mechanism of toxicity is not well understood, they induce aryl hydrocarbon (AH) hydroxylases and bind to the AH receptor, which is believed to mediate toxicity (Schiestl et al. 1997).

The family of Dioxins was actually discovered (Campbell and Friedman 1966; Sanger et al. 1958; Schmittle et al. 1958) several years before they were named. During the late 1950s, a serious and lethal disease outbreak appeared among poultry flocks, involving the formation of edema within the heart sac, or pericardium (Sanger et al. 1958; Schmittle et al. 1958). The name given to this unknown factor known to be present in the oil of the feed was 'chick edema factor'. A previously observed condition known as X Disease in cattle also was likely due to the same group of chemicals.

It was not until this same group of chemicals was found (Courtney et al. 1970) as a contaminant of the herbicide, 2,4,5-trichlorophenoxyacetic acid (2,4,5-T), that the chemical structure and nomenclature were finally established. It was this herbicide, with the code name Agent Orange, which was used as a defoliant during the Vietnam War that drew the attention of public health authorities to the serious toxicity of these compounds (Whitlock 1987; Courtney et al. 1970; Whiteside 1970).

The previously discovered chick edema factor was present in the oil component of the chicken feed (Sanger et al. 1958; Schmittle et al. 1958), perhaps arising either from the use of the herbicide on the crops used to make the oil or from the bleaching of the oil with chloro-phenols during processing. The solubility of Dioxin in oil is a fundamental property of these organic, aromatic and halogenated type compounds. Dioxins were first discovered in the emissions of trash incinerators (Olie et al. 1977). In 1987, Hagenmaier and co-workers in Germany reported that the levels of PCDDs and PCDFs in the fly ash collected from medical waste incinerators could be two orders of magnitude higher than the levels found in the fly ash in municipal waste incinerators (Hagenmaier et al. 1987). In September 1994, the USEPA published a draft report which examined the known sources of PCDDs in the U.S., and concluded that medical waste incineration was the largest identified source: 5100 grams out of a grand total of 9300 grams of PCDD toxic equivalents per year. Medical waste incinerators produce more PCDDs and PCDFs, per ton of waste burned, than municipal waste incinerators (USEPA 1994a).

TCDD induces a wide spectrum of biological effects including enzyme induction and vitamin A depletion. Not all of these effects are observed in any single animal species. The most characteristic toxic effects observed in all laboratory animals are body weight loss, thymus atrophy, and immunotoxicity teratogenicity, reproductive effects, and carcinogenicity (Nau et al. 2006). Chloracne and related dermal lesions are the most frequently noted signs of 2,3,7,8-TCDD toxicosis in humans; dermal lesions are also observed in rhesus monkeys, hairless mice, and rabbits. In contrast, most rodents do not develop chloracne and related dermal toxic lesions after exposure to 2,3,7,8-TCDD. Many of the toxic lesions are noted primarily in epithelial tissues.

For occupational and accidental exposures to PCDDs and PCDFs, inspite of many clinical and follow-up studies, no clear-cut persistent systemic effects have been delineated except for chloracne (Baccarelli et al. 2006). Other effects have been noted, but, apart from chloracne and perhaps minor functional disorders, none have been persistent. In the Seveso accident, the only clear-cut adverse health effect recorded was chloracne. Chloracne (193 cases) occurred in 1976 and 1977, and 20 of those individuals still had active chloracne in 1984. Many studies were performed to find possible links between exposure to Agent Orange and health effects in civilians or military personnel in Vietnam. However, the information available to date does not allow definite conclusions to be drawn with regard to effects on human reproduction or any other significant health effects.

The only documented intoxications with PCDFs in humans are the two instances of contamination of rice oil with PCDFs, that is, Yusho in Japan, 1968, and Yu-cheng in Taiwan, 1979. In total, several thousand people were acutely intoxicated. From the data it appears most likely that the causative agent was the PCDFs. The general symptoms were similar to that seen in intoxications with TCDD, with the differences reflecting the intensity of exposure and the ages and sex of those exposed (Todaka et al. 2006). The average daily intake of 2,3,7,8-substituted PCDFs by Yusho patients was estimated to be 0.1–0.2 μg/kg body weight for a period of several months, while the lowest dose causing disease was estimated to be 0.05–0.1 μg/kg body weight per day over a period of 30 days.

Although Dioxins are encountered in both the vapor and particulate phases, it has been suggested that ingestion results in 90 percent of the human exposure (Charnley and Doull 2005). Atmospheric Dioxins deposit on vegetation which form animal feed. Humans then ingest crops, fish, meat and dairy products that could substantially increase total risk. Octachlorodibenzo-p-dioxin (OCDD) is an experimental teratogen and an irritant to the eye. Ingestion of this chemical results in poisoning. These solvents are fat-soluble and therefore accumulate in the tissues of animals and humans in the food chain. Humans are typically exposed to these chemicals through the consumption of fish, meat, and milk. Exposure to Dioxins results in a drop in sperm count, an increase in testicular and prostate cancer, endometriosis, and an increased risk of developing breast cancer.

Toxic Equivalents or TEQs, are used to report the toxicity-weighted masses of mixtures of Dioxins/Furans. The TEQ method of Dioxin reporting is more meaningful than simply reporting the total number of grams of a mixture of variously toxic compounds because the TEQ method offers toxicity information about the mixture. Each Dioxins/Furans compound is assigned a Toxic Equivalency Factor or TEF. This factor denotes a given Dioxin compound's toxicity relative to 2,3,7,8-TCDD, which is assigned the maximum toxicity designation of one. Other PCDD compounds are given equal or lower numbers, with each number roughly proportional to its toxicity relative to that of 2,3,7,8-TCDD. Developed by the World Health Organization, TEFs are used extensively by scientists and governments around the world (Van den Berg et al. 2006). The EPA uses units of grams-TEQ to report emissions of Dioxins/Furans from known sources to the open environment in its Inventory of Sources of Dioxin in the United States. To obtain the number of grams-TEQ of a Dioxins/Furans mixture, one simply multiplies the mass of each compound in the mixture by its TEF and then totals them.

1.3 ANALYTICAL METHODS

The exceptionally high toxicity of 2,3,7,8-TCDD and other congeners of Dioxins and Furans and bioaccumulation due to hydrophobic nature necessitates very low detection limits. The characteristics of methods for determination of PCDDs and PCDFs are derived from several factors related to their chemical, physical, and toxicological properties. An understanding of the environmental levels, transport and fate of these compounds is important to understand the significance of these compounds as well as the low detection limits and broad range of detection required.

In the case of Dioxins and Furans, ultimate sensitivity and selectivity is required while both speed (analysis time) and cost are sacrificed. Another factor in method development and application is the large amount of litigation that occurred because of the detection of these compounds in humans as well as the environment. Methods of analysis are required to generate data that could withstand scrutiny in a court of law, which led to the necessity of developing methods based on sophisticated techniques. The earliest reported method used to detect 2,3,7,8-TCDD was a rabbit skin test (Adams et al. 1941). Test samples were applied to the inner surface of the ear and to the shaven belly of albino rabbits, and inflammatory responses were observed. Subsequently, Jones and Krizek (1962) developed a test based on the recovery and weight of the keratin formed on the rabbit ear after application of a sample. These biological methods were non-specific as to isomers and not sufficiently sensitive to detect low levels of contamination. In the late 1960s and early 1970s, gas chromatographic methods were used for the quantification mainly of 2,3,7,8-TCDD in commercial 2,4,5-T formulations. The detection level was normally in the range of μg/g. These analyses were not isomer-specific and the results could not be confirmed. Ryhage (1964) solved the problem of combining a gas chromatograph with a mass spectrometer PCDDs and PCDFs. During the 1970s and 1980s, various types of mass spectrometer and gas chromatograph/mass spectrometer combinations were used in analytical work. Use of these more sophisticated instruments allowed for the development of isomer-specific and validated analyses for the TCDDs in the very late 1970s and for the other Dioxins and Furans in the early 1980s.

A large number of the individual Dioxins have been synthesized by various methods and characterized, mainly by GC-MS (Rappe et al. 1985; Taylor et al. 1985;

Buser and Rappe 1980) but also by using nuclear magnetic resonance (NMR) or ultraviolet (UV), infrared (IR) (Kende et al. 1974; Pohland and Yang 1972), or X-ray analyses (Slonecker et al. 1983; Boer et al. 1973). Paasivirta et al. (1977) have shown that 2,3,7,8-TCDD can be detected down to the pg level using a glass capillary column and a ^{63}Ni electron-capture detector. Combined with efficient clean-up procedures, this method has shown to be useful down to a level of 9 ppt (Niemann et al. 1983), although positive samples need confirmation by mass spectroscopy. Several review articles discussing methods of analyzing PCDDs and PCDFs have appeared (Crummett et al. 1985; Tiernan 1983; Karasek and Anuska 1982; Harless and Lewis 1982; Esposito et al. 1980; Rappe and Buser 1980). Other techniques, such as enzyme induction and radioimmunoassay have been described and discussed by Firestone (1978). McKinney et al. (1982) used the radioimmunoassay method for determining 2,3,7,8-TCDD in human fat, and found the reliable sensitivity at 95% confidence interval to be 100 pg per sample. An analytical method based on the keratonization response of epithelial cells in an *in vitro* system has been described by Gierthy and Crane (1985). This method can be an assay for Dioxin-like activity in environmental and biological samples. A positive response was found for 2,3,7,8-TCDD at a concentration of 10–11 mol/litre.

The latest technique for the analysis of Dioxins/Furans uses classical extraction techniques such as Soxhlet, liquid-liquid extraction (USEPA 1994b), solid-phase extraction (SPE) (Pujadas et al. 2001; Taylor et al. 1995), or the more recent accelerated solvent extraction (Dionex 1999; Richter 1996). Once the extract has been transferred to a suitable solvent, a three-stage (silica, alumina and carbon) open-column clean-up is followed by GC-MS. Because of the large number of isomers and congeners, and due to the extreme toxicity of some Dioxins/Furans isomers, highly sensitive and specific analytical techniques are required for the measurements. Detection limits for the analysis of environmental samples should be lower than the usual detection levels required for pesticide analysis. A detection level of one pg or less might be required to measure 2,3,7,8-TCDD and the other toxic isomers in a 1 gm environmental sample. Analyses at such low levels are complicated by the presence of a multitude of other interfering compounds for which intensive clean-up procedures are required. The mono-, di-, and trichloro congeners are not usually included in analyses. Such compounds are considered to be much less toxic than the higher chlorinated congeners and are also much more volatile and losses may occur during clean-up. The level of sophistication needed in the analyses for Dioxins/Furans will depend upon the objectives thereof. In cases where the objectives were primarily to screen samples to identify groups of Dioxins/Furans (in a qualitative or semi-quantitative manner), routine assays are adequate. In other instances, where the objective of the analysis is to quantify accurately specific Dioxins/Furans isomers in the samples, sophisticated analytical procedures are required. Clearly, both types of analyses can be useful, depending on the purpose for which the analytical results are to be used. Many analytical methods have been developed in recent years for the analysis of trace amounts of PCDDs and PCDFs in environmental samples, especially for 2,3,7,8-TCDD. The most specific of these methods are based on MS. There are many requirements to be met by such an analytical method, including representative sampling and appropriate storage, efficient extraction, high selectivity in the clean-up, high specificity in the gas chromatography, high sensitivity in the detection, safe and reliable quantification, good reproducibility, and useful confirmatory information.

A variety of analytical techniques have been developed over the past few years to speed up analysis. Fast GC (Snow 2005; Boden et al. 2002; van Ysacker 1995; Giddings 1962), Dual Column GC analysis, analyte specific GC phases (MacPherson 2003), comprehensive 2 Dimensional GC (2DGC) (Pani and Górecki 2006) and time of flight (TOF) mass spectrometry have been used to reduce analysis times, combine analyte scans, and reduce analytical costs without significant losses in sensitivity and selectivity (de Wit 2002).

Need for measuring of 2,3,7,8-substituted congeners—PCDDs and PCDFs are considered to be very stable and persistent, as illustrated by the half life of TCDD in soil of 10–12 years. This persistence, combined with high partition coefficients (up to 8.20 for OCDD) provides the necessary conditions for these compounds to bio-concentrate in organisms. Bio-concentration factors of 26707 has been reported in rainbow trout (Salmo gairdneri) exposed to 2,3,7,8-TCDD. The chemical properties of PCDDs and PCDFs (low water solubility, high stability, and semi-volatility) favor their long range transport. As with most other organochlorines, food is a major source of PCDDs and PCDFs in the general population, with food of animal origin contributing the most to human body burdens. In a survey of Dioxins in US food, total PCDD/Fs ranged from 0.42 ppt to 61.8 ppt (wet weight) (total TEQ range: 0.02 ppt to 1.5 ppt). The estimated daily intake for adults ranged from 0.3 to 3.0 pg TEQs/kg body weight, and for breast fed infants the range was 35.3 to 52.6 pg TEQs/kg body weight. Recent estimates of adult average daily intake for Canada, Germany and the Netherlands are 1.52, 2 and 1 pg TEQs/kg bodyweight, respectively. These are below the TDI of 4 pg/kg body weight/day for lifetime exposure estimated by WHO. Table 1.1 gives the PCDDs and PCDFs congeners substituted with chlorine in the 2,3,7,8-positions. As these are the most toxic congeners, these congeners are generally determined. The detection, quantification, and confirmation of different congeners of Dioxins and Furans are usually performed by mass spectrometry. MS allows the determination of the number and type of halogens present from characteristic isotope distribution patterns (Buser 1991).

Many analytical methods have been developed in recent years for the analysis of trace amounts of Dioxins and Furans in environmental samples, especially for

Table 1.1: 2,3,7,8-chlorine substituted PCDDs and PCDFs congeners.

Compound	**Homologue Name**	**2,3,7,8-chlorinated Congeners**
TCDD	Tetrachlorodibenzo-p-dioxin	1
PeCDD	Pentachlorodibenzo-p-dioxin	1
HxCDD	Hexachlorodibenzo-p-dioxin	3
HpCDD	Heptachlorodibenzo-p-dioxin	1
OCDD	Octachlorodibenzo-p-dioxin	1
TCDF	Tetrachlorodibenzofuran	1
PeCDF	Pentachlorodibenzofuran	2
HxCDF	Hexachlorodibenzofuran	4
HpCDF	Heptachlorodibenzofuran	2
OCDF	Octachlorodibenzofuran	1

2,3,7,8-TCDD. The most specific of these methods are based on MS. There are many requirements to be met by such an analytical method, including representative sampling and appropriate storage, efficient extraction, high selectivity in the clean-up, high specificity in the gas chromatography, high sensitivity in the detection, safe and reliable quantification, good reproducibility, and useful confirmatory information. Great precautions should be taken in handling the samples. The large number of isomers in some homologous groups (Table 1.2) makes the separation and quantification of individual congeners difficult. The toxic equivalent factors for the seventeen toxic congeners are given in (Table 1.3).

Table 1.2: Homologues and congeners of PCDDs and PCDFs.

Compound	Homologue Name	No. of Possible Congeners		Elemental Composition
		Total	2,3,7,8-chlorinated Congeners	
MCDD	Monochlorodibenzo-p-dioxin	2	0	$C_{12}H_7O_2Cl$
DiCDD	Dichlorodibenzo-p-dioxin	10	0	$C_{12}H_6O_2Cl_2$
TrCDD	Trichlorodibenzo-p-dioxin	14	0	$C_{12}H_5O_2Cl_3$
TCDD	Tetrachlorodibenzo-p-dioxin	22	1	$C_{12}H_4O_2Cl_4$
PeCDD	Pentachlorodibenzo-p-dioxin	14	1	$C_{12}H_3O_2Cl_5$
HxCDD	Hexachlorodibenzo-p-dioxin	10	3	$C_{12}H_2O_2Cl_6$
HpCDD	Heptachlorodibenzo-p-dioxin	2	1	$C_{12}HO_2Cl_7$
OCDD	Octachlorodibenzo-p-dioxin	1	1	$C_{12}O_2Cl_8$
MCDF	Monochlorodibenzofuran	4	0	$C_{12}H_7OCl$
DiCDF	Dichlorodibenzofuran	16	0	$C_{12}H_6OCl_2$
TrCDF	Trichlorodibenzofuran	28	0	$C_{12}H_5OCl_3$
TCDF	Tetrachlorodibenzofuran	38	1	$C_{12}H_4OCl_4$
PeCDF	Pentachlorodibenzofuran	28	2	$C_{12}H_3OCl_5$
HxCDF	Hexachlorodibenzofuran	16	4	$C_{12}H_2OCl_6$
HpCDF	Heptachlorodibenzofuran	4	2	$C_{12}HOCl_7$
OCDF	Octachlorodibenzofuran	1	1	$C_{12}OCl_8$

Table 1.3: Toxic equivalent factors for the 17 toxic congeners (WHO 2005).

Dioxins	Factor	Furans	Factor
2,3,7,8-TCDD	1	2,3,7,8-TCDF	0.1
1,2,3,7,8-PeCDD	1	2,3,4,7,8-PeCDF 1,2,3,7,8-PeCDF	0.3 0.03
1,2,3,4,7,8-HxCDD 1,2,3,6,7,8-HxCDD 1,2,3,7,8,9-HxCDD	0.1	1,2,3,4,7,8-HxCDF 1,2,3,7,8,9-HxCDF 1,2,3,6,7,8-HxCDF 2,3,4,6,7,8-HxCDF	0.1
1,2,3,4,6,7,8-HpCDD	0.01	1,2,3,4,6,7,8-HpCDF 1,2,3,4,7,8,9-HpCDF	0.01
OCDD	0.0003	OCDF	0.0003

1.4 PHYSICO-CHEMICAL PROPERTIES OF DIOXINS AND FURANS

The physico-chemical properties of Dioxins are given in Tables 1.4 and 1.5.

1.5 ENVIRONMENTAL FATE

Chlorinated dibenzodioxins and dibenzofurans have been found in all environmental media including water, soil, air, and sediment (USEPA 2000). In general, these compounds are resistant to abiotic and biotic degradation. Consequently, Dioxin-like compounds released to the atmosphere can travel long distances before they are deposited onto water, soil and vegetation causing a widespread occurrence of such compounds.

Table 1.4: Physico-chemical properties of Dioxins.

	Mono-CDD	**Di-CDD**	**Tri-CDD**	**Tetra-CDD**
Molecular mass	218.6	253.1	287.5	322
Density (g/dm^3)	No data	No data	No data	1.827
Melting point (°C)	89.0–105.5	114–210	128–163	175–306
Boiling point (°C)	No data	No data	374	446.5
Soil sorption coefficient (log K_{oc})	No data	No data	No data	No data
Octanol/Water Partition (log K_{ow})	4.52–5.45	5.86–6.39	6.86–7.45	6.6–8.7
Solubility in water in 25°C (mg/dm^3)	0.278–0.417	3.75×10^{-3} -1.67×10^{-2}	4.75×10^{-3} -8.41×10^{-3}	7.9×10^{-6} -6.3×10^{-4}
Vapor pressure at 25°C (mm Hg)	9.0×10^{-5} -1.3×10^{-4}	9.0×10^{-7} -2.9×10^{-6}	6.46×10^{-8} -7.5×10^{-7}	7.4×10^{-10} -4×10^{-3}
Henry's law coefficient (K_H) atm m^3/mol	82.7×10^{-6} -146.26×10^{-6}	21.02×10^{-6} -80.04×10^{-6}	37.9×10^{-6}	7.01×10^{-6} -101.7×10^{-6}
	Penta-CDD	**Hexa-CDD**	**Hepta-CDD**	**Octa-CDD**
Molecular mass	356.4	390.9	425.3	459.8
Density (g/dm^3)	No data	No data	No data	No data
Melting point (°C)	195–206	238–286	265	330–332
Boiling point (°C)	No data	No data	507.2	485–510
Soil sorption coefficient (log K_{oc})	No data	No data	No data	No data
Octanol/Water Partition (log K_{ow})	8.64–9.48	9.19–10.4	9.69–11.38	8.78–13.37
Solubility in water in 25°C (mg/dm^3)	1.18×10^{-4}	4.42×10^{-6}	2.4×10^{-6} -1.9×10^{-3}	2.27×10^{-9} -7.4×10^{-8}
Vapor pressure at 25°C (mm Hg)	6.6×10^{-10}	3.8×10^{-11}	5.6×10^{-12} -7.4×10^{-8}	8.25×10^{-13}
Henry's law coefficient (K_H) atm m^3/mol	2.6×10^{-6}	44.6×10^{-6}	1.31×10^{-6} -2.18×10^{-5}	6.74×10^{-6}

Table 1.5: Physico-chemical properties of Furans.

	1,3,7,8-TetraCDF	**2,3,6,8-TetraCDF**	**2,3,7,8-TetraCDF**	**1,2,3,4,8-PentaCDF**
Molecular mass	305.96	305.96	305.96	340.42
Density (g/dm^3)	No data	No data	No data	No data
Melting point (°C)	No data	197–198	219–221	177–178
Boiling point (°C)	No data	No data	No data	No data
Soil sorption coefficient (log K_{oc})	No data	No data	5.61 (estimated)	No data
Octanol/Water Partition (log K_{ow})	No data	No data	5.82	6.79
Solubility in water in 25°C (mg/dm^3)	No data	No data	4.2×10^{-4}	No data
Vapor pressure at 25°C (mm Hg)	No data	No data	9.21×10^{-7}	No data
Henry's law coefficient (K_H) atm m^3/mol	1.48×10^{-5}	1.48×10^{-5}	1.48×10^{-5}	2.63×10^{-5}
	1,2,3,7,8-PentaCDF	**1,2,3,7,8-PentaCDF**	**1,2,3,4,7,8-HexaCDF**	**1,2,3,6,7,8-HexaCDF**
Molecular mass	340.42	340.42	374.87	374.87
Density (g/dm^3)	No data	No data	No data	No data
Melting point (°C)	225–227	196–196.5	225.5–226.5	232–234
Boiling point (°C)	No data	No data	No data	No data
Soil sorption coefficient (log K_{oc})	No data	No data	No data	No data
Octanol/Water Partition (log K_{ow})	6.79	6.79	No data	No data
Solubility in water in 25°C (mg/dm^3)	No data	2.4×10^{-4}	8×10^{-6}	1.8×10^{-5}
Vapor pressure at 25°C (mm Hg)	2.73×10^{-7}	1.63×10^{-7}	6.7×10^{-8}	6.7×10^{-8}
Henry's law coefficient (K_H) atm m^3/mol	2.63×10^{-5}	2.63×10^{-5}	2.78×10^{-5}	2.78×10^{-5}

1.5.1 Air

PCDDs and PCDFs are released into the air in emissions from municipal solid waste and industrial incinerators. The transport of these compounds from stacks and other stationary point sources, as well as from waste disposal sites and other area sources, can be predicted from dispersion modeling (SAI 1980). Exhaust from vehicles powered with leaded and unleaded gasoline and diesel fuel also release PCDDs to the air. Other sources of PCDDs in air include: emissions from oil- or coal-fired power plants, burning of chlorinated compounds such as PCBs, and cigarette smoke (Lohman and Seigneur 2001). PCDDs formed during combustion processes are associated with small particles in the air, such as ash. The larger particles will be deposited close to the emission source, while very small particles may be transported longer distances. Some of the lower chlorinated Dioxins and Furans may vaporize from the particles (and soil or water surfaces) and be transported long distances in the atmosphere, even around the globe. It has been estimated that 20% to 60% of 2,3,7,8-TCDD in the air is in the vapor phase. Sunlight and atmospheric chemicals will break down a very small portion of the PCDDs, but most PCDDs will be deposited on land or water.

PCDDs occur as a contaminant in the manufacture of various chlorinated pesticides and herbicides, and releases to the environment have occurred during the use of these chemicals (Thibodeaux 1983). Because PCDDs remain in the environment for a long time, contamination from past pesticide and herbicide use may still be of concern. In addition, improper storage or disposal of these pesticides and waste generated during their production can lead to PCDDs contamination of soil and water.

1.5.2 Water

The solubility of 2,3,7,8-TCDD in water has been extensively studied, but much less data are available for the other Dioxins and Furans. However, data from microbiological experiments indicate that 2,3,7,8-TCDD is highly adsorbed to sediments and biota. Matsumura et al. (1983) suggested that more than 90% of the 2,3,7,8-TCDD in an aquatic medium could be present in the adsorbed state. Even though PCDD/PCDFs have very low vapor pressures, they can volatilize from water. However, volatilization is not expected to be a significant loss mechanism for the tetra- and higher chlorinated PCDD/PCDFs from the water column under most non-spill scenarios. Podoll et al. (1986) calculated volatilization half-lives of 15 days and 32 days for 2,3,7,8-TCDD in rivers and ponds/lakes, respectively. Dioxins and Furans are released in waste waters from pulp and paper mills that use chlorine or chlorine-containing chemicals in the bleaching process. Some of the PCDDs deposited on or near the water surface will be broken down by sunlight. A very small portion of the total PCDDs in water will evaporate to air. Because PCDDs do not dissolve easily in water, most of the PCDDs in water will attach strongly to small particles of soil or organic matter and eventually settle to the bottom. PCDDs may also attach to microscopic plants and animals (plankton) which are eaten by larger animals that are in turn eaten by even larger animals. This is called a food chain. Concentrations of chemicals such as the most toxic, 2,3,7,8-chlorine substituted PCDDs, which are difficult for the animals to break down, usually increase at each step in the food chain. This process, called biomagnification, is the reason why undetectable levels of PCDDs in water can result in measurable concentrations in aquatic animals. The food chain is the main route by which PCDDs concentrations build up in larger fish, although some fish may accumulate PCDDs by eating particles containing PCDDs directly off the bottom.

1.5.3 Soil

PCDDs deposited on land from combustion sources or from herbicide or pesticide applications bind strongly to the soil, and therefore are not likely to contaminate groundwater by moving deeper into the soil. However, the presence of other chemical pollutants in contaminated soils, such as those found at hazardous waste sites or associated with chemical spills (for example, oil spills), may dissolve PCDDs, making it easier for PCDDs to move through the soil. The movement of chemical waste containing PCDDs through soil has resulted in contamination of groundwater. After PCDD/PCDFs have been deposited onto soil or plant surfaces, there can be an initial loss of PCDD/PCDFs due to photodegradation and/or volatilization. Factors affecting the extent of the initial loss include climatic factors, soil characteristics, and the concentration and physical form of the deposited congener (Nicholsan et al. 1993; Paustenbach et al. 1992; Freeman and Schroy 1989). Soil erosion and surface runoff can also transport PCDDs into surface waters. A very small amount of PCDDs at the soil surface will evaporate into air. Certain types of soil bacteria and fungus can break PCDDs down, but the process is very slow. In fact, PCDDs can exist in soil for many years. Paustenbach et al. (1992) reviewed many major published studies on Dioxin persistence in soil and concluded that 2,3,7,8-TCDD probably has a half-life of 25 to 100 years in subsurface soil and 9 to 15 years at the soil surface (i.e., the top 0.1 cm). Plants take up only very small amounts of PCDDs by their roots. Most of the PCDDs found on the parts of plants above the ground probably come from air and dust and/or previous use of PCDD-containing pesticides or herbicides. Animals (such as cattle) feeding on the plants may accumulate PCDDs in their body tissues (meat) and milk.

Study were carried out to obtain experimental evidence for the fate and transport of 2,3,7,8-TCDD in natural soils using its nontoxic isomers. The results from the batch experiments (Fan et al. 2006) indicated a high sorption affinity of all the TCDD isomers to soils and a strong correlation to organic matter content. TCDDs were more tightly bound to the soil with high organic matter than to the soil with low organic matter; however, longer contact time was required to approach sorption equilibrium of TCDDs in the soil with high organic matter. Miscible-displacement breakthrough curves indicated chemical non-equilibrium transport, where there was a rate-limited or kinetic sorption that was likely caused by organic matter. Combustion analyses of extracted soil from the soil columns showed that most TCDDs were adsorbed in the top 1–5 cm of the column. These column combustion results also showed that sorption was correlated to specific surface and soil depth, which suggested the possibility of colloidal transport.

1.6 REMOVAL TECHNIQUES

1.6.1 Adsorption

Crosby et al. (1971) and Windal et al. (1999) studied the effects of various treatment methods for the removal and control of emissions of Dioxins. Activated carbon treatment is the most effective method for removal of any organic. Dioxin removal by activated carbon have been studied by using a fixed bed type column and 1,2,3,4-TCDD. Effects of running time, TCDD concentration of inlet gas, valid particle diameter and amount of activated carbon used, gas temperature, and superficial gas velocity in the column on TCDD removal are analyzed. The effects are explained by the model that TCDD removal is controlled by mass transfer rate in a boundary film and in the pores on activated carbon particles. TCDD removal is little affected by the existence of 1,2,3-trichlorobenzene and sulfur dioxide in the gas phase, and is enhanced by the existence of hydrochloride in the gas phase (Furubayashi et al. 2001).

High performance Dioxin removal device sucks the Dioxin included in the exhaust gas, which is mainly discharged from waste incinerators. Such Dioxin is sucked into the activated carbon charge layers on two floors in a vertical tower and so the Dioxin density is be lowered in a stable condition even if the Dioxin production quantity is sharply changed. Dioxin removal rate is 95% or more (Furukawa Co., Ltd., 6-1, Marunouchi 2-chome, Chiyoda-ku, Tokyo 100-8370 Japan). The WKV Dioxins Removal System from WKV GmbH (Germany) introduced by NKK corporation, Japan removes Dioxins from flue gas along with heavy metals such as mercury. The system adopts activated carbon base adsorbent packed in a specially designed adsorber to remove Dioxins and heavy metals from flue gas. Owing to the efficient contact of flue gas with the adsorbent, the removal efficiency is high (NKK Corporation 1-1-2, Marunouchi, Chiyoda-ku, Tokyo 100-8202 Japan).

1.6.2 Photodegradation

Photolysis and photocatalytic processes for 2,3,7,8-tetrachloro-dibenzo-p-dioxin (2,3,7,8-TCDD) and 1,2,3,6,7,8-hexachlorodibenzo-p-dioxin (1,2,3,6,7,8-HxCDD) have been studied. The photocatalytic process was found to be faster than direct photolysis for the chlorinated PCDDs, and the rate decreased with increasing PCDDs quantity. The photocatalytic rate of the PCDDs decreased with increasing chlorination extent (Wu et al. 2005).

The advanced technique, developed by Kubota Corp., can degrade PCDDs in water to the detection limit using a photochemical degradation process by a combination of ozone and ultra violet (UV) light. It has been known that PCDDs in water can be degraded by ultra violet alone. However, the rate of degradation is too low for practical use. A photochemical degradation process by a combination of ozone and ultra violet light improves degradation at considerably high rate. The UV light converts dissolved ozone into highly reactive hydroxyl radicals. PCDDs degradation is conducted by a combination of dechlorination by ultra violet and fission of double bonds by hydroxyl radicals. This photochemical degradation process can degrade PCDDs to carbon dioxide, water, and no harmful compounds. It has high Dioxin degradation ability. The UV light with ozone produces powerful photochemical degradable potential which can break down Dioxins in water under 0.1 pg-TEQ/L. This system is applicable to every kind of water, which contains Dioxins at a wide range of concentration, landfill leachate, pond water around Dioxin emission site, waste water from scrubber, leachate from contaminated soils, and groundwater and industrial effluent (Kubota Corporation Ltd., 1-3, Nihonbashi-muromachi 3-chome, Chuo-ku, Tokyo 103-8310 Japan).

1.6.2.1 In Water

Numerous studies have demonstrated that Dioxins and Furans undergo photolysis following first order kinetics in a solution. Photolysis is slow in water but increases dramatically when solvents serving as hydrogen donors are present such as hexane, benzene, methanol, acetonitrile, isooctane, and acetonitrile/water (Koester and Hites 1992; Friesen et al. 1990a; Choudry and Webster 1989; Dulin et al. 1986; Crosby et al. 1971; Dobbs and Grant 1979). Natural waters have differing quantities and types of suspended particulates and dissolved organic material that could either retard or enhance the photolysis of Dioxins and Furans. For example, Choudry and Webster (1989) reported that photolysis of 1,3,6,8-TCDD was slower in a pond water matrix than was predicted from a laboratory solution. Conversely, Friesen et al. (1990b) reported that photolysis of PeCDD, HpCDD, TCDF, and PeCDF proceeds much faster in a pond or lake water matrix than was predicted from or measured in a laboratory solution.

Dobbs and Grant (1979) investigated the photolysis of a series of hexa-, hepta-, and octa-CDDs in hexane. Photolysis half-lives ranged from 0.4 days to 2 days. Meta- and para-substituted congeners were degraded more rapidly than ortho-substituted congeners. Dulin et al. (1986) studied the photolysis of 2,3,7,8-TCDD in various solutions under sunlight and artificial light. Using the results obtained in a water:acetonitrile solution (1:1, v/v) under sunlight conditions, Dulin et al. (1986) calculated the half-life of 2,3,7,8-TCDD in surface water to be 4.6 days in summer at 40° north latitude. The quantum yield for photodegradation of 2,3,7,8-TCDD in water was three times greater under artificial light at 313 nm than under sunlight, and the artificial light photolysis quantum yield for hexane, a good hydrogen donor, was 20 times greater than for the water:acetonitrile solution, a poor hydrogen donor.

Choudry and Webster (1989) studied the photolytic behavior under 313 nm light of a series of PCDDs in a water:acetonitrile solution (2:3, v/v). Assuming that the quantum yields observed in these studies are the same as would be observed in natural waters, they estimated the mid-summer half-life values at 40° north latitude in clear near-surface water to be as follows: 1,2,3,7-TCDD (1.8 days); 1,3,6,8-TCDD (0.3 days); 1,2,3,4,7-PeCDD (15 days); 1,2,3,4,7,8-HxCDD (6.3 days); 1,2,3,4,6,7,8-HpCDD (47 days); and OCDD (18 days). In addition, the authors also experimentally determined the sunlight photolysis half-life of 1,3,6,8-TCDD in pond water to be 3.5 days (i.e., ten times greater than the half-life predicted by laboratory experiments). A study by Friesen et al. (1990a) examined the photolytic behavior of 1,2,3,4,7-PeCDD and 1,2,3,4,6,7,8-HpCDD in water:acetonitrile (2:3, v/v) and in pond water under sunlight conditions at 50° north latitude. The observed that the half-lives of these two compounds in the acetonitrile solution were 12 and 37 days, respectively, and 0.94 and 2.5 days in pond water, respectively. Crosby et al. (1971) reported that polychlorinated dibenzofurans undergo photolytic dechlorination in the presence of a hydrogen donor, with more highly chlorinated congeners being more stable. In contrast, Hutzinger (1973) and Buser and Bosshardt (1976) reported that the more highly chlorinated congeners undergo photodegradation at a rate similar to that of lower chlorinated CDFs. Hutzinger (1973) found that both 2,8-DCDF and OCDF photolyze rapidly in methanol and hexane.

Buser (1988) studied the photolytic decomposition rates of 2,3,7,8-TCDF, 1,2,3,4-TCDF, and 1,2,7,8-TCDF in dilute isooctane solutions under sunlight and artificial laboratory illumination (fluorescent lights). When the solutions were illuminated with sunlight, the estimated half-lives were 0.2 days for a solution containing 3 ng/μL of 2,3,7,8-TCDF, 0.1 days for a solution containing 2 ng/μL of 1,2,3,4-TCDF, and 0.4 days for a solution containing 0.3 ng/μL of 1,2,7,8-TCDF. For the same solutions illuminated with artificial light, the half-lives were greater than 28 days. Friesen et al. (1996) studied the photodegradation of 2,3,7,8-TCDF and 2,3,4,7,8-PeCDF using water:acetonitrile (2:3, v/v) and lake water. The observed half-lives of the TCDF and PeCDF in the acetonitrile solution were 6.5 and 46 days, respectively, and 1.2 and 0.19 days in lake water, respectively.

1.6.2.2 In Soil

Photodegradation of PCDD/PCDFs is limited only to the soil surface. Below the top few millimeters of soil, photodegradation is not a significant process (Puri et al. 1989; Yanders et al. 1989). Substantial research on the environmental persistence of 2,3,7,8-TCDD has been performed as part of the decontamination of the area around the ICMESA chemical plant in Seveso, Italy. The levels of Dioxin in the soil decreased substantially during the first six months (DiDomenico et al. 1982). An experiment was conducted at this site to determine the effectiveness of photolysis in decontaminating surface deposits on foliage. Test plots were sprayed with olive oil to act as a hydrogen donor, and the levels of Dioxin on grass were found to be reduced by over 80 percent within nine days (Crosby 1971). The 2,3,7,8-TCDD in contaminated soil was also found to be photolabile in sunlight when the soil was suspended in an aqueous solution of a surfactant.

Buser (1988) studied the photolytic decomposition rates of 2,3,7,8-TCDF, 1,2,3,4-TCDF, and 1,2,7,8-TCDF dried as thin films on quartz vials. When exposed to sunlight, the substances slowly degraded with reported half-lives of 5 days, 4 days, and 1.5 days, respectively. Koester and Hites (1992) studied the photodegradation of a series of tetra- through octa-chlorinated PCDDs and PCDFs on silica gel. In general, the PCDFs degraded much more rapidly than the PCDDs, and half-lives increased with increasing level of chlorination (1,2,7,8-TCDF excluded). The half-lives for PCDDs ranged from 3.7 days for 1,2,3,4-TCDD to 11.2 days for OCDD. The half-lives for PCDFs ranged from 0.1 day for 1,2,3,8,9-PeCDF to 0.4 days for OCDF. Sunlight or ultraviolet light irradiation of PCDDs and PCDFs in the presence of vegetable oil offers a potential method for the cleanup of contaminated soil. In this study, the effects of extra virgin olive oil on the photochemical degradation of 1,2,3,4,6,7,8-heptachlorodibenzofuran and heptachlorodibenzo-p-dioxin (1,2,3,4,6,7,8-HpCDF/HpCDD) have been investigated (Isosaari et al. 2005). OCDD is a common contaminant of the widely used biocide pentachlorophenol occurring at a concentration of approximately 1000 μg per gm in the commercial product. Photolytic reductive dechlorination of OCDD using solar irradiation has been studied (Dobbs et al. 1979).

Photodegradation of 2,3,7,8-TCDD have been achieved in field soils, and on any other material contaminated by this substance, providing a layer of a substance consisting of a hydrogen donor is added followed by irradiation by sunlight or high-pressure mercury lamps. A suitable layer is obtained by spraying surfaces with a 1:1 xylene-ethyl oleate solution into which TCDD dissolves. The photodecomposition rate is affected by the intensity of radiation, the medium upon which 2,3,7,8-tetrachloro dibenzo-p-dioxin is dispersed and temperature. Radiations from high-pressure mercury lamps can be used to decontaminate both indoors and outdoors of buildings, while natural summer solar radiation has been found to be effective in the decontamination of floors, walls, and soil. Decontamination occurs to a certain extent also beneath the soil surface (Liberti et al. 1978).

1.6.3 Solar Radiations

Dioxins/Furans have a strong UV absorption at wavelengths shorter than 270 nm; however, their absorption band extends into the wavelengths longer than 290 nm which makes is possible to use sunlight as a radiation source for PCDD/F photolysis (Choi et al. 2000; Wagenaar et al. 1995; Tysklind et al. 1993; Choundry and Webster 1989; Dulin et al. 1986). Upon absorption of light, excitation of the PCDD/Fs molecule leads to the cleavage of either the C-Cl bond or the C-O bond. Cleavage of the C-Cl bond, that is, dechlorination, accounts for less than 30% of the photodegradation products in isooctane and in aqueous solutions (Rayne et al. 2002; Kim and O'Keefe 2000; Kieatiwong et al. 1990). In hexane, dechlorination yields have typically been higher: 50% from 2,3,7,8-TCDF (Dung and O'Keefe 1994), 5 to 32% from OCDD and 35 to 68% from OCDF (Konstantinov and Bunce 1996; Wagenaar et al. 1995). The relative yield of dechlorination products decreases with an increasing conversion of the original compound (Rayne et al. 2002; Konstantinov and Bunce 1996).

Dioxins and Furans are formed in a multitude of industrial and non-industrial processes in which sources of chlorine, organic matter, and sufficient energy are incorporated (Fiedler 1998; Tuppurainen et al. 1998; Rappe 1993; Bumb et al. 1980). Minor formation takes place in forest fires and other natural processes (Prange et al. 2003; Ferrario et al. 2000). However, most of the Dioxins and Furans are of anthropogenic origin: humans bear responsibility for their formation through combustion processes and the manufacture of chlorinated chemicals (Breivik et al. 2004; Green et al. 2001; Alcock and Jones 1996). From these sources, Dioxins and Furans are released into air, land, and water. Despite their poor mobility, their long half-live does facilitate transportation and redistribution between environmental compartments (Sinkkonen and Paasivirta 2000; Cole et al. 1999; Lohmann and Jones 1998). In certain human populations, the exposure to Dioxins and Furans is close to the concentrations that potentially can cause adverse health effects (Büchert et al. 2001; WHO 2000; Alaluusua et al. 1999; Johnson 1995). Significant contaminations in water have been attributed to accidental fires and treatment and disposal practices of municipal solid waste, combustion at old incineration plants in particular (QuaB et al. 2000a; Wenborn et al. 1999; Dyke et al. 1997).

Dioxin-like compounds may induce a wide spectrum of biological responses at the biochemical, cellular, and tissue levels (ATSDR 1998). In humans, a wide variety of health effects have been linked to high exposure to Dioxins, including mood alterations, reduced cognitive performance, diabetes, changes in white blood cells, dental defects, endometriosis, decreased male/female ratio of births, decreased testosterone, and (in neonates) elevated thyroxin levels. Presently the effects have been proven only in the case of chloracne. The effect that has caused the greatest public concern is cancer. IARC (2000) classified TCDD as a human carcinogen. A cancer potency factor from a meta-analysis of human data from three occupational coherts of 1×10^{-3} pg/kg bw/day has been established by USEPA (2000). Therefore, it is essential to identify the main sources of PCDDs and PCDFs releases in the various environmental matrices. The collection of baseline data for any contaminant at trace levels in the environmental samples definitely depends upon the available analytical methodology. The available analytical methods for Dioxins and Furans have limitations due to their presence in picogram and femtograms levels and resemblance in structural formulas. It has been proposed to develop and study the analytical techniques for the separation, identification, and quantification of specific numbers of Dioxins and Furans which can be carried at ease.

In India, there is lack of capability for providing a cheap and fast analytical service for PCDDs and PCDFs analysis. Thus, the main aim of the study was to develop a cost effective an alternative method for the identification and measurement of identified toxic PCDDs and PCDFs congeners namely, 2,3,7,8-TCDD, 2,3,7,8-TCDF, OCDD, and OCDF in water and soil samples.

The study aims to study techniques for their removal in contaminated water and soil matrices.

Analysis usually requires transfer to liquid or gas phase. Most methods are based on liquid-solid extraction. Thus, a classical Soxhlet extraction method and alternative rotary shaker extraction method were used for extraction of PCDDs and PCDFs in soil samples. Liquid-liquid extraction method was used for extraction of aqueous samples. Clean-up was carried out using silica, alumina, and celite/carbopack columns. A number of laboratory exercises were conducted to establish the method for identification of Dioxins and Furans in water and soil matrices. Simulated samples were prepared for the analysis of 2,3,7,8-TCDD, 2,3,7,8-TCDF, OCDD and OCDF in soil and water. A specific set of samples in duplicate were processed using various organic solvents with different extraction methods. The concentrated extracts of both soil and water were subjected to specific clean-up processes with chromatographic columns and subjected to GC-MS for the separation and quantification. The various conditions of samples extraction, clean-up process, and GC-ECD and GC-MS conditions were optimized for identification and quantification of selected PCDDs and PCDFs congeners, namely, 2,3,7,8-TCDD, 2,3,7,8-TCDF, OCDD, and OCDF. The data was evaluated to arrive at the efficient, easy, and economic method for extraction of PCDDs and PCDFs in water and soil samples.

1.7 INHOUSE REMOVAL TECHNIQUES FOR CONTROL OF DIOXINS/FURANS

The laboratory studies on removal of Dioxins/Furans viz. 2,3,7,8-TCDD, 2,3,7,8-TCDF, OCDD, and OCDF in the water and soil samples have been carried out using two different methods.

1.7.1 Water

The adsorption techniques using imported and indigenous granular activated carbon were studied for the removal of 2,3,7,8-TCDD, 2,3,7,8-TCDF, OCDD, and OCDF in simulated samples. The removal efficiency of GAC was evaluated with respect to contact time, carbon dose, pH, and varying analyte concentrations.

For the photochemical method, the effectiveness of UV irradiations on removal of 2,3,7,8-TCDD, 2,3,7,8-TCDF, OCDD, and OCDF in simulated samples with respect to contact time, pH, and varying analyte concentrations has been studied.

Using solar energy, the effect of solar irradiations on the degradation of 2,3,7,8-TCDD, 2,3,7,8-TCDF, OCDD, and OCDF in simulated samples has been evaluated with respect to contact period in conjunction with titanium dioxide (TiO_2) as a catalyst.

1.7.2 Soil

The effectiveness of UV irradiations on the removal of 2,3,7,8-TCDD, 2,3,7,8-TCDF, OCDD, and OCDF using simulated samples were studied with respect to contact period and analyte concentrations using a photochemical method.

The effectiveness of solar irradiations on the degradation of 2,3,7,8-TCDD, 2,3,7,8-TCDF, OCDD, and OCDF in simulated samples was evaluated with respect to contact period by exposing the simulated samples to sunlight and sunlight in conjunction with groundnut oil.

CHAPTER 2

Sources and Formation of Dioxins and Furans

2.1 INTRODUCTION

There are about 210 Dioxins and related compounds called Furans. They are classified into two classes of chemical compounds: The class of the polychlorinated dibenzo-p-dioxins and the class of the polychlorinated dibenzofurans comprising 75 Dioxins (PCDDs) and 135 related compounds called Furans (PCDFs). All differ from one another in terms of location and number of chlorine atoms attached to the molecule, and degree of toxicity. Only 17 out of 210 Dioxins and Furans have four lateral chlorine atoms and are of concern. The one which is most often referred to is the 'Seveso Dioxin' or 2,3,7,8-tetrachlorodibenzo-p-dioxin (2,3,7,8-TCDD), and is the most toxic.

2.2 STOCKHOLM CONVENTION ON POPs

Technological and industrial advances of the 20th century have resulted in enormous amounts of hazardous wastes. These wastes have been created primarily in the last half of the century for industrial or commercial use, or are a byproduct from industrial processes. Of particular concern is a group of chemicals termed persistent organic pollutants (POPs), which includes 12 chemicals. Dioxins and Furans belong to the group of POPs and are the most toxic of all the chemicals known. The Stockholm Convention is an important milestone in international law and is a global treaty to protect human health and the environment from POPs that remain intact in the environment for long periods, become widely distributed geographically, accumulate in the fatty tissue of living organisms, and are toxic to humans and wildlife. A major impetus to the Stockholm Convention was the finding of POPs contamination in relatively pristine Arctic regions—thousands of miles from any known source. Much of the evidence for long-range transport of airborne gaseous and particulate substances to the United States focuses on dust or smoke because they are visible in satellite images. Tracing the movement of most POPs in the environment is complex because these compounds can exist in different phases (e.g., as a gas or attached to airborne particles) and can be exchanged among environmental media. For example, some

POPs can be carried for many miles when they evaporate from water or land surfaces into the air, or when they adsorb to airborne particles. Then, they can return to Earth on dust particles or in snow, rain, or mist. POPs also travel through oceans, rivers, lakes, and, to a lesser extent, even through animal carriers, such as migratory species. Over 150 countries signed the Convention out of which 83 ratified it and the convention came into effect on 17 May 2004 and became legally binding. Taking cognizance of growing scientific evidence and public awareness around the world concerning the harm that is caused by these toxic pollutants; and noting special concerns about their accumulation in food and in human body tissues, the Stockholm Convention on Persistent Organic Pollutants (May 2001) focuses on reducing and eliminating release of 12 POPs (coined the "Dirty Dozen" by the United Nations Environment Programme (UNEP)). These 12 chemicals include eight pesticides (aldrin, chlordane, DDT, dieldrin, endrin, heptachlor, mirex, and toxaphene); two industrial chemicals (polychlorinated biphenyls and hexachlorobenzene); and two unintended by-products, Dioxins and Furans. Many of the POPs included in the Stockholm Convention are no longer produced in this country. However, humans and animals can still be at risk from POPs that have persisted in the environment from unintentionally produced POPs that have been released elsewhere and then transported to other places. They can spread over long distances (e.g., through migratory species, or long range air transport) and may therefore have an impact far away from their original source. These intrinsic properties of POPs create a dangerous combination that makes it practically impossible to effectively control them, once they are released into the environment. Although most developed nations have taken strong action to control POPs, a great number of developing nations have only recently begun to restrict their production, use, and release.

Recognizing the serious and long lasting injury to ecosystems and human health that POPs (Dioxins and Furans) can cause in communities that immediately surround their source locations, and also in far distant regions, a treaty was signed at Stockholm, Sweden on 22 May 2001. The convention outlines the goal of elimination of POPs, sets out a process to add new POPs to the global ban, and supports the need for the destruction of POPs' wastes at their source and in the stockpiles. The convention also establishes a financial and technical assistance mechanism for the developing countries and countries in economic transitions that would facilitate implementation of the convention. In support of this last objective, the United Nations Industrial Development Organization (UNIDO) and the Global Environment Facility (GEF) have initiated a program for POPs destruction in developing countries.

2.3 ADDRESSING POPs GLOBALLY

The United States (US) has taken a leading role to reduce and/or eliminate POPs and their release at a regional and global basis. Some highlights of US efforts:

- Canada and the US signed an agreement for the Virtual Elimination of Persistent Toxic Substances in the Great Lakes. The strategy sets long-term goals to promote emission reduction of toxic substances.
- The United States, Canada, and Mexico established the Commission for Environmental Cooperation (CEC) under the North American Agreement on

Environmental Cooperation (NAAEC), which developed a regional initiative on the sound management of chemicals. Under this initiative, the CEC developed Regional Action Plans, which recognize activities that reduce or eliminate risks from chemicals of concern.

- The US signed the legally binding regional protocol with other member nations (including European countries, Canada, and Russia) of the United Nations Economic Commission for Europe (UNECE) on POPs under the Convention on Long-Range Transboundary Air Pollution (LRTAP). This agreement seeks to eliminate production and reduction of emissions of POPs in the UNECE region. The original agreement addressed the 12 Stockholm Convention POPs and four additional chemicals (hexachlorocyclohexanes, hexabromobiphenyl, chlordecone, and polycyclic aromatic hydrocarbons), but, like the Stockholm Convention, included a mechanism for adding additional substances to the agreement. Elements from the LRTAP POPs Protocol were used in negotiations for the Stockholm Convention.
- Other international work has addressed trade in hazardous substances, some of which are POPs. The US, along with 71 other countries and the European Community, have signed the Rotterdam Convention on the Prior Informed Consent (PIC) Procedure for Certain Hazardous Chemicals and Pesticides in International Trade, building on a 10-year-old voluntary program. The PIC Convention identifies pesticides and industrial chemicals of concern, facilitates information sharing about their risks, and provides countries with an opportunity to make informed decisions about whether they should be imported. Some of the POP substances are already on the PIC list.
- The US has also provided technical and financial assistance for POPs-related activities to a variety of countries and regions, including Mexico, Central and South America, Russia, Asia, and Africa. Examples of this assistance include development of Dioxin and Furan release inventories in Russia and Asia, the Chemicals Information Exchange and Networking Project for chemicals managers in targeted countries in Africa and Central America, the destruction of pesticide stockpiles in Africa and Russia, and the reduction of PCB sources in Russia, which reduced emissions of PCBs and enabled Russia to meet the requirements of both the Stockholm Convention and the LRTAP POPs Protocol.
- The United States is also an observer of the Basel Convention, which was designed to reduce cross-border movements of hazardous waste. The Convention focuses on improving controls on the movement of waste, including some POPs waste, preventing illegal traffic, and ensuring that waste is disposed of as close as possible to its source.

2.4 PARTIES TO STOCKHOLM CONVENTION

2.4.1 United States

The US has taken a leading role to reduce and/or eliminate POPs and their release on a regional and global basis (Table 2.1). Canada and the United States signed an agreement for the Virtual Elimination of Persistent Toxic Substances in the Great Lakes.

Table 2.1: Status of POPs-Dioxins/Furans.

Country	**Global Historical Use/Source**	**Status**
USA	Unintentionally produced during most forms of combustion, including burning of municipal and medical wastes, backyard burning of trash, and industrial processes. Also can be found as trace contaminants in certain herbicides, wood preservatives, and in PCB mixtures.	Regulated as hazardous air pollutants (CAA). Dioxin in the form of 2,3,7,8-TCDD is a priority toxic pollutant (CWA).
United Kingdom		- 'Dioxins and dioxin-like PCBs in the UK environment'—government consultation document published in 2002. - Research Priorities for Dioxins and Polychlorinated Biphenyls (PCBs) (2003). - Development of UK Cost Curves for Abatement of Dioxin Emissions to Air (2003).
		Substantial decrease in the past 10 years in emissions to the environment and concentrations in food: - Approx 60% reduction in emissions of dioxins to air. - Approx 75% reductions in emissions of PCBs to air. - Approx 70% reduction of dioxins and dioxin-like PCBs in food.
		Increased controls on: - Industrial processes. - MWI, metal processing, power stations, and chemical manufacturing. - Open agricultural burning. - Marketing and use. - Vehicular emissions. - Safe disposal of PCBs and phasing out of remaining identifiable PCBs.
Australia	**Dioxins**-Produced unintentionally due to incomplete combustion as well as the manufacture of some pesticides and other chlorinated substances. Emitted from a range of sources including bushfires, uncontrolled burning of waste and metal processing. **Furans**-Produced unintentionally from many of the same processes that produce dioxins. Furans are structurally similar to dioxins and share many of their toxic effects.	No federal emission standards but most states have some regulations. Reporting under National Pollutant Inventory.
		Management of POPs in Australia is being addressed through measures related to releases of dioxins.

Table 2.1 contd. ...

...Table 2.1 contd.

Country	Global Historical Use/Source	Status
China		The monitoring system for dioxins and furans in China has not yet been developed.
South Korea	Generated as waste itself, or during waste treatment, and from synthesis and production of products.	Due to governmental regulatory action, during the most recent five years, emissions of dioxins have been significantly reduced. In response to the Stockholm Convention, periodic monitoring has been conducted in South Korea.
India	Unintentional Production	Specific standards available.
Austria	Waste incinerators, including co-incinerators of municipal, hazardous, or medical waste or of sewage sludge.	- Adaptation to BAT necessary for PCDDs/Fs. - Determination of concentrations of PCDD/F and relevant precursors especially in bleached (Kraft-)pulp (imported and domestic production), papers (packaging papers, paper boards, paper made from recovered fibers), colors and inks, and deinking sludge.
European Union (EU)		Provisions on the reduction of emissions of unintentionally produced POPs.

2.4.2 United Kingdom

The requirements of the Stockholm Convention are implemented in UK principally by the UK Persistent Organic Pollutants Regulations 2007. The Regulations contain provisions regarding production, placing on the market and use of chemicals, management of stockpiles and wastes, and measures to reduce unintentional releases of POPs. The Dioxin Action Plan in UK comprise evaluation of release inventories for Dioxins, PCBs and HCB; effectiveness of current legislation and policies; development of future strategies; promotion of education, training, and awareness; and implementation and strategy review mechanism.

Emissions of Dioxins and Furans have declined significantly following measures taken to control industrial releases. However, unintentional releases from diffuse sources, such as backyard burning, continue to present a challenge in achieving further reductions. Since 2007, the UK has been affected by one major incident relating to Dioxins in food. This originated in the Republic of Ireland and was associated with highly contaminated feed which was supplied mainly to pig farms in the Republic of Ireland. Some potentially contaminated pork and pork products were exported to other Member States, including the UK, and were withdrawn from the food chain as a precaution. However, some of the highly contaminated feed was also supplied to a small number of beef and dairy farms in Northern Ireland. As a consequence, hundreds of thousand liters of milk were discarded, several hundred tons of meat were withdrawn and destroyed, and about 5,000 contaminated cattle were culled in order to prevent the further entry of contaminated meat into the food chain. Although further product withdrawals have been instigated as a result of contamination incidents in other Member States, there have been no other significant incidents originating in the UK.

2.4.3 Australia

Australia ratified the Stockholm Convention on 20 May 2004 and become a Party on 18 August 2004. In relation to the 12 original POPs, Australia is well advanced in meeting the measures agreed under the Convention. Australia's NIP and aims to:

- Assist in the implementation of the NIP;
- provide an information conduit for non-government organizations with an interest in POPs and to act as a resource for the sharing of knowledge and experience between organizations; and
- assist with the development of educational and information material relevant to POPs.

2.4.4 China

The environmental monitoring of POPs is very limited in China and the nationwide pollution status of POPs is still unclear. The monitoring system for Dioxins and Furans in China has not yet been developed. However, the further development of qualified supervisory capabilities to monitor the unintentional by-products, and the exact pollution sources and pollution status of Dioxins and Furans is encouraged.

2.4.5 Russia

POPs are a problem of national nature demanding immediate measures under the obligation of Stockholm Convention. For ensuring the chemical and biological safety of Russia, the Federal Assembly of the Russian Federation was recommended to ratify the Stockholm Convention on POPs. The Government of the Russian Federation was also recommended to prepare and bring it to the State Duma for ratification.

2.4.6 South Korea

The south Korean government has followed actively the needs of Stockholm Convention; (1) regulation in use, distribution, and treatment of POPs or POPs containing wastes, (2) development and application of BAT and BEP to reduce POPs release, and (3) monitoring and investigation on POPs contamination nationwide. Due to governmental regulatory action, during the most recent five years, emissions of Dioxins have been significantly reduced. However, the annual rate per unit area was still extremely high in relation to other countries. Consequently, environmental concentrations and human exposure levels in the source area frequently exceeded criteria suggested to protect environments and human. Thus to decrease human and ecological exposure, more reductions in emissions is required.

2.4.7 India

India actively participated in the International Negotiation Committee meeting leading to the drafting and acceptance of the Stockholm Convention. India ratified the Stockholm Convention on January 13, 2006 and the Convention entered into force on April 12, 2006.

2.5 PRODUCTION, USE AND DISPOSAL

Production

Dioxins are not manufactured commercially in the US except on a small scale for use in chemical and toxicological research. Dioxins are unique among the large number of organochlorine compounds of environmental interest in that they were never intentionally produced as desired commercial end products (Zook and Rappe 1994). Typically, Dioxins are unintentionally produced during various uncontrolled chemical reactions involving the use of chlorine (EPA 1990c) and during various combustion and incineration processes (Zook and Rappe 1994).

In general, there are two conventional methods for the preparation of Dioxins for research purposes: condensation of a polychlorophenol and direct halogenation of the parent dibenzo-p-dioxin or a monochloro-derivative. For example, 2,3,7,8-tetrachlorodibenzo-p-dioxin (2,3,7,8-TCDD) is generally synthesized by the condensation of two molecules of 2,4,5-trichlorophenol in the presence of a base at high temperatures or by chlorination of dibenzo-p-dioxin in chloroform in the presence of iodine and ferric chloride (EPA 1987k; IARC 1977). Other methods of 2,3,7,8-TCDD synthesis include the following: pyrolysis of sodium α-(2,4,5-trichlorophenoxy) propionate at 500°C for 5 hrs; reaction of dichlorocatechol salts with o-chlorobenzene by refluxing in alkaline dimethyl sulfoxide; ultraviolet irradiation of CDDs of high chlorine content; Ullman reaction of chlorinated phenolates at 180–400°C; pyrolysis of chlorinated phenolates and chlorinated phenols; and heating 1,2,4-trichloro-5-nitrobenzene, and 4,5-dichlorocatechol in the presence of a base (EPA 1984a; IARC 1977).

Use

The only reported use of Dioxins/Furans is as research chemicals (NTP 1989). A diversified group of researchers use various Dioxins in studies of toxicology, environmental fate, transformation, and transport, and in residue analysis of a variety of contaminated media. Dioxins have been tested for use in flame-proofing polymers such as polyesters and against insects and wood-destroying fungi; however, there are no data reporting its commercial production or use for these purposes (IARC 1977).

Disposal

Several treatment or disposal methods for Dioxins and Dioxin-contaminated materials have been investigated, including land disposal, thermal destruction, and chemical and biological degradation. Each of these methods has limitations regarding economics, technical feasibility, and acceptability (HSDB 1995). Land disposal of Dioxin-containing wastes is prohibited (EPA 1986f, 1988f).

Thermal destruction technologies offer the most straightforward approach to treating or disposing of CDD-contaminated materials because under the appropriate conditions, the breakdown of the CDDs is assured (U.S. Congress 1991). Incineration, involving the high-temperature oxidation of Dioxin molecules, is the most extensively tested method for disposal of Dioxins. EPA has classified Dioxins such as TCDD, PeCDD, and HxCDD as Principal Organic Hazardous Constituents and are required to be incinerated under conditions that achieve a destruction and removal efficiency of

99.99 percent (EPA 1990b; Sedman and Esparza 1991). To ensure complete destruction, a temperature of at least 1,500–2,600°F in incinerator operating conditions is required adequate for destruction of 2,3,7,8-TCDD and most other chlorinated organics, with residence times of at least 30 minutes (although 1.5 hours is a more common residence time) (EPA 1990a).

2.6 SOURCES OF DIOXINS AND FURANS

Dioxins and Furans were never produced intentionally but occur as trace contaminants in a variety of industrial and thermal processes. Dioxins are ubiquitous in the environment (Podoll et al. 1986). Although all of the sources or processes that contribute to Dioxins in the environment have not been identified, Dioxins are known to be formed in the manufacture of chlorinated intermediates and pesticides, during smelting of metals (EPA 1998j), in the incineration of municipal, medical, and industrial wastes (Podoll et al. 1986), and from the production of bleached wood pulp and paper (Fletcher and McKay 1993). Dioxins are also found in emissions from the combustion of various other sources, including coal-fired or oil-fired power plants, wood burning, and home heating systems (Chiu et al. 1983; Czuczwa and Hites 1984; EPA 1998j; Gizzi et al. 1982; Thoma 1988). Generally, the more highly chlorinated CDDs are the most abundant congeners present in the emissions from these combustion sources and from the production of bleached wood pulp and paper (Fletcher and McKay 1993). Dioxins also occur in other combustion products (e.g., cigarette smoke) (Bumb et al. 1980; Lofroth and Zebuhr 1992; Muto and Takizawa 1989), automobile exhaust from cars running on leaded gasoline with chlorine scavengers and to a lesser extent from cars running on unleaded gasoline (Birmingham et al. 1989; Marklund et al. 1987, 1990), and diesel exhaust (Jones 1995; Cirnies-Ross et al. 1996). Dioxins and Furans can form during the synthesis and combustion of chlorine-containing materials, such as polyvinylchloride, in the presence of naturally occurring phenols, vegetation treated with phenoxy acetic acid herbicides, paper and wood treated with chlorophenols, and pesticide-treated wastes (Arthur and Frea 1989).

2.6.1 Primary Sources

2.6.1.1 Industrial-Chemical Processes

Primary sources of environmental contamination with PCDD/PCDF in the past were due to production and use of chloroorganic chemicals, including the pulp and paper industry. In wet-chemical processes, the propensity to generate PCDD/PCDF during synthesis of chemical compounds decreases in the following order (Fiedler et al. 1990; Hutzinger and Fiedler 1993): chlorophenols > chlorobenzenes > aliphatic chlorinated compounds > inorganic chlorinated compounds. 2,3,7,8-TCDD is a by-product formed in the manufacture of 2,4,5-trichlorophenol (2,4,5-TCP) (Arthur and Frea 1989). 2,4,5-TCP was used to produce the bactericide, hexachlorophene, and the chlorophenoxyherbicide, 2,4,5-trichlorophenoxy acid (2,4,5-T). Trichlorophenol-based herbicides have been used extensively for weed control on crops, rangelands, roadways, right-of-ways, etc. Various formulations of 2,4-dichlorophenoxy acetic acid (2,4-D) contaminated mainly with higher chlorinated CDDs/CDFs and 2,4,5-T contaminated mainly with 2,3,7,8-TCDD were used extensively for defoliation and crop destruction by the American military during the Vietnam War.

Factors favorable for the formation of PCDD/PCDF are high temperatures, alkaline media, presence of UV-light, and presence of radicals in the reaction mixture/chemical process (Hutzinger and Fiedler 1993). Emissions of PCDD/PCDF into the environment via water and to soils occur from kraft pulp and paper mills. The US-EPA inventory estimates annual emissions from kraft pulp and paper mills in the range of 20 g I-TEQ. In addition, PCDD/PCDF were detected in the final product (pulp, paper) as well as in the pulp and paper sludges. With advanced bleaching technology, the PCDD/PCDF contamination in effluents, products, and sludges was reduced.

CDDs have been detected in chimney soot samples from various home heating systems using unleaded heating oil, coal, and wood in Germany (Thoma 1988). A Canadian study of wood-burning stoves detected only OCDD in particulates from the stack emissions (Wang et al. 1983). Open-air burning of PCP-treated wood produced levels of CDDs ranging from 2 ppb (TCDD) to 187 ppb (OCDD) (Chiu et al. 1983). Combustion of untreated wood also produces CDDs (TCDD, PeCDD, HxCDD, HpCDD, OCDD) (Clement et al. 1985).

2.6.1.2 Thermal Processes

The EPA has recently identified stationary source categories that release 2,3,7,8-TCDD TEQ to the atmosphere (EPA 1998j). The percentage contribution of the five highest source categories are: 68 percent from municipal waste incineration, 12.3 percent from medical waste incineration, 8.9 percent from Portland cement manufacture hazardous waste kilns, 3.5 percent from secondary aluminum smelting, and 3.0 percent from other biological incineration. These five source categories account for 95.9 percent of all stationary emissions of 2,3,7,8-TCDD TEQ to the air. The "Trace Chemistries of Fire Hypothesis" suggests that CDDs and CDFs can also form during a variety of combustion processes including natural ones, such as forest fires and volcanic eruptions (Crummett 1982). However, there is very limited evidence suggesting that such natural processes could be minor sources of these compounds in the environment. Only data from one study were found that directly measured CDD/CDFs in actual emissions from forest fires. Tashiro et al. (1990) detected the concentration of total CDD/CDFs in air ranging from 15 to 400 pg/m^3.

Dioxins have been detected in chimney soot samples from various home heating systems using unleaded heating oil, coal, and wood in Germany (Thoma 1988). A Canadian study of wood-burning stoves detected only OCDD in particulates from the stack emissions (Wang et al. 1983). Open-air burning of PCP-treated wood produced levels of CDDs ranging from 2 ppb (TCDD) to 187 ppb (OCDD) (Chiu et al. 1983). Combustion of untreated wood also produces CDDs (TCDD, PeCDD, HxCDD, HpCDD, OCDD) (Clement et al. 1985). The presence of PCDD/PCDF in the emissions and residues from municipal solid waste incinerators were detected first in 1997 in the MSWI in Amsterdam (Olie et al. 1977). Some key-parameters are briefly summarized, which have been identified to influence the formation of PCDD/PCDF in combustion processes (Fiedler 1998).

Role of Temperature: Some early experiments were performed at high temperatures, when, for example, Rubey et al. studied the thermal stability of PCB and the formation of PCDF. The experiments clearly showed that PCB (here: 2,3,4,4,5-pentachlorobiphenyl

= 2,3,4,4′,5-CB) are stable up to temperatures around 700°C. With increasing temperatures, there is a decrease in the PCB concentration and an increase in PCDF formation. Preferentially, lower chlorinated PCDF (Cl_4DF) were formed with a maximum at about 750°C. Further increase of the temperatures results in destruction of the newly formed PCDF (Rubey et al. 1985).

Role of Temperature and Residence Time: For gas-phase reactions, temperature is not the single limiting factor and the combination of two, for example, temperature and residence, is an important parameter for determining the efficiency of how organic substances are being destroyed. As a general rule: higher temperatures need shorter residence times of the gaseous molecules. Thus, it is an engineering question how to build and operate a plant.

Role of Precursors: From experiments to condense pentachlorophenol (PCP) over fly ash by Karasek and Dickson (1987) in a temperature range from 250 to 350°C the precursor theory was established. The authors concluded that metallic constituents in the fly ash act as catalysts for the formation of PCDD. In more recent works, Milligan and Altwicker found that gas-phase 2,3,4,6-tetrachlorophenol was the most efficient precursor in PCDD formation (Milligan and Altwicker 1996).

Role of Sulfur/Chlorine Ratio: In 1986, Griffin established a hypothesis to explain the formation of PCDD/PCDF as a result of the sulfur-to-chlorine ratio in the feed (Griffin 1986). It is well known that combustion of fossil fuels like coal generates much less PCDD/PCDF than combustion of municipal solid waste. The hypothesis states that in coal there is a sulfur-to-chlorine ratio of 5/1 whereas in municipal waste the ratio S/Cl is 1/3. The latter ratio allows formation of molecular chlorine according to the Deacon process catalytically driven by metals, for example, copper. The molecular chlorine is considered to be responsible for the *de novo* Dioxin formation (Griffin 1986).

Role of Chlorine Species: The influence of the chlorine species can be summarized in that chlorination of aromatic compounds readily occurs in the presence of Cl_2. Such substitution reactions do occur in the presence of fly ash (heterogeneous phase, probably surface-catalyzed) as well as in the gas phase (homogeneous phase). At temperatures up to 250°C, HCl does chlorinate chlorine-free dibenzodioxin, 1,2,3,4-Cl_4DD or toluene when adsorbed to fly ash. Without fly ash, Cl_2 was 4-times more efficient than HCl in chlorinating these compounds (Gullett et al. 1994). Gaseous chlorine (Cl_2) was found to be the most efficient chlorinating agent (Gullett et al. 1990).

Role of Oxygen: From laboratory, pilot-scale, and large-scale experiments it was concluded that increasing oxygen concentrations from 0 to 10 percent resulted in increasing formation of PCDD/PCDF. The O_2 content pushes the Deacon reaction towards Cl_2-production and subsequently to formation of organochlorine compounds (Vogg et al. 1987). Under pyrolytic conditions (oxygen deficiency), dechlorination of PCDD/PCDF occurs at temperatures above 300°C.

Role of Metals: When testing the efficiency of metals to catalyze formation of Dioxins and Furans, copper was found to be the most efficient compound (Stieglitz 1989). Further studies have shown that small hydrocarbons such as acetylene and ethylene are readily chlorinated in the presence of cupric chloride or cupric oxide and HCl (Froese and Hutzinger 1993, 1996). The mechanism to reduce Cu(II) to Cu(I) and the oxychlorination of the newly formed Cu(I)Cl to reconvert to Cu(II)Cl_2 completes the catalytic cycle. The mechanism is very similar to the copper catalyzed Deacon reaction that converts HCl into Cl_2. With acetylene, however, the reaction is accelerated as the activation energy for the formation of Cu(I)Cl is reduced.

Role of Deposits and Other Parameters: Kanters and Louw (1996) showed that in the absence of fly ash, deposits in the cooler ends of a municipal solid waste incinerator favor the formation of PCDD/PCDF and other PICs (products of incomplete combustion).

2.6.2 Secondary Sources

Dioxin reservoirs are those matrices where Dioxins and Furans are already present, either in the environment or as products. The PCDD/PCDF found in these reservoirs are not newly generated but concentrated from other sources. A characteristic of the reservoir sources is that they have the potential to allow re-entrainment of Dioxins and Furans into the environment. Product reservoirs include PCP-treated wood, PCB-containing transformers and sewage sludge, compost, and liquid manure, which can be used as fertilizers in agriculture and gardens. Reservoirs in the environment are, for example, landfills and waste dumps, contaminated soils (mainly from former chemical production or handling sites), and contaminated sediments (especially in harbors and rivers with industries discharging directly to the waterways).

2.6.3 Natural Sources

Biological formation of Dioxins and Furans from chlorinated precursors was discussed for compost and sewage sludge and questions on the possibility of a biogenic formation did arise for sediments and soils (especially forest soils). Based on the results of Öberg and Rappe 1992 the turnover to convert pentachlorophenol (PCP, the most suitable precursor) to Dioxins is in the low ppm-range (Wagner et al. 1990; Öberg and Rappe 1992; Öberg et al. 1992, 1990). Consequently, a chlorinated precursor present in an environmental matrix, such as soil or sediment, at ppm-concentrations should be converted to not more than ppt levels of high-chlorinated Dioxins (Cl_7DD and Cl_8DD). In other words, ppm-concentrations of chlorophenols would generate ppt-levels of HxDD and OCDD or ppq-concentrations in TEQ. Thus, based on present knowledge, biological formation of Dioxins from chlorinated phenols under environmental conditions is negligible (Fiedler 1995).

2.7 ACCIDENTAL RELEASE OF DIOXINS AND FURANS

Dioxin first came to widespread public notice during the Vietnam War, when it was identified as a component of the defoliant Agent Orange (Hay 1982). Previously, campaigns on behalf of agricultural and forestry workers had been mounted to have TCP banned because of its alleged toxic effects on humans. These frequently met with scientific disapproval, partly because the evidence was only "anecdotal." The United Kingdom's regulatory system was particularly unsympathetic to such claims (Wynne 1989).

2.7.1 Seveso Disaster

Around midday on Saturday 10 July 1976, an explosion occurred in a TCP (2,4,5-trichlorophenol) reactor of the ICMESA chemical plant on the outskirts of Meda, a small town about 20 kilometres north of Milan, Italy. A toxic cloud containing TCDD (2,3,7,8-tetrachlorodibenzo-p-dioxin), then widely believed to be one of the most toxic man-made chemicals (Mocarelli et al. 1991), was accidentally released into the atmosphere. The Dioxin cloud contaminated a densely populated area about six km long and one km wide, lying downwind from the site. This event became internationally known as the Seveso disaster, after the name of a neighbouring municipality that was most severely affected (Hay 1982; Pocchiari et al. 1987). Eleven communities in the rolling countryside between Milan and Lake Como were directly involved in the toxic release and its aftermath. The four most impacted municipalities included Seveso (1976 population 17,000), Meda (19,000), Desio (33,000), and Cesano Maderno (34,000). Two other municipalities were subject to post accident restrictions: Barlassina (6,000) and Bovisio Masciago (11,000). Health monitoring was extended to a further five municipalities.

2.7.2 Agent Orange

Another type of disaster was due to "Agent Orange" which was extensively used as a defoliant during the Vietnam War between 1961 and 1970. Agent orange is a disaster with a protracted latency period, that is, the possible effects are not manifested until years or decades after exposure. The US military, in order to clear vegetation, weeds, and leaves on trees to reduce sites for enemy troops to hide, applied Agent Orange by airplanes, helicopters, trucks, and backpack sprayers. The exposure to Agent Orange caused delayed health effects. Agent Orange contained minute amounts of highly toxic compound TCDD. Dioxin contaminants of Agent Orange have persisted in the environment in Vietnam for over 30 years.

2.7.3 Times Beach Crisis

The chemical contamination and subsequent evacuation of Times Beach, Missouri was an environmental disaster which affected over 2,000 residents of low socioeconomic status. By 1970, the Times Beach, Missouri had a population of over 1,200 people. Times Beach had mostly dirt roads and could not afford to pave them. The roads were sprayed with oil in an attempt to control dust problems. In 1972 and 1973 the town's roads were sprayed by a waste oil hauler. The small town had unpaved roads, so a private contractor was hired in the 1970s to repeatedly spray the streets with oil in order to control the dust. Unbeknownst to those in Times Beach, the contractor was using waste oil that he was being paid to dispose off from a chemical factory. This waste oil was heavily contaminated with Dioxins, chemicals found in the herbicide Agent Orange which was used during the Vietnam War. The roads turned purple and animals began dying.

In 1982 a leaked EPA document was published and alerted Times Beach officials that their town may be contaminated with hazardous substances. Under pressure, the EPA conducted an analysis on the soil. In December 1982, the town was devastated by heavy flooding shortly before the EPA's results confirmed toxic levels of Dioxin. By the 23rd of December, the residents were advised to leave town as the EPA confirmed

Times Beach was contaminated. The CDC recommended the town not be reinhabited, and the EPA announced they would be using Superfund money to buy-out all the houses. Eventually an incinerator was constructed and the contaminated materials were burned and buried. Today the area has been renovated into the Route 66 State Park, and further EPA testing has declared the park safe.

2.7.4 Yusho History

A strange disease was first reported on October 10, 1968 in western Japan, even though several patients with peculiar symptoms had been recognized by alert clinicians several months prior (Kuratsune et al. 1996; Katsuki 1969; Yoshimura 2003). This strange disease—characterized by acne-like eruptions, pigmentation of the skin, and eye discharge—was named Yusho ("oil disease") (Kuratsune et al. 1996; Goto and Higuchi 1969).

In 1968, a mass food poisoning (yusho) occurred in western Japan involving more than 1,850 people, the majority of whom were residents of Fukuoka and Nagasaki prefectures. The poisoning is now understood to have been caused by ingestion of a commercial brand of rice oil contaminated with polychlorinated derivatives of biphenyls, dibenzofurans, quaterphenyls, and some other related compounds. At its outbreak in 1968, Yusho patients suffered severe skin symptoms. Although the blood concentrations of PCBs and Dioxins, especially highly toxic 2,3,4,7,8-pentachlorinated dibenzofuran (2,3,4,7,8-PeCDF) remain high in these patients. The number of deaths seen among 1,761 victims (887 males and 874 females) from the date of official registration as yusho up to the end of 1983 was compared with the expected number of deaths which was calculated on the basis of the national age, sex, and cause-specific death rates. Neither significantly increased nor significantly decreased mortality was seen among overall causes of death in males and females. A significant excess mortality was seen for malignant neoplasms at all sites in males but not in females. Neither significantly increased nor decreased mortality was seen for cancer of the esophagus, stomach, rectum and colon, pancreas, breast, and uterus. For cancer of the liver, however, a considerably increased mortality was seen in both males and females but the excess was statistically significant only in males. It was also notable that such increased mortality due to liver cancer was seen mainly among the patients living in Fukuoka prefecture but not at all among those in Nagasaki prefecture. Deaths from chronic liver diseases and liver cirrhosis were also found to be increased in both sexes but the increase was not statistically significant.

2.7.5 Dioxin Leakage in Central Tokyo

The Clean Association of Tokyo 23, the public organization who is responsible for the intermediate treatment (separation, fragmentation, and incineration of collected wastes), announced on 13th June 2011 that one of the incineration plants that they are taking charge of had stopped because of the high Dioxin concentration level in the working area of the plant. It was Setagaya Incineration Plant located in the midst of residential area of Setagaya-ward, Tokyo. The announcement of the accident was made two weeks after the accident happened in the plant. The citizens of Setagaya-ward and other Tokyo Metropolitan had learnt about this accident by the newspaper on 14th June. The Clean Association of Tokyo 23 announced that "There is no impact to the surrounding environment of the plant by this accident, but we are monitoring the ambient air Dioxin level on the roof top of the plant" (http://eritokyo.jp/independent/ikeda-col1018...html).

2.7.6 Belgian Dioxin Scare 1999

The Belgium 'Dioxin crisis' of 1999 provides a salutary lesson. The Belgian food industry was badly damaged when high levels of Dioxin were discovered in eggs and chickens and traced back to Dioxin contaminated animal feed. Import bans by countries worldwide included chicken, eggs, meat, and any products containing eggs or milk. The Belgian government estimated the cost of the crisis at €465 million.

2.7.7 Dioxin Contamination of Animal Feed in Europe

BRUSSELS, Belgium, November 5, 2004 (Environmental News Service 2004): Fear of Dioxin contamination in European foods spread today on reports that Dutch potato by-products tainted with the cancer causing chemical had been sold to farmers in the Netherlands, Belgium, and Germany. More than 160 farms were closed in the Netherlands and Belgium after Dioxin was found in dairy products. All of those farms reportedly had received shipments of animal feed which contained potato by-products from Canadian frozen potato chip fries manufacturer McCain that were contaminated with Dioxin (http://www.chaseireland.org/inthemedia/ens-05-11-04.htm).

CHAPTER 3

Global Contamination and Movement of Dioxins and Furans in the Environment

3.1 GLOBAL TRANSPORT OF DIOXINS AND FURANS

When Dioxins/Furans are used in the environment or produced, they are likely to get distributed among water, air, soil, and biota. This is linked to their intrinsic physico-chemical properties. There are two major ways by which they can reach surface and ground water-runoff and leaching. Runoff will occur if the chemical does not adsorb onto soil. Leaching occurs when the chemical is weakly adsorbed by soil and can easily move through the soil. Weak acid pesticides are bound weakly to soil so they can easily move downward to ground water. Organic carbon and global temperature differences have been established as key drivers of Dioxins/Furans global distribution. However, to accurately predict the global fate of Dioxins/Furans, further research is required to understand the various bio-geochemical and geo-physical cycles. The movement of Dioxins/Furans in the environmental compartments is illustrated in the Fig. 3.1. In 1974, it was first suggested that POPs may migrate through the atmosphere as gases and aerosols and condense in low-temperature regions. Individual Dioxins/Furans separate in the atmosphere in a fashion similar to the familiar fractionation processes (such as fractional distillation). Dioxins/Furans may be fractionated during their journey toward the poles because they migrate at different velocities (Wania 1996). There have been many measurements of Dioxins/Furans concentrations in the global environment, particularly in arctic regions and the oceans. Other important advances in understanding the global distribution of Dioxins/Furans have occurred. It is thus timely to evaluate, modify, and extend some of these concepts, especially in the light of recently announced international intentions (UNEP 1999) to control such substances. Evidence indicates that most Dioxins/Furans are volatile enough to evaporate and deposit, among air, water, and soil at ordinary environmental temperatures. Warm temperatures favor deposition from the Earth's surface in tropical and subtropical regions. Cool temperatures at higher latitudes favor deposition from the atmosphere, for instance, favour greater adsorption of these compounds to atmospheric particulate matter, which then deposit (Wania and Mackay 1996; Mackay and Wania 1995) onto soil and water. Several factors are involved in POPs' tendency

Fig. 3.1: Movement of Dioxins/Furans in the environment.

to condense, deposit, and accumulate in cold ecosystems. Natural decomposition is very slow in the cold, thus allowing Dioxins/Furans to remain integral for longer time and become more persistent. Likewise, cold slows the evaporation of Dioxins/Furans from water and promotes their condensation and movement, or "partitioning," from the atmosphere to the surface. These compounds do not actually condense in the sense that the atmosphere becomes saturated or supersaturated. Conditions are always below saturation limits. Rather Dioxins/Furans merely partition from a gaseous to a non-gaseous phase. Dioxins/Furans can migrate to higher latitudes. The compounds migrate, rest, and again migrate with seasonal temperature changes at mid-latitudes.

3.2 SOIL AND WATER CONTAMINATION DUE TO DIOXINS AND FURANS

Although many of the primary sources are not yet known, Dioxins and Furans in very small concentrations are ubiquitous in the environment. They have been found over the entire globe, even in remote areas. Dioxins and Furans are adsorbed on airborne particulate matter or found in industrial effluent which deposited on soil and eventually bind to other organic substances and bottom sediments in lakes and rivers. Although Dioxins/Furans are encountered in both the vapor and particulate phases, it has been suggested that ingestion results in 90 percent of the human exposure (Charnley and Doull 2005; Taioli et al. 2005). Atmospheric Dioxins deposit on vegetation which becomes the food for animals. Humans then ingest crops, fish, meat, and dairy products and thus accrue a body burden of dioxin. Secondary exposure, due to such soil and water pollution, may be as significant as atmospheric exposure and could substantially increase total risk.

The following section compiles present knowledge of Dioxins/Furans sources and concentrations in various environmental matrices. Dioxins/Furans emission inventories have utilized an integrated approach to estimate emissions to the air, land, and soil. Direct releases to water have only been qualitatively estimated (Quaβ et al. 2000a; Wenborn et al. 1999). The inventories suggest that emissions of Dioxins/Furans to land, including landfills, may be as high as or even higher than emissions into the air (Hansen and Hansen 2003; Haynes and Marnane 2000; Wenborn et al. 1999; Dyke et al. 1997). The greatest releases are attributed to the production and the use of organochlorine pesticides (Wenborn et al. 1999). Pesticides and other chlorinated chemicals have an especially high potential of penetrating into water systems. There are many chlorinated chemicals that may contain Dioxins and Furans impurities, which makes it extremely difficult for researchers to estimate the historical and current emissions.

3.2.1 Distribution

The processes such as sedimentation, adsorption, and volatilization are responsible for the distribution of Dioxins/Furans in the environment. They can then be degraded by chemical (oxidation, reduction, hydrolysis, and photolysis) and/or biological processes. Biological degradation in soil and living organisms utilize hydrolysis, oxidation, reduction, and conjugation to degrade chemicals. The process of degradation is largely governed by the medium in which the Dioxins/Furans are distributed.

3.2.1.1 Soil/Sediment Adsorption

Adsorption of Dioxins/Furans on soils or sediments is a major route for their transportation and eventual degradation. Most of the Dioxins/Furans are non-polar and hydrophobic. Non-polar chemicals tend to be pushed out of water and onto soils which contain non-polar carbon material. Soils vary in the amount of organic carbon content, which mainly determines the amount of Dioxins/Furans adsorbed. The factors which are responsible for soil adsorption include the following:

Organic Carbon Content: The amount of carbon present in soil is probably the greatest factor influencing the amount of POP adsorbed. This is because organic matter is non-polar and has a relatively light negative charge. Most Dioxins/Furans are non-polar and will be attracted to the lightly charged surface. As the organic content of soil increases so does the amount of Dioxins/Furans adsorbed.

Polarity: Polarity greatly determines the partitioning of the Dioxins/Furans. A polar pesticide will be very water soluble and tend not to be adsorbed onto soil. If the pesticide is non-polar it will tend to leave water and be adsorbed onto soil. Once adsorbed, non-polar molecules will strongly bind to soils than polar ones, thus polar molecules will be more mobile in soil. This process occurs when water is present because the chemical is able to dissolve in water and move through soil. The chemical moves by dissolving and being deposited on another portion of soil and the process continues. This movement is what allows Dioxins/Furans to reach ground water.

Organic matter in solution: Non-polar chemicals can be adsorbed by this organic matter in water which can then settle to the bottom or be transported away.

3.2.1.2 Bioconcentration Factor

Bioconcentration factor (BCF) is an important indicator to show how much a Dioxins/Furans chemical will accumulate in living organisms such as fish. Once absorbed into an organism, chemicals move through the food chain. The increase in concentration of a chemical as it moves through a food chain is called biomagnification or bioaccumulation. However, BCF is different from biomagnification in that it is a measure of the absorption of a chemical from an aqueous environment by a single organism. Factors that influence BCF include:

Polarity: Polar molecules are soluble in water (polar) and not very soluble in tissues (non-polar). Polar chemicals tend not to bioconcentrate in biota because of this low solubility, whereas non-polar chemicals will accumulate in fatty tissues.

Solubility: Dioxins and Furans are non-water soluble chemicals which are non-polar and are more likely to accumulate.

Lipid content: Every living organism has a different lipid content so each will absorb different amounts of a chemical accordingly. If the lipid content is high, the organism will tend to absorb more.

Metabolism: When an animal absorbs a chemical, the biological system of that animal has mechanisms for getting the chemical out of the system. This is done by changing part of the chemical in order to make it water soluble so that it can be expelled through the urine.

3.2.1.3 Volatilization

The volatilization rates of a chemical is best indicator of the method of transportation by which the Dioxins/Furans can be dispersed in the environment. Volatilization is one of the main transport pathways by which Dioxins/Furans move from water and soil surfaces into the atmosphere. The volatilization rate provides information about how much chemical is present in air *vs* soil and water. A non-volatile chemical that migrates and accumulates at the surface can be transported to other areas by water runoff. The amount that is transported depends on the volatility of the Dioxin and Furans congeners. A chemical that is extremely volatile is quickly spread over a large area by wind. A non-volatile congener can accumulate on the soil or water surface and be transported down through the soil layer to ground water. The factors influencing volatilization include:

Temperature: An increase in temperature generally increases the vapor pressure and volatilization rate. An increase in temperature will increase water and chemical evaporation rates. If all of the water evaporates off the soil, non-polar compounds will be strongly bound to the soil, so volatilization will decrease with dry soil.

Chemical Properties: If the chemical is non-polar it will tend to move onto the soil and water surface. Some readily volatilize while others don't. If a chemical (non-polar) is not soluble in water it will have a tendency to be pushed onto soil or the water surface which increase the volatilization. Chemicals that are polar will be very soluble in water so the volatility will generally not be high.

Molecular Properties: As the molecular weight increases, the volatilization generally slows down because larger molecules move slower and they are not moving fast enough to go into the gas phase. A chemicals shape and structure also affect the volatilization rate since they can influence how strongly the chemical bonds to soil or water.

Concentration: Higher chemical concentrations mean that more chemical is present in the water, so more will reach the surface and be volatilized.

Vapor pressure: Vapor pressure can be used to calculate Henry's Law Constant which is an indicator of the volatilization of a chemical. POPs volatilize from soils differently according to the adsorbing surfaces.

3.2.1.4 Photolysis

Photochemical and free radical reactions are major degradation pathways in the atmosphere. The products formed by photolysis may or may not be more toxic than the parent compound. Once a POP chemical has been degraded, a major removal process for chemicals is to precipitate out of air and return to the earth's surface. The most important factor in determining the rate of photolysis is the amount of light (photons) present. Large quantities of Dioxins/Furans are lost by volatilization into the atmosphere. In the atmosphere, there are two major degradation pathways that can occur, photochemical reactions caused by sunlight and the free radical reactions. The products formed may or may not be more toxic than the parent chemical. Photochemical reactions can take place in air or water when sunlight is present. The factors influencing photolysis include:

Time of day: The rate determining step in photolysis is the amount of photons present that can react with the molecules. Photochemical degradation processes increase with temperature, so the maximum degradation rates will occur at midday.

Weather: Clouds can prevent light from reacting with molecules to form radicals, thus slow down the degradation process.

Radicals Present: The amount of ozone and other free radicals present can greatly influence the rates. Increases in free radical concentration can speed up the degradation process.

Particulates: Some chemicals are not degraded by light because the wavelength needed is not present. The particles in air can scatter light, preventing photons from reaching molecules. This will decrease the amount of radicals formed and the amount of pesticide that is degraded by light directly.

Water Depth: Photochemical reactions decrease with water depth due to a decrease in the amount of light present.

3.3 MICROBIAL METABOLISM

Microbial metabolism of Dioxins/Furans is an important degradation process in water and soil. The process can take several steps and the end goal is to mineralize the chemical into the basic components of CO_2, H_2O, and mineral salts. Organisms, like fish, are able to metabolize compounds but can't mineralize them. There are four types of microbes: bacteria, fungi, protozoa, and algae. Bacteria and fungi are the most abundant in nature so they are the most important in biological transformation processes. Bacteria account for 65 percent of the total bio-mass in soil, so they generally account for most degradation processes. Whether aerobic or anaerobic metabolism will occur in the degradation of Dioxins/Furans will be determined by the surrounding conditions. The factors influencing microbial metabolism include:

Water Depth: at different water depths different microbes exist. Dioxins/Furans may or may not be metabolized depending on the microbes present at that depth. The rate of metabolism is slower because of the lower temperature at lower depths.

Mobility: For biodegradation to occur, Dioxins/Furans needs to be absorbed through a microbes' membrane so that the enzymes can do their work. If Dioxins/Furans are strongly bound to soil the biodegredation process cannot occur. Most microbes move more freely in water than in soil so chemicals in water will be degraded faster in water than in soil. Thus Dioxins/Furans are biodegraded to a lesser extent as they are strongly bound to the soil.

Primary/Secondary Metabolism: Primary metabolism occurs when microbes are able to live by deriving energy from the metabolism of a chemical. When a chemical is applied to soil, the microbe population increases because it is food for the microbes. When the chemical is depleted, the microbial population decreases.

Conditions: The rate of metabolism is influenced by pH, soil moisture, amount of oxygen present, temperature, and salinity.

3.4 ENVIRONMENTAL FATE OF DIOXINS

Dioxins and Furans are highly persistent compounds and have been detected in air, water, soil, sediments, animals, and foods. These POPs are released to the environment during combustion processes (e.g., municipal solid waste, medical waste, and industrial hazardous waste incineration, and fossil fuel and wood combustion); during the production, use, and disposal of certain chemicals (e.g., PCBs, chlorinated benzenes, chlorinated pesticides); during the production of bleached pulp by pulp and paper mills; and during the production and recycling of several metals (Buser et al. 1985; Czuczwa and Hites 1986a, 1986b; Oehme et al. 1987, 1989; Zook and Rappe 1994).

Dioxins generated by combustion may be transported long distances (as vapors or associated with particulates) in the atmosphere (Czuczwa and Hites 1986a, 1986b; Tysklind et al. 1993). They may eventually be deposited on soils, surface waters, or plant vegetation as a result of dry or wet deposition. Dioxins (primarily MCDD, DCDD, TrCDD) will slowly volatilize from the water column, while the more highly chlorinated Dioxins will adsorb to suspended particulate material in the water column and be transported to the sediment (Fletcher and McKay 1993; Muir et al. 1992). Dioxins deposited on soils will strongly adsorb to organic matter. Dioxins are unlikely to leach to underlying groundwater but may enter the atmosphere on soil dust particles or enter surface waters on soil particles in surface runoff. Low water solubilities and high lipophilicity indicate that Dioxins will bioconcentrate in aquatic organisms, although as a result of their binding to suspended organic matter the actual uptake by such organisms may be less than predicted. This is also true of uptake and bioconcentration by plants, although foliar deposition and adherence may be significant.

3.4.1 Atmospheric Fate

Combustion processes appear to have contributed to the ubiquity of Dioxins in the environment (Hites and Harless 1991; Tysklind et al. 1993). Dioxins have relatively long residence times in the atmosphere, and Dioxins associated with particulates become distributed over large areas (Tysklind et al. 1993). During transport in the atmosphere, Dioxins are partitioned between the vapor phase and particle-bound phase (Hites and Harless 1991). However, because of the very low vapor pressure of Dioxins, the amount present in the vapor phase generally is negligible as compared to the amount adsorbed to particulates (Paustenbach et al. 1991). The two environmental factors controlling the phase in which the congener is found are the vapor pressure and the atmospheric temperature (Hites and Harless 1991). Congeners with vapor pressure $< 10^{-8}$ mm Hg will be primarily associated with particulate matter while congeners with a vapor pressure $> 10^{-4}$ mm Hg will exist primarily in the vapor phase (Eisenreich et al. 1981). With a reported vapor pressure ranging from 7.4×10^{-10} to 3.4×10^{-5} mm Hg, 2,3,7,8-TCDD falls into the intermediate category.

The detection of Dioxins in sediments from Siskiwit Lake, Isle Royale, suggests that they can be transported to great distances in air (Czuczwa and Hites 1986a, 1986b). Similar amounts of Dioxins were also found in Lake Huron and Lake Michigan sediments, which indicates that atmospheric transport is a source of Dioxins found on these Great Lake sites (Czuczwa and Hites 1986a, 1986b; Hutzinger et al. 1985). Atmospheric deposition of TCDD to Lake Erie may contribute up to 2 percent of the annual input of TCDD to the lake (Kelly et al. 1991). Dioxins are physically removed from the atmosphere via wet deposition (scavenging by precipitation), particle dry deposition (gravitational settling of particles), and gas-phase dry deposition (sorption of Dioxins in the vapor phase onto plant surfaces) (Rippen and Wesp 1993; Welschpausch et al. 1995). Precipitation (rain, sleet, snow) is very effective in removing particle-bound Dioxins from the atmosphere (Hites and Harless 1991; Koester and Hites 1992).

Dioxins can be found in both the vapor and particle-bound phases (Eitzer and Hites 1989a; Hites and Harless 1991), with the low vapor pressure of OCDD resulting in its enrichment in the particulate phase in the atmosphere. When this particulate matter is deposited on water, OCDD-enriched sediments will result (Eitzer 1993). The less chlorinated Dioxin congeners, TCDD and PeCDD occur in higher proportion in the vapor and dissolved phases of air and rain, whereas the more chlorinated congeners, HpCDD and OCDD are associated with the particulate-bound phases (EPA 1991d). Data from one study of Dioxins in the ambient atmosphere of Bloomington, IN, found that vapor-to-particle ratios for individual Dioxins ranged from 0.01 to 30 and were dependent on the ambient temperature and the compound's vapor pressure (Eitzer and Hites 1989b). Since the less-chlorinated Dioxins have higher vapor pressures, they are found to a greater extent in the vapor phase (Eitzer and Hites 1989a). Photodegradation of the Dioxins in vapor-phase occurs and they are lost more readily than the particulate-bound Dioxins as air moves. Vapor-phase Dioxins are not likely to be removed from the atmosphere by wet or dry deposition (Atkinson 1991), although this is a primary removal process for particulate-bound Dioxins. Wet or dry deposition could result in greater concentrations of the more chlorinated Dioxins reaching soil or water surfaces and eventually sediment (EPA 1991d). All Dioxins are found to some extent in both the vapor phase and bound to particulates. At warmer temperatures (28ºC), Dioxins, particularly the MCDDs, DCDDs, TrCDDs, and TCDDs will have a greater tendency to exist in the vapor phase. At cooler temperatures (16–20ºC and < 3ºC), all Dioxins will have less propensity to exist in the vapor phase and greater tendency to adsorb to particulates. At a constant temperature, there is a positive relationship between increasing numbers of chlorine atoms on the molecule and decreased tendency to exist in the vapor phase relative to particulate adsorption (Eitzer and Hites 1989b; Paustenbach et al. 1991).

The primary transformation reaction for Dioxins in the atmosphere depends on whether the Dioxin is in the vapor or particulate phase. Vapor-phase Dioxins are not likely to undergo reactions with atmospheric ozone, nitrate, or hydroperoxy radicals; however, reactions with hydroxyl radicals may be significant, particularly for the less-chlorinated congeners (MCDD through TCDD) (Atkinson 1991). Based on the photolysis lifetime of Dioxins in solution, it is expected that vapor-phase Dioxins will also undergo photolysis in the atmosphere, although reactions with hydroxyl radicals will predominate. A half-life of 8.3 days was estimated for the gas-phase reaction of 2,3,7,8-TCDD with photochemically produced hydroxyl radicals in the atmosphere (Podoll et al. 1986). Using the gas-phase hydroxyl radical reaction rate constant of 1×10^{-11} cm^3-molecule^{-1} sec^{-1} and an average 12-hour daytime hydroxyl radical concentration of 1.5×10^6 molecules cm^{-3}, the atmospheric lifetimes of Dioxins are estimated to range from 0.5 days for MCDD to 9.6 days for OCDD, with TCDD having a lifetime of 0.8–2 days (Atkinson 1991). Particulate-bound Dioxins are removed by wet or dry deposition with an atmospheric lifetime of 10 days (Atkinson 1991) and, to a lesser extent, by photolysis. Miller et al. (1987) measured photolysis of 2,3,7,8-TCDD sorbed onto small-diameter fly ash particulates suspended in air. The results indicated that fly ash confers photostability to the adsorbed 2,3,7,8-TCDD. It was reported that little (8%) to no loss of 2,3,7,8-TCDD on the fly ash samples after 40 hours of illumination in simulated sunlight. The photodegradation of Dioxins naturally adsorbed to five fly ash samples collected from coal-fired plants, municipal incinerators, and from a hospital incinerator were studied (Koester and Hites 1992). The photodegradation experiments conducted for 2 to 6 days showed no significant degradation.

3.4.2 Aquatic Fate

Dioxins are removed from the water column by volatilization to the atmosphere to a minor extent, and by binding to particulates and sediment, or bioaccumulation in aquatic organisms to a larger extent (Fletcher and McKay 1993; Muir et al. 1992; Paustenbach et al. 1992). Henry's Law Constants ranging for Dioxins from 1.31×10^{-6} to 146×10^{-6} atm-m^3/mol (Shiu et al. 1988) indicate that volatilization from water is likely to be very slow, with the transfer rate controlled by the gas-phase resistance Lyman et al. 1982; Shiu et al. 1988). The more chlorinated congeners, Tetra-CDD, Penta-CDD, Hexa-CDD, Hepta-CDD, Octa-CDD have lower Henry's law constant values than the less chlorinated Mono-CDD, Di-CDD, Tri-CDD. Thus volatilization from the water bodies is not very significant for the Tetra-CDD through Octa-CDD congeners as compared to adsorption to particulates. With increasing chlorine number the Henry's law constants decrease as a result of the decrease in vapor pressure and water solubility (Shiu et al. 1988). Podoll calculated volatilization half-lives for 2,3,7,8-TCDD for ponds and lakes (32 days) and for rivers (16 days) (Podoll et al. 1986). The main removal mechanism for Dioxins from the water column is by sedimentation, with 70–80 percent of the Dioxins associated with the particulate phase (Muir et al. 1992). The remainder was associated with dissolved organic substances. Dioxins bound to sediment particles may be resuspended in the water column if the sediments are disturbed. This could increase both the transport and availability of the Dioxins for uptake by aquatic biota (Fletcher and McKay 1993). Generally, Dioxins strongly adsorb to soil and their vertical mobility in the terrestrial environment is low which is characterized by low vapor pressure, low aqueous solubility, and high hydrophobicity (Eduljee 1987).

Photolysis is the primary route of Dioxins disappearance in aqueous solutions (Hutzinger et al. 1985). While photolysis is a relatively slow process in water, under certain conditions Dioxins are rapidly photolyzed, when exposed to ultraviolet light at appropriate wavelength and in the presence of an organic hydrogen donor. These hydrogen donors can be expected to be present in chlorophenol pesticides either as formulation solvents (e.g., xylene or petroleum hydrocarbons), as active constituents of the formulation (e.g., the alkyl esters of 2,4-D and 2,4,5-T), or as natural organic films on soils (Crosby et al. 1973). The photolytic behavior of CDDs in an organic solvent or in a watcr-organic solvent, however, may not accurately reflect the photolytic behavior of these compounds in natural waters (Hutzinger et al. 1985). For example, Choudry and Webster (1989) reported that photolysis of 1,3,6,8-TCDD was slower in natural pond-water solutions than was predicted from studies with laboratory solutions. Conversely, Friesen et al. (1990) reported that photolysis of PeCDD and HpCDD proceeds faster in a pond or lake-water solutions than was predicted or measured in a laboratory solution. In general, however, lower chlorinated CDDs are degraded faster than higher chlorinated congeners. Chlorine atoms in the lateral positions (e.g., 2, 3, 7, 8) are also more susceptible to photolysis than are chlorine atoms in the para positions (e.g., 1, 4, 6, 9) (Choudhry and Hutzinger 1982; Crosby et al. 1973; Hutzinger et al. 1985). Photolysis half-lives for dissolved 2,3,7,8-TCDD in sunlight range from 118 hours in winter, to 51 hours in fall, to 27 hours in spring, to 21 hours in summer (Podoll et al. 1986).

3.4.3 Terrestrial Fate

The transport of Dioxins is primarily affected by adsorption. The organic carbon fraction of the soil governs the degree of adsorption of hydrophobic organic contaminants. Dioxins adsorb more strongly to soils that have higher organic carbon content than to soils that have low organic carbon content (Yousefi and Walters 1987). Because of their very less water solubilities and vapor pressures, Dioxins found below the top few mm of soil are strongly adsorbed and show little vertical migration, particularly in soil with high organic carbon content (Yanders et al. 1989). Vertical movement of Dioxins in soil may result from the saturation of sorption sites of the soil matrix, migration of organic solvents, or human or animal activity (Hutzinger et al. 1985). Des Rosiers (1989) studied the adsorption/desorption of 2,3,7,8-TCDD in contaminated soils. The K_{oc} range indicates that 2,3,7,8-TCDD is immobile in soil (Swann et al. 1983). Podoll et al. pointed out that if organic co-solvents that can solubilize 2,3,7,8-TCDD are present in the soil, the mobility of 2,3,7,8-TCDD in soil will increase (Podoll et al. 1986). In a study carried out by Orazio et al. (1992), after 15 months, only 1.5 percent of the Dioxins applied to soil surfaces had leached to a depth of 2.5 cm below the soil surface. The leaching through the soil was principally associated with carriers such as petroleum oil.

Most of the Dioxins entering surface waters are coupled with particulate and eroded soil particulates contaminated with Dioxins (Hallett and Brooksbank 1986). In the aquatic environment, significant partitioning of Dioxins may occur from the water column to sediment and suspended particulate organic matter. Dissolved Dioxins will partition to suspended solids and dissolved organic matter (detritus, humic substances) and are likely to remain sorbed once in the aquatic environment. Hallett and Brooksbank (1986) put forward that Dioxins were strongly associated with suspended sediment collected from the Niagara River on the New York-Canada border.

K_{ow} values for Dioxins are found to increase relative to increasing chlorination (range from 10^4 to 10^{12} for MCDD through OCDD). Because of these physico-chemical properties, Dioxins are expected to adsorb to embedded and suspended sediments and also result in bioaccumulation and bioconcentration in aquatic organisms. Although as a result of their strong binding to suspended organic matter, actual uptake by these organisms may be less than predicted. Dioxins may be transported long distances in the atmosphere. Dioxins deposited on soils will strongly adsorb to organic matter. They are unlikely to leach to underlying groundwater, but may enter the atmosphere on soil or dust particles or enter surface water in runoff.

Degradation

Photolysis of 2,3,7,8-TCDD on soils is comparatively a slow process than in an aqueous media (Kieatiwong et al. 1990). 2,3,7,8-TCDD applied to soil or a solid surface seems to be extremely resistant to the action of sunlight and decomposes very slowly (Plimmer et al. 1973). A methanol solution of 2,3,7,8-TCDD (2.4 ppm) when applied to glass plates coated with soil and illuminated for 96 hrs with a fluorescent UV lamp remained unchanged (Plimmer et al. 1973). McPeters and Overcash studied the effect of organic solvents on degree of photolysis of TCDD. They found that when organic solvents are added to the soil, the degree of photolysis was enhanced. Use of a solvent mixture of tetradecane and 1-butanol to TCDD-treated soil, combined with exposure to sunlight, resulted in 61–85 percent photodegradation of TCDD after 60 days. Principally the solvent was effective in transporting TCDD from deeper in the soil column (60 cm) to the soil surface via evaporation where photodegradation could occur. TCDD concentrations at 60 cm decreased from 23.8 ng/g to 7.1 ng/g after 60 days (McPeters and Overcash 1993).

Photolysis of OCDD (10 mg/kg) on soils was found to produce lower chlorinated CDDs, notably 2,3,7,8-TCDD, 1,2,3,7,8-PeCDD, three HxCDD isomers substituted at the 2,3,7,8-positions, and 1,2,3,4,6,7,8-HpCDD. Photolysis of OCDD occurred in mean soil depths between 0.06 and 0.13 mm (Miller et al. 1989). After five days of irradiation approximately 30–45 percent of OCDD was lost; no further significant loss of OCDD was observed following 10 additional days of irradiation. Photolysis of OCDD occurred at shallow soil depths and the conversion of OCDD to the more toxic isomers, TCDD, PeCDD, and HxCDD homologues was small (0.5–1%) compared with the photodechlorination to HpCDD (67%) (Miller et al. 1989).

The loss of 2,3,7,8-TCDD in contaminated soil was studied under natural conditions in experimental plots at the Dioxin Research Facility, Times Beach, Missouri (Yanders et al. 1989). The 2,3,7,8-TCDD concentration profiles of sample cores taken at Times Beach in 1988 were virtually the same as those in cores taken in 1984. In the four years of the study, the loss of 2,3,7,8-TCDD due to photolysis at Times Beach was minimal (Yanders et al. 1989). Estimates of the half-life of TCDD on the soil surface range from 9 to 15 years, whereas the half-life in subsurface soil may range from 25 to 100 years (Paustenbach et al. 1992).

3.5 ENVIRONMENTAL FATE OF FURANS

3.5.1 Atmospheric Fate

Furans are present in the atmosphere both in the vapor and particulate phase (Hites 1990). With increasing temperature the ratio of the vapor to particulate phase Furans in air increases. The ratio in Bloomington, Indiana was as high as 2 during the warm summer months and < 0.5 in the winter. However, the distribution of Furans between the vapor and particulate phase depends on the amount and nature of the particulate matter in the atmosphere and the temperature (Hites 1990). The vapor to particle ratio is also different for the different congeners. In the air, a higher proportion of tetra-CDF congeners are present in the vapor phase, whereas hepta-CDF and octa-CDF congeners are found predominantly in the particulate phase (Hites 1990). The transport of atmospheric Furans to soil and water occurs by dry and wet deposition.

The loss of vapor phase Furans by reactions with HO_2 radicals, NO_3 radicals and ozone has been estimated to be of negligible importance in the troposphere (Atkinson 1991). The estimated rate constants for the reactions of vapor phase Furans with OH radicals are as follows (10–12 cm^3/molecule-sec): tetra-CDFs, 1.4–8.3; penta-CDFs, 1.0–4.3; hexa-CDFs, 0.74–2.6; hepta-CDFs, 0.53–0.92; and octa-CDFs, 0.39. Using a 12-hour average daytime OH radical concentration of $1.5 \times 106/cm^3$, the estimated tropospheric lifetimes of tetra-, penta-, hexa-, hepta-, and octaCDF are 1.9–11, 3.6–15, 5.9–22, 17–31, and 39 days, respectively. The vapor phase reaction of Furans with OH radicals is the dominant loss process and this loss process is more important for the lower, than the higher, chlorinated congeners, because the lifetimes due to this reaction are shorter for lower chlorinated congeners and the vapor phase concentrations of lower chlorinated congeners are higher. Based on the available information, the reactions of OH radicals with particulate phase Furans are insignificant and the principal air removal mechanism for Furans is wet and dry deposition. Photodegradation of Furans bound to atmospheric particles is not a significant process in removing these compounds from air (Koester and Hites 1992). No data regarding vapor phase photolysis of Furans were located. In the absence of data, the half-lives of these compounds in the vapor phase have been estimated from aqueous phase photolysis data and it was concluded that photolysis is relatively unimportant, even when compared to reaction with hydroxyl radicals (with the possible exception of 1,3,6,8-tetra-CDF) (Atkinson 1991).

3.5.2 Aquatic Fate

The volatilization of Furans from water, depends on their Henry's Law Constants ($< 1.48 \times 10^{-5}$ atm-m^3/mol). The rate of volatilization of these Furans thus is slow and is controlled by slow diffusion through air (Thomas 1981). The volatilization rates are moreover declined as the Furans present in water are mostly in the adsorbed states. However, no experimental data pertaining to the volatilization of Furans from water have been reported. The adsorption of Furans to suspended solids and sediment in water depends on their K_{oc} values. The estimated log K_{oc} values for 2,3,7,8-TCDD and octa-CDF are 5.61 and 8.57, respectively, due to which these compounds strongly adsorb to suspended Furans in sediment and not in water. The Furans concentration in water is so low that they are rarely measured. Thus, sediments are the ultimate environmental sinks for Furans (Czuczwa and Hites 1986b).

Degradation

The loss of Furans in water by abiotic processes such as hydrolysis and oxidation is not likely to be significant (EPA 1986a). The photolysis of Furans occurs in hydrogen donating solvents and is faster in methanol than in hexane. Photolysis in these solvents however, proceeds with quick dechlorination and eventual formation of unidentified resinous polymeric products (Hutzinger et al. 1973). At shorter wavelengths (254 nm), photolysis may proceed at a much faster rate than are available from sunlight (> 290 nm). Muto and Takizawa concluded that the higher chlorinated congeners of Furans photodegrade faster than lower chlorinated congeners and rate of photolysis in hexane is faster for Furans than Dioxins (Muto and Takizawa 1991). It has been reported that the rates of photolysis of 2,3,7,8-substituted congeners in solution are faster compared to the rates of non-2,3,7,8-substituted congeners (Tysklind and Rappe 1991). The photolysis of octaCDF in dioxane under xenon lamp, resulted in formation of hexa- and pentaCDFs as major products, with small amounts of hepta- and tetra-CDFs (Koshioka et al. 1987).

The rate of photolysis is much slower in the sorbed phase compared to solution phase (the estimated lifetimes data of Atkinson (1991) is based on solution phase photolysis) (Tysklind and Rappe 1991). If the photolysis rates of Furans are assumed to be faster than Dioxins (Muto and Takizawa 1991), the photolysis lifetimes of Furans are expected to be shorter than those for Dioxins. However, the persistence of Furans in natural water (based on a half-life of 1 year for Dioxins in a model aquatic ecosystem) (EPA 1986a), contradicts the estimated photolytic lifetimes in natural water. This is possibly due to the fact that Dioxins/Furans in natural water are present predominantly in particulate-sorbed phase.

No data in the literature indicate that biodegradation of Furans in water is significant. Biodegradation studies in sediments of a lake water indicated that 2,3,7,8-TCDD is resistant to biodegradation (EPA 1986a). Therefore, biodegradation of Furans in aquatic medium may also be insignificant.

3.5.3 Terrestrial Fate

The transport of Furans from soil to air is possible via volatilization and by wind blown dusts. The very low vapor pressures and high soil sorption coefficients of those Furans for which data are available indicate that volatilization of these compounds from soil is insignificant (Hutzinger et al. 1985b). As no loss of 2,3,7,8-TCDD, from the contaminated soil at Times Beach, Missouri, occurred in four years (Yanders et al. 1989), it strongly suggests that volatilization for Furans is also insignificant. No evidence of appreciable loss of Furans due to volatilization was found in contaminated soils during a period of eight years (Hagenmaier et al. 1992). Furans may be transported from soil to water via leaching and runoff. Soil leaching experiments conducted by Carsch et al. (1986) indicated that Furans remain strongly adsorbed even in sandy soil and leaching of these compounds from soil by rainwater is not significant. Hagenmaier et al. (1992) found that the vertical movement of Furans was very slow and > 90 percent of Furans were found in the top 10 cm after three years. Therefore, transport of Furans from landfill soil to adjacent land or surface water by runoff water is more likely than leaching. Leaching or vertical movement of Furans in soil can occur under special conditions, such as saturation of the sorption sites of the soil matrix, presence of organic solvents in the soil facilitating co-solvent action, cracks in the soil, or burrowing activity of animals (Hagenmaier et al. 1992; Hutzinger et al. 1985b).

Degradation

The photodegradation of thin film of fly ash bound Furans under sunlight was much slower than solution phase photolysis (Hutzinger et al. 1973; Tysklind and Rappe 1991). Direct evidence of sunlight initiated photolysis of Furans in soil was not located. Given the fact that sunlight cannot penetrate beyond the surface layer of soil and the lack of photolysis of Furans adsorbed to fly ash (Koester and Hites 1992; Tysklind and Rappe 1991), the photolysis of Furans in soil and sediment may not be significant. It may be significant for airborne particles. No direct evidence suggesting significant biodegradation of Furans in soil and sediments are reported in the literature. The lack of biodegradation of Dioxins in soil and sediments (although a few microbes degraded 2,3,7,8-TCDD at a slow rate) (EPA 1986a) and the lack of evidence for any degradation of Furans in dated lake sediments (Czuczwa et al. 1985; Czuczwa and Hites 1986) indirectly suggest that biodegradation of Furans in soil or sediments is not significant.

3.6 CONTAMINATION TRENDS OF DIOXINS

Dioxins in very small concentration are ubiquitous in the environment and it is likely that many of the primary sources are not yet known. They have been found worldwide, even in remote areas. Dioxins adsorbed on airborne particulate or in industrial effluent are deposited on soil and eventually bind to other organic substances and bottom sediments in lakes and rivers. Various studies at informal e-waste recycling sites (EWRSs) in Asian developing countries found the soil contamination levels of Dioxin/Furans from tens to ten thousand picogram TCDD-equivalents (TEQ) per gram (Nguyen Minh Tue et al. 2013). Although these are encountered in both the vapor and particulate phases, it has been suggested that ingestion results in 90 percent of the human exposure (Charnley and Doull 2005; Taioli et al. 2005). Atmospheric Dioxins deposit on vegetation which form animal feed. Humans then ingest crops, fish, meat and dairy products and thus accrue, due to such soil and water pollution, may significantly as atmospheric exposure and could substantially increase total risk. Dietary intake is the most important source of exposure to Dioxins for the general population. This pathway contributes more than 90 percent of the daily intake for the general population of Korea (Seung-Kyu Kim et al. 2007). Dioxin/Furans emission inventories have utilized an integrated approach to estimate emissions to the air, land, and soil. Direct releases to water have only been qualitatively estimated (Quaβ et al. 2000a; Wenborn et al. 1999). The inventories suggest that emissions of Dioxins/Furans to land, including landfills, may be higher than emissions into the air (Hansen and Hansen 2003; Haynes and Marnane 2000; Wenborn et al. 1999; Dyke et al. 1997). There are many chlorinated chemicals that may contain Dioxin/Furans impurities, which makes it extremely difficult to estimate the historical and current emissions.

3.6.1 Levels in Air

Indoor household dust samples gathered by a vacuum cleaner from rooms with furniture treated with a wood-preserving formulation were analyzed for Dioxins (Christmann et al. 1989b). The wood-preserving formulation contained PCP, which is known to be contaminated with Dioxins, particularly HxCDD, HpCDD, and OCDD. OCDD was the most abundant congener found in the dust samples at an average concentration of 191 μg/kg (ppb), followed by HpCDD (20 μg/kg), HxCDD (2.5 μg/kg), PeCDD (0.9 μg/kg), and TCDD (0.2 μg/kg) (Christmann et al. 1989b). Indoor air concentrations of Dioxins/Furans were measured in kindergarten classrooms in West Germany to evaluate releases from wood preservatives (PCP) that may have been used in building materials (Päpke et al. 1989a). Measured indoor air concentrations of total Dioxins/Furans ranged from 1.46 to 4.27 pg/m^3, while measured outdoor air concentrations ranged from 0.61 to 78.97 pg/m^3. The 2,3,7,8-substituted congeners predominated with mean concentrations as: OCDD (131.5 pg/m^3), 1,2,3,4,6,7,8-HpCDD (77 pg/m^3), 1,2,3,4,6,7,8-HpCDF (51 pg/m^3), and OCDF (25.3 pg/m^3).

Measured indoor air samples collected in an office building in Binghamton, New York, 2 years after a fire in an electrical transformer that contained PCBs and tri- and tetra-chlorobenzenes were reported to have 2,3,7,8-TCDD concentrations ranging from 0.23 to 0.47 pg/m^3 (Smith et al. 1986a). The 2,3,7,8-TCDD isomer constituted 23–30 percent of the 1.0–1.3 pg/m^3 total TCDDs (LOD) = 0.003 pg/m^3 (Smith et al. 1986a). Reed et al. (1990) measured the background levels of Dioxins in air in a semirural location in Elk River, Minnesota, located 25 miles northwest of Minneapolis-St. Paul. No major industrial or commercial activity occurred in the area at the time of the study. Ambient air samples were collected in the winter and summer of 1988. 2,3,7,8-TCDD was not detected in any of the ambient air samples taken in the summer (LOD for 2,3,7,8-TCDD ranged from 0.005 to 0.065 pg/m^3). 2,3,7,8-TCDD was noted in a wintertime sample at concentrations of 0.015 pg/m^3 and 0.019 pg/m^3. Detection limits in the remaining wintertime samples for 2,3,7,8-TCDD ranged from 0.005 to 0.01 pg/m^3. Winter time Dioxin concentrations were greater than those observed for summer time. The winter time Dioxins congener profile showed increasing concentrations with increasing chlorine substitutions. Average ambient air concentrations in winter time of HpCDD and OCDD ranged from approximately 0.5 to 4.1 pg/m^3 and 0.74 to 8.2 pg/m^3, respectively (Reed et al. 1990). Average ambient air concentrations of HpCDD and OCDD in summer time ranged from approximately 0.204 to 0.246 pg/m^3 and 0.018 to 0.024 pg/m^3, respectively (Reed et al. 1990). In general, the highly chlorinated congeners were found to present at higher concentrations than the less chlorinated congeners.

Dioxins have been found in urban air particulates from Washington, D.C., and St. Louis, Missouri; OCDD was the predominant congener at concentrations of 200 ppb and 170 ppb for Washington and St. Louis, respectively (Czuczwa and Hites 1986a). Combustion of municipal and chemical wastes was the most likely source of these compounds. Dioxins were detected in air samples from Albany, Binghamton, Utica, and Niagara Falls, NY (Smith et al. 1990b). Concentrations of Dioxin congener groups for all four cities were as follows: total TCDD, not detected ($<$ 0.21 pg/m^3); total PeCDD, $<$ 0.04–0.62 pg/m^3; total HxCDD, 0.10–2.4 pg/m^3; total HpCDD, $<$ 0.21–4.4 pg/m^3; and OCDD $<$ 0.54–4.6 pg/m^3 (Smith et al. 1990b). In 1988–89, total Dioxins measured downwind from an industrial source in Niagara Falls, NY, ranged from 0.3 pg/m^3 to 133 pg/m^3 and were approximately 2.5 times higher than upwind concentrations (Smith et al. 1990b). Between 1986 and 1990, total Dioxin concentrations averaged 2.3 pg/m^3, of which 65 percent was OCDD (Smith et al. 1992).

An extensive multi-year monitoring program for Dioxins/Furans was conducted at eight sampling locations in the Los Angeles South Coast Air Basin from 1987 to 1989 (Hunt and Maisel 1992). The monitoring network, which monitored for both vapor and particulates, included several sites situated in residential areas as well as sites in the vicinity of suspected CDD/CDF sources. Monitoring results indicated that 2,3,7,8-TCDD was virtually undetected. The most commonly detected 2,3,7,8-substituted congener was OCDD followed by 1,2,3,4,6,7,8-HpCDD. The predominance of 1,2,3,4,6,7,8-HpCDD as the most persistent congener is associated with stationary or mobile combustion source emissions. 1,2,3,4,6,7,8-HpCDD was found at all seven sampling sessions at a mean concentration of 1.140 pg/m^3. OCDD also was found at all seven sampling sessions at a mean concentration of 2.883 pg/m^3. The mean total TCDD concentration was 0.114 pg/m^3 and was measured during only three sampling sessions (Hunt and Maisel 1992).

A mixture of Dioxins (TCDD, PeCDD, HxCDD, HpCDD, OCDD) has been found in emissions from the combustion of various sources, including municipal incinerators, power plants, wood burning, house-heating systems, and petroleum-refining operations (Chiu et al. 1983; Clement et al. 1985; Thoma 1988; Thompson et al. 1990). Dioxins were found in stack and fly ash samples from the following combustion sources (ranges given): municipal incinerator, 8 ppb (OCDD) to 390 ppb (HxCDD) (TCDD was found at 10 ppb); open-air burning of PCP-treated wood, 2 ppb (TCDD) to 187 ppb (OCDD); coal-fired power plant, 1 ppb (TCDD) to 6 ppb (PeCDD and HxCDD); hydroelectric power plant, 0.5 ppb (OCDD) to 5.2 ppb (TCDD) (Chiu et al. 1983); and petroleum refining, 0.8 (OCDD) to 3.4 ng/m^3 (PeCDD) (Thompson et al. 1990). Samples of ash from wood-burning stoves, a fireplace, and open-air wood burning contained detectable levels of CDDs ranging from 0.3 to 33 ppb (Clement et al. 1985). The open-air burning ash contained the highest total CDD concentration (33 ppb), with HpCDD being the most abundant homologue (11 ppb). The total CDD concentrations in four samples from wood-burning stoves ranged from 0.3 to 15 ppb, with the relative amounts of each homologue varying for each sample. Ash samples from the fireplace contained total CDD concentrations ranging from 3.1 to 5.4 ppb, with HxCDD (0.3–1.7 ppb) and OCDD (0.4–3.1 ppb) being the most abundant homologues present (Clement et al. 1985). TCDD was present in ash samples from open-air burning (0.8 ppb) and was detected in ash from the fireplace. 2,3,7,8-TCDD has been detected in air samples (concentrations unspecified) collected at 9 of the 91 NPL hazardous waste sites where it has been detected in some environmental media (HazDat 1998). CDDs have been detected in air samples (concentrations unspecified) collected at 10 of the 126 NPL sites where they have been detected in some environmental media.

Thus, most of the measurements of CDDs in air tend to be very close to current detection limits. Dioxins are found at the greatest concentrations in urban air with OCDD being the most prevalent congener (up to 0.100 ppq). Concentrations of all Dioxins are highest in the air near industrial areas. Rural areas usually have very low or unquantifiable levels of all CDDs. In urban and suburban areas, concentrations of CDDs may be greater during colder months of the year when furnaces and wood stoves are used for home heating.

3.6.2 Levels in Water

The environmental levels of Dioxins in water are given in Tables 3.1 and 3.2. Significant emissions to water have been attributed to accidental fires and treatment and disposal practices of municipal solid waste, combustion at old incineration plants in particular (Marek Zielinski et al. 2014; Quab et al. 2000a; Wenborn et al. 1999; Dyke et al. 1997).

In Latvia, catchment of Baltic Sea, potentially Dioxins/Furans releasing industrial processes include municipal and hospital waste incineration and illegal processing of non-ferrous scrap and iron and steel production (Jensen 2003; Quab et al. 2000a). A release estimate of 0.28 g to 0.6 g Nordic-TEQ/year from the Swedish chloralkali plants to water was reported in 1993 (UNEP Chemicals 1999). The application of chlorine gas for pulp bleaching used to be a significant source of PCDD/Fs before its replacement with less hazardous forms of chlorine. In the 1980's, the releases of 2,3,7,8-TCDF from a single pulp mill were claimed to be as high as 60 g WHO-TEQ/year (MacDonald et al. 1998). The estimated releases from pulp and paper industry in 1994 were almost the same in Finland and Sweden, 1.2 to 83 gI-TEQ/year to land and 1.5 to 5.3 g I-TEQ/year to water (lowest to highest estimates) (Wenborn et al. 1999).

Table 3.1: Environmental levels of Dioxins in water.

Dioxin	**Conc. Range (ppq)**	**Conc. Mean**	**Location**	**Sampling Year**
2,3,7,8-TCDD	ND (0.7)	NA	Lockport, New York raw surface drinking water	88
	ND (0.02–0.024)	NA	Eman River, Sweden raw surface drinking water	NR
TCDDs	ND-40	2.70	Ontario, Canada raw surface drinking water	83–89
	ND (0.4–2.6)	NA	New York State treated surface drinking water	86–88
	1.7	1.7	Lockport, New York raw surface drinking water	88
	0.05–0.084	0.067	Eman River, Sweden raw surface drinking water	NR
1,2,3,7,8-PeCDD	ND (1.0)	NA	Lockport, New York raw surface drinking water	88
	ND (0.025–0.039)	NA	Eman River, raw surface drinking water Sweden	NR
PeCDDs	ND (1.0)	NA	Lockport, York New raw surface drinking water	88
	ND (1.2–7.4)	NA	New York State treated surface drinking water	86–88
	0.067–0.12	0.094	Eman River, Sweden raw surface drinking water	NR
1,2,3,4,7,8-HxCDD	ND (1.8)	NA	Lockport, New York raw surface drinking water	88
	ND-0.054	0.027	Eman River, Sweden raw surface drinking water	NR
1,2,3,6,7,8-HxCDD	ND (1.5)	NA	Lockport, New York, raw surface drinking water	88
	ND-0.12	0.06	Eman River, Sweden raw surface drinking water	NR
1,2,3,7,8,9-HxCDD	ND (1.5)	NA	Lockport, New York raw surface drinking water	88
	ND-0.075	0.038	Eman River, Sweden raw surface drinking water	NR
HxCDDs	ND (1.5)	NA	Lockport, New York raw surface drinking water	88
	ND (0.4–4.7)	NA	New York State treated surface drinking water	86–88
	0.13–0.67	0.4	Eman River, Sweden raw surface drinking water	NR
1,2,3,4,6,7,8-HpCDD	ND (2.8)	NA	Lockport, New York raw surface drinking water	88
	0.15–0.30	0.22	Eman River, Sweden raw surface drinking water	NR

Table 3.1 contd. ...

...Table 3.1 contd.

Dioxin	Conc. Range (ppq)	Conc. Mean	Location	Sampling Year
HDioxins	ND (2.8)	NA	Lockport, New York raw surface drinking water	88
	ND (0.4–6.8)	NA	New York State treated surface drinking water	86–88
	0.17–0.64	0.40	Eman River, Sweden raw surface drinking water	NR
OCDD	ND-175	10.6	Ontario, Canada raw surface drinking water	83–89
	ND-46	3.16	Ontario, Canada treated surface drinking water	83–89

NA = not applicable;
ND = non-detected (limit of detection);
NR = not reported;
Sources: 1. Jobb et al. 1990; 2. Meyer et al. 1989; 3. Rappe et al. 1989b

Table 3.2: Dioxins and Furans emissions to water from the EU countries in 1994.

Source	Release to Water Potential (pg I –TEQ)
Production and use of chemicals	
Pesticide production	High
Pesticide use	High
Chemical production	High
Paper and pulp production	Medium (8–27)
Textile treatment	Medium
Combustion	
Accidental fires	High
Combustion of wood	
Waste treatment and disposal	
Incineration of MSW	
Disposal of MSW to landfill	High
Incineration of industrial waste	Medium (0.3–16)
Waste oil disposal	Medium (1.8–7.9)
Metal industry	
Secondary Pb production	
Secondary Cu production	
Electric furnace steel plant	
Secondary Al production	
Metal industry	Medium

Adapted from: Wenborn 1999

Paustenbach et al. (1996) reported that stormwater runoff from 15 sites in the San Francisco area contained PCDD/F I-TEQ at levels ranging from 0.01 to 65 pg/L; however, most of the samples contained less than 15 pg/L Dioxins/Furans.

A bleached sulfate integrated pulp and paper mill producing printing and writing paper from mixed tropical hardwood and bamboo was studied. The mill uses a "conventional bleaching sequence", C-E-H_1-H_2, with an average chlorine consumption of 80 kg per ton of air-dried pulp (ADP). The content of Furans and Dioxins in the bleaching filtrate in term of the Nordic Toxicity Equivalent (N-TEQ) was 33.5, 1.15, 0.56 and 0.014 pg/L for the E-, C-, H_1 and H_2-bleaching stages, respectively. The corresponding Furans and Dioxins loads in ng/tons ADP were in the same ranking, that is, 670, 69, 11.2, and 0.28, respectively (*www.environmental-expert.com/events/r2000/r2000.htm*). A study carried out by Wang suspected TCDD in wastewaters from pesticides industries manufacturing raw materials such as 2,4,5-TCP and 1,2,4,5-tetra chlorobenzene which are characteristic precursors of TCDD. TCDD level as high as 111 mg/L was found in drums of waste from pesticides 2,4,5-T (Wang 2004). Dyke and Amendola (2007) carried out intensive study of on-site releases of Dioxins and Furans from 19 chemical production facilities in the US producing, or using, large volumes of chlorine. Releases to water varied widely from essentially non-detectable amounts to more than 5 g of PCDD/F (as I-TEQ) annually.

During 1986, a survey of 20 community water systems throughout the state of New York was conducted to evaluate CDD/CDF concentrations (Meyer et al. 1989). The sampling sites selected were representative of major surface water sources in the state used to obtain drinking water. The sites included surface water sources receiving industrial discharges and those known to contain CDD-contaminated fish, as well as water sources from more remote areas. Raw water sampled at the Lockport, NY, facility contained concentrations of TCDDs (1.7 ppq) as well as concentrations of TCDFs to OCDFs (18, 27, 85, 210, and 230 ppq, respectively). These data show that the CDF congener group concentrations increased with increasing chlorine numbers. TCDFs were also detected in finished water sampled at the Lockport facility (duplicate samples contained 2.1 and 2.6 ppq). Except for a trace of OCDF detected at one other location, no other Dioxins/Furans were detected in finished water at any of the other 19 community water systems surveyed.

2,3,7,8-TCDD has been detected in surface water samples collected at nine sites of the 91 NPL hazardous waste sites where it has been detected in some environmental media (HazDat 1998). Dioxins have been detected in surface water samples collected at 14 of the 126 NPL sites where they have been detected in some environmental media. TCDDs, PeCDDs, HxCDDs, HDioxins, and OCDD have been detected in surface water samples at 10, 1, 4, 4, and 6 sites of the 105, 34, 43, 49, and 53 NPL sites, respectively, where they have been detected in some environmental media. Groundwater in the vicinity of an abandoned wood treatment facility was sampled from monitoring wells constructed at depths ranging from 6.1 to 30.5 meters and was analyzed for Dioxins in January 1984 (Pereira et al. 1985). Concentrations of HxCDD, HpCDD, and OCDD in groundwater samples taken from wells at a depth of 6.1 meters were 61 ppt, 1,500 ppt, and 3,900 ppt, respectively. The authors noted that the high concentrations of Dioxins in the sample from a depth of 6.1 meters probably resulted from the presence of micro-emulsions of oil that were difficult to separate from the sample. Groundwater samples collected from deeper wells (12.2–30.5 meters) contained HxCDD, HpCDD, and OCDD at concentration ranges of not detected to 21 ppt, ND to 34 ppt, and ND to 539 ppt, respectively (Pereira et al. 1985). 2,3,7,8-TCDD were detected in groundwater samples collected at 15 sites of the 91 NPL hazardous waste sites where it has been detected in some environmental media (HazDat 1998). Dioxins have been detected in groundwater samples collected at 32 of the 126 NPL sites where they have been detected in some environmental media. TCDDs, PeCDDs, HxCDDs, HDioxins, and OCDD have been detected in groundwater samples at 21, 3, 10, 14, and 16 sites of the 105, 34, 43, 49, and 53 NPL sites, respectively, where they have been detected in some environmental media.

Dioxins are rarely detected in drinking water at ppq levels or higher. Raw water samples generally have higher concentrations of Dioxins (9–175 ppq) than finished drinking water samples (19–46 ppq) because conventional water treatment processes remove the Dioxins along with the particulates from raw water. Dioxins have been detected in treated effluent samples collected at pulp and paper mills using the bleach kraft or sulfite bleaching process. In groundwater samples collected near industrial sites, Dioxins have been detected at concentrations up to 3,900 ppt.

3.6.3 Levels in Soil and Sediment

The occurrence of exceptionally high levels of Dioxins in soils and sediments is usually attributable to the manufacture and use of chlorinated chemicals. For example, in the 1970's and 1980's, releases from the manufacture of 2,4,5-TCP and 2,4,5-T contaminated soil in Seveso (Italy), Hamburg (Germany), and New Jersey (USA) (Exner 1987). Dioxins were found in at least 110 of 1428 National Priorities List sites identified by the USEPA (Pohl et al. 1998). In Japan, the annual emissions of Dioxins from the herbicides pentachlorophenol and chloronitrofen (2,4,6-trichlorophenyl-4′-nitrophenyl ether) to agricultural land were more than 10 kg I-TEQ/year during the 1960's and early 1970's (Masunaga and Nakanishi 1999). In Poland, the production of 2,4,5-T, sodium pentachlorophenate, pentachlorophenol and 2,4,5-trichlorophenol was started in 1969 in Neratovice. Dumping of chemical wastes around the production facility led to soil contamination with 2,3,7,8-TCDD levels up to 29800 pg/g were reported (Holoubek et al. 2000). Severe contamination of sawmill sites were reported in Arkhangelsk district in Northern Russia, with concentrations up to 2,24,000 pg I-TEQ/g d.w. (Troyanskaya et al. 2003). Direct inputs of Dioxins/Furans to land typically lead to local contamination in relatively small areas. Volatilization, erosion, leakages to groundwater, and transportation with surface runoff may result in a wider distribution in terrestrial, atmospheric, and aqueous compartments. Paepke et al. (2004) reported environmental levels measured in different countries for soil and for sediment in Europe. For comparison, these data are given in Tables 3.3 and 3.4. Background values for Dioxins in sediments and in soil samples in the European Union were found between 1 and 20 ng TEQ/kg. The typical congener distribution was similar to the pattern found in deposited particulate matter/ambient air samples: increasing concentration with increasing number of chlorines for Dioxins and decreasing concentration with increasing number of chlorines for Furans.

Dioxins and Furans contaminated sediments have been found in the vicinity of many industrial sites all around the globe. The highest reported quantity emanates from magnesium production in Norway, where concentrations up to 18000 pg Nordic-TEQ/g d.w. were analyzed in the sediments of Frierfjord, giving rise to a total load of approximately 50 to 100 kg I-TEQ (Knutzen and Oehme 1989). The manufacture and use of chlorine and chlorinated chemicals were considered as the major source of severe sediment contamination in Arkhangelsk district in Russia (Troyanskaya et al. 2003), Tokyo Bay in Japan (Yao et al. 2002; Sakurai et al. 2000), Lake Baikal in Hungary (Mamontov et al. 2000); Newark Bay (Passaic River) in New Jersey, USA

Table 3.3: PCDD/PCDF concentrations (ng TEQ/kg dry matter) in sediment from the EU states.

EU States	Background	Urban	Contaminated
Finland	0.7–100	–	80000
Germany	1.2–19	12–73	> 1500
Italy	0.1–10	0.5–23	570
Luxembourg	–	2.4–16	–
Netherlands	1–10	–	4000
Spain	–	–	0.2–57
Sweden	0.8–207	–	1692
U.K.	–	2–123	7410

Source: (Paepke et al. 2004)

Table 3.4: PCDD/PCDF concentrations (ng TEQ/kg dry matter) in soil from the EU states.

EU States	Any Type	Pasture/Rural	Contaminated
Austria	–	1.6–14	332
Belgium	2.7–8.9	2.1–2.7	–
Finland	–	–	85000
Germany	0.1–42	0.1–30	38000
Greece	2–45		1144
Ireland	0.2–8.6	0.8–13	–
Italy	0.1	0.1–43	–
Luxembourg	1.8–20	1.4	–
Netherlands	2–55	2.2–17	98000
Spain	0.6–8.4	0.1–8.4	–
Sweden	–	0.1	11446
U.K.	0.8–87	0.8–20	1585

Source: (Paepke et al. 2004)

(Bopp et al. 1991), Venice Lagoon in Italy (Green et al. 1999), Hamburg harbour in Germany (Götz et al. 1990), and Thunder Bay in Ontario, Canada (McKee et al. 1990).

The concentrations of Dioxins and Furans were determined in soils from dumping sites in the Philippines, Cambodia, India, and Vietnam (Nguyen et al. 2003). Residue concentrations of Dioxins in dumping site soils were apparently greater than in soils collected in agricultural or urban areas far away from dumping sites. Elevated fluxes of Dioxins to soils in dumping sites were encountered in the Philippines, Cambodia, India, and Vietnam-Hanoi, and these levels were higher than those reported for other countries. Considerable loading rates of Dioxins and Furans ranging from 20 to 3900 mg/yr (0.12–35 mg TEQ/yr) were observed in the dumping sites of these countries. Dioxins and Furans concentrations in some soil samples from the Philippines, Cambodia, India, and Vietnam-Hanoi exceeded environmental guideline values.

In 1994, the mean concentration in 16 sediment samples from the estuary of the Warnow River, Bodden area, Mecklenburg Bight, Oder Bight, and Arkona Basin was 8.53 ± 5.40 pg TEQ/g d.w. Concentrations of Dioxins were typically much higher than those of Furans, and the predominating congener was OCDD. Dannenberg and Lerz (1999) found 13–2991 pg/g d.w. of Dioxins and 2.5–820 pg/g d.w. of Furans in sediment samples collected from approximately the same areas. Unidentified sources have also been reported of Dioxins contamination in soil and sediment samples from different regions and of different ages. For example, Hashimoto et al. (1995) found a Dioxin concentration as high as 2000 pg/g d.w. in sediment slices that were one to ten million years old. The Dioxins/Furans concentration analyzed from Australian sediment cores that represented the last 350 years was 2500 pg/g d.w. at its highest (Gaus et al. 2001). In rural lake sediments from the 1940s, a concentration of 100 pg/g d.w. was analysed in the UK (Green et al. 2001). Müller et al. (2002) reported Dioxins/Furans concentrations ranging from 3 to 33 pg I-TEQ/g d.w. in Hong Kong sediments and Mai Po marshes. One characteristic of all these samples was a congener profile that was dominated by highly chlorinated Dioxins. The profile was different from known industrial sources.

The soil samples collected in the vicinity of the incinerator of Hsinchu during winter 2001 showed concentration ranging from 1.29 pg-TEQ/g d.m. to 5.02 pg-TEQ/g d.m. with an average value of 3.03 pg-TEQ/g d.m. (Pai-Sheng et al. 2003). Whereas the soil sample for surveying the Dioxins/Furans level collected in summer 2001 showed a range from 0.52 to 1.97 pg-TEQ/g d.m. with an average value of 1.19 pg-TEQ/g d.m. Not much variation in the Dioxins/Furans concentrations in soil sample was observed even with the seasonal variation and site of sampling variation. 2,3,7,8-TCDD was detected in 35 residential surface soil samples at Paritutu, New Plymouth (0–75 mm) at concentrations ranging from 0.71 to 92 ng/kg. The majority of the results (31 out of 35) had TCDD concentrations less than 10 ng/kg and 23 results were less than 5 ng/kg. Based on a correlation between 2,3,7,8-TCDD and TEQ from the full Dioxin profile analysis, the TEQ range for the surface samples was estimated to be 2.7 to 99 ng/kg. The concentration of 2,3,7,8-TCDD in deeper samples (75–150 mm) ranged from 0.71 to 17 ng/kg. Concentrations of2,3,7,8-TCDD in six soil samples taken from gardens ranged from 2 to 7.3 ng/kg (Table 3.4) (Pattle Delamore Partners Ltd. 2004). In an earlier study, soil concentrations of 2,3,7,8-TCDD were measured in industrialized areas of a group of mid-western and mid-Atlantic states (Illinois, Michigan, New York, Ohio, Pennsylvania, Tennessee, Virginia, and West Virginia) (Table 3.5) (Nestrick et al. 1986). Many of the samples were taken within one mile of major steel, automotive, or chemical manufacturing facilities or of municipal solid-waste incinerators. The data show that in these typical industrialized areas, 2,3,7,8-TCDD soil concentrations were below 0.01 ppb (range, ND–9.4 ppt). The widespread

Table 3.5: Concentration of 2,3,7,8-TCDD in surface and garden soil samples of Paritutu, New Plymouth.

Site No.	2,3,7,8-TCDD (ng/kg dry weight)		Garden Type
	Surface (0-75 mm)	Garden	
03	5.8	4.5	Vegetable
04	7.4	4.9	Ornamental, raised, old
09	17	2.8	Terraced, ornamental, from filled area
12	2.9	2	Ornamental lawn border
14	8.0	7.3	Terraced, from natural ground level
23	0.71	1.3	Ornamental lawn border, slightly raised

Source: Pattle Delamore Partners Ltd. 2004

occurrence of 2,3,7,8-TCDD in U.S. urban soils at levels of 0.001–0.01 ppb suggested that local combustion sources, including industrial and municipal waste incinerators, are the probable sources of the trace level 2,3,7,8-TCDD soil concentrations in those locations (Nestrick et al. 1986). Soil samples collected in the vicinity of a sewage sludge incinerator were compared with soil samples from rural and urban sites in Ontario, Canada (Pearson et al. 1990). Soil in the vicinity of the incinerator showed a general increase in Dioxin concentration with increasing degrees of chlorination. Of the Furans measured, only OCDF was detected (mean concentration, 43 ppt). Rural woodlot soil samples contained only OCDD (mean concentration, 30 ppt). Soil samples from undisturbed urban parkland revealed only concentrations of HDioxins and OCDD, but all Furan congener groups from TCDF to OCDF were present. The parkland samples showed an increase in concentrations from the HDioxins to OCDD and PeCDFs to OCDF. The TCDFs were found at the highest concentration (mean, 29 ppt) of all the Furan congener groups.

3.6.4 Levels in Humans

The highest exposure to Dioxin-contaminated breast milk reported was associated with the widespread use of Agent Orange as a defoliant during the Vietnam War. Human milk samples collected from Ho Chi Minh City and Song Be Province in South Vietnam had lower 2,3,7,8-TCDD values in the late 1980s (7.1 and 17 ppt lipid basis) (TEQ values of 18.5 and 31.7 ppt), respectively, than they did in the 1970s when Agent Orange spraying occurred (Schecter et al. 1989e). A mean value for 2,3,7,8-TCDD in human milk in South Vietnam was reported to be 484.9 ppt (range, not detectable to 1,450 ppt) (Baughman and Meselson 1973; Schecter et al. 1986a). These values serve as reference values for the highest levels of 2,3,7,8-TCDD documented in human milk (Schecter et al. 1989e). Estimated daily intakes of TEQs by nursing infants from Vietnam have been reported (Schecter and Gasiewicz 1987a). The estimated daily intake by nursing infants in southern Vietnam in 1970 was 908 pg TEQs/kg body weight/day, whereas the daily intakes in southern and northern Vietnam in 1984 were 88.7 and 5.1 pg TEQs/kg body weight/day, respectively. Analysis of nine milk samples from women living in northern Vietnam showed no detectable concentrations of 2,3,7,8-TCDD (detection limit 2 ppt) (Schecter and Gasiewicz 1987a). Very extensive residential contamination by 2,3,7,8-TCDD occurred in Seveso, Italy, when a 2,4,5-TCP reactor exploded in 1976 (Mocarelli et al. 1991). The contaminated area was divided into three zones based on the concentration of 2,3,7,8-TCDD in the soil. Families in zone A, the most heavily contaminated area based on soil 2,3,7,8-TCDD levels, were evacuated within 20 days of the explosion and measures were taken to minimize exposure of residents in nearby zones. A recent analysis of 19 blood samples from residents of zone A, which were collected and stored shortly after the accident, showed serum lipid levels of 2,3,7,8-TCDD that ranged from 828 to 56,000 ppt. These serum lipid levels were among the highest ever reported for humans (Mocarelli et al. 1991).

In a study conducted in Missouri, 2,3,7,8-TCDD was measured in the adipose tissue of 39 volunteers with a history of residential, recreational, or occupational exposure (14 years post-exposure) and in 57 persons in a control group (Patterson et al. 1986a). Based on questionnaire responses, the eligible exposed group for this study consisted of people who were exposed either to areas with 2,3,7,8-TCDD concentrations in soil between 20 and 100 ppb for 2 or more years or to 2,3,7,8-TCDD concentrations > 100 ppb for at least 6 months. Persons who met these criteria were classified as having one of three types of exposure: residential (either living in close proximity to areas with 2,3,7,8-TCDD-contaminated soil or having evidence of contamination inside the home), recreational (riding or caring for horses in 2,3,7,8-TCDD contaminated stable arenas at least one time per week), or occupational (working either in a hexachlorophene production facility or at truck terminals where the grounds had been sprayed with 2,3,7,8-TCDD-contaminated waste oil). All study participants had detectable levels of 2,3,7,8-TCDD in their adipose tissue, but the group with known previous exposures had significantly higher levels than controls. Nineteen (49%) of the 39 exposed persons had levels higher than the highest 2,3,7,8-TCDD concentration (20.2 ppt) detected in the 57 controls. Six (15%) of the 39 exposed persons had 2,3,7,8-TCDD concentrations > 100 ppt. Five of the 6 values > 100 ppt were from persons exposed to 2,3,7,8-TCDD during the production of hexachlorophene. The other high value (577 ppt) was found in a man exposed to 2,3,7,8-TCDD while horseback riding in a contaminated arena. 2,3,7,8-TCDD concentrations measured in the occupational group (average 136.2 ppt; range 3.5–750 ppt) were, in general, higher than those in the residential group (average 21.1 ppt; range 2.8–59.1 ppt), the recreational group (average 90.8 ppt; range 5.0–577 ppt), and the control group (average 7.4 ppt; range 1.4–20.2 ppt) (Patterson et al. 1986a).

2,3,7,8-TCDD has been detected at concentrations ranging from 20 to 173 ppt in adipose tissue from 3 Vietnam veterans reported to have been heavily exposed to Agent Orange (Gross et al. 1984). Except for these few men, however, 2,3,7,8-TCDD concentrations in American Vietnam and non-Vietnam veterans were nearly identical with mean serum levels of approximately 4 ppt (CDC 1988). Concentrations of 2,3,7,8-TCDD in the controls (those who never served in Vietnam) ranged from not detected (4 ppt) to 20 ppt. The veterans had served in Vietnam in 1967 and 1968 in areas where Agent Orange had been heavily used (CDC 1988). In another study, 2,3,7,8-TCDD was detected in adipose tissue of 14 Vietnam veterans and 3 control patients at levels ranging from not detected (2–13 ppt) to 15 ppt. No significant differences in the tissue levels of Vietnam veterans and the controls were found in this study (Weerasinghe et al. 1986). Air Force personnel associated with Operation Ranch Hand (spraying of Agent Orange) in Vietnam from 1962 to 1971 had serum Dioxin levels up to 10 ppt (521 persons). A correlation was found between Dioxin concentrations and increased body fat (USAF 1991). The median half-life of 2,3,7,8-TCDD in 36 veterans was estimated to be 7.1 years (Pirke et al. 1989). In 1987, many of the exposed Air Force personnel had serum Dioxin concentrations > 50 ppt and several had concentrations exceeding 300 ppt (CDC 1987). Wolfe et al. (1994) reported a half-life value of 11.3 years for Air Force personnel involved in Operation Ranch Hand. Using individuals from two rice oil poisoning episodes (Yusho [Japan] and Yu-Cheng [Taiwan]), Ryan et al. (1993a) have shown that the elimination of related Furans is not constant, but variable, with faster clearance at higher doses followed by a slowing down in the rate of loss as body burden decreases. By analogy, the same may be true for Dioxins. It is also likely that individual congeners or those with the same degree of chlorination are excreted from the body at rates that differ from those estimated for 2,3,7,8-TCDD. Because the rate of clearance is not constant, uncertainty in determining the half-life measurement may result, especially for estimates of the changing body burden of total CDDs measured as TEQs. Ayotte et al. (1997) measured concentrations of Dioxins/

Table 3.6: 2,3,7,8-TCDD levels measured in soil samples collected in 1984 from industrialized areas of US cities.

Sample Location[a]	2,3,7,8-TCDD (ppq)
Lansingh, MI	3000 (700)[b] ND (800)
Gayford, MI	ND (200)
Detroit, MI	3600 (700) 2100 (400)
Chicago, Il	9400
Middletown, OH	4200 ND (300)
Barberton, oH	ND (300)
Akron, OH	5600
Nashville, TN	6300
Pittsburgh, PA	800 (300)
Marcus hook, PA	2600 (500)
Philadelphia, PA	400 (300)
Clifton Heights, PA	900 (300)
Brooklyn, NY	2600 (400)
South Carolina, WV	ND (400)
Arlington, VA	ND (400)
Newport News, VA	400 (300)

[a] Post Office state Abbreviations used
[b] Values in parentheses show the detection limit, 2.5 times noise, when the experimental result is less than 10 times the measured detection limit
ND = not detected; ppq = parts per quadrillion;
Source: Nestrick et al. 1986

Furans and PCBs in plasma of adult Inuit living in Arctic Quebec, Canada. The Inuit consume large amounts of fish and marine mammal tissue. The mean concentration of 2,3,7,8-TCDD was 8.4 ppt (range 2.5 to 36.0 ppt) in the Inuit population and < 2 ppt (range < 2) for the control population in Southern Quebec. The TEQ values for all Dioxins/Furans was 39.6 ppt (range 17.1 to 81.8 ppt) in the Inuit population and 14.6 ppt (range 11.5 to 18.9 ppt) for the control population. When PCBs and Dioxins/Furans are considered together, the mean TEQ values for all Dioxin-like compounds was 184.2 ppt in the Inuit population (range 55.8 to 446.7 ppt) and 26.1 ppt (range 20.1 to 31.7 ppt) for the control population.

A large-scale environmental survey was conducted by the Dow Chemical Company to determine soil levels of 2,3,7,8-TCDD on the Dow Midland Plant site and in the city of Midland, Michigan (Nestrick et al. 1986). The Dow Midland Plant site manufactures a variety of chlorophenolic compounds. Soil samples were taken from three different types of areas: locations known to be directly associated with current or historic chlorophenolic production and handling, locations known to be associated with incineration of chemical and conventional wastes and with ash storage, and locations away from established 2,3,7,8-TCDD sources that provide a measure of general background levels of 2,3,7,8-TCDD surface soil within the Dow property. Soil samples taken from chlorophenolic production areas showed a range of 2,3,7,8-TCDD concentrations from 0.041 to 52 ppb. Two localized areas of elevated

Table 3.7: Dioxins, Furans and Dioxin Toxicity Equivalencies (TEQs) in US food (ppt, wet weight).

Food Type	**Total CDDs/CDFs**		
	CDD	**CDF**	**TEQ**
Fish			
Haddock	0.75	0.14	0.03
Haddock fillet	0.35	0.07	0.02
Crunchy Haddock	2.91	0.51	0.13
Perch	1.55	1.14	0.02
Cod	0.82	0.09	0.02
Meat			
Ground beef	4.1	7.0	1.5
Beef rib sirloin tip	0.6	0.2	0.04
Beef rib steak	30.7	4.6	0.3
Pork Ham	59.3	2.5	0.3
Cook ham	29.3	2.5	0.3
Lamb sirloin	8.95	0.85	0.4
Bologna	3.7	0.4	0.12
Chicken drumstick	0.95	0.14	0.03
Dairy			
Cottage cheese	0.6	0.3	0.04
Soft blue cheese	14.0	5.0	0.7
Heavy cream	5.0	2.0	0.4
Soft cream cheese	4.0	2.0	0.3
American cheese slices	4.0	2.0	0.3

Source: Schecter et al. 1994d

concentrations (above 5 ppb) were identified with peaks at 34 ppb and 52 ppb. All other samples taken around this area had 2,3,7,8-TCDD soil concentrations below 1 ppb. Two of 10 surface soil samples with 2,3,7,8-TCDD concentrations above 1 ppb (2.0 and 4.3 ppb) were found near the waste incinerator. The concentrations observed there (0.018–4.3 ppb) closely matched the 2,3,7,8-TCDD content of the ash produced by the incinerator, which ranged up to 10 ppb. The background levels of 2,3,7,8-TCDD (0.0065–0.59 ppb) within the Dow Midland Plant site were well below 1 ppb. Soil samples taken within the city of Midland showed 2,3,7,8-TCDD soil concentrations below the 1 ppb concern level established by the U.S. Public Health Service, Center for Disease Control and Prevention (CDC), for residential areas (Kimbrough et al. 1984). 2,3,7,8-TCDD soil concentrations in the city of Midland (0.6–450 ppt) were higher in areas nearer the Dow Chemical Company Midland Plant site (22–450 ppt) (Nestrick et al. 1986). This gradient suggests that operations on the Midland Plant site are associated with the appearance of the trace levels of 2,3,7,8-TCDD in the nearby environment.

Table 3.8: Dioxins/Furans concentrations in food samples of animal origin (ppt, lipid base).

Congener	Dairy		Meat					Fish		
	Cows Milk	Butter	Pork	Cattle	Sheep	Chicken	Eggs	Herring	Cod	Redfish
2,3,7,8-TCDF	0.7	0.15	0.11	0.3	0.6	2.1	1.1	57.0	98.0	78.0
2,3,7,8-TCDD	0.2	0.08	0.03	0.6	0.01	0.3	0.2	4.7	23.0	2.8
1,2,3,7,8-PeCDF	0.2	0.09	0.01	0.01	0.01	0.01	0.6	16.0	48.0	31.0
2,3,4,7,8-PeCDF	1.4	0.45	0.08	1.5	0.9	1.5	0.8	29.0	3.1	25.0
1,2,3,7,8,-PeCDD	0.7	0.41	0.12	0.8	0.5	0.7	0.4	12.0	1.3	6.5
1,2,3,4,7,8-HxCDF	0.9	0.43	0.15	0.8	0.9	0.6	0.4	3.0	6.9	3.5
1,2,3,6,7,8-HxCDF	0.8	0.44	0.07	0.6	1.2	0.4	0.3	4.2	13.0	6.0
2,3,4,6,7,8-HxCDF	0.7	0.31	0.05	1.3	1.5	0.3	1.7	3.6	8.2	7.2
1,2,3,4,7,8-HxCDD	0.3	0.15	0.21	0.6	0.3	0.5	1.3	1.2	0.01	0.5
1,2,3,6,7,8-HxCDD	1.1	0.95	0.29	1.9	1.5	2.8	1.4	5.8	17.0	8.4
1,2,3,7,8,9-HxCDF	0.4	0.26	0.06	0.6	0.4	0.6	5.0	1.0	5.2	1.3
1,2,3,4, 6,7,8-HpCDF	0.5	0.34	1.1	2.2	8.1	0.8	0.6	1.6	10.0	1.5
1,2,3,4,6,7,8-HpCDD	2.0	1.5	2.1	18.0	15.0	6.0	0.4	3.6	10.0	3.0
OCDF	1.0	0.25	0.41	0.2	0.3	0.6	0.2	1.4	2.1	0.3
OCDD	10.0	3.4	19.0	25.0	68.0	52.0	12.0	19.0	83.0	11.0
TEQ	0.86	0.43	0.14	1.31	0.52	1.16	0.80	21.3	39.7	20.0

Source: Beck et al. 1989a
TEQ = Toxic equivalency

Several studies have analyzed soil samples in the State of Missouri for 2,3,7,8-TCDD contamination and all reported values are comparable. Concentrations of 2,3,7,8-TCDD in soil samples from contaminated sites throughout Missouri ranged from 30 to 1,750 ppb and concentrations in Times Beach, MO, a heavily contaminated site ranged from 4.4 to 317 ppb (Tiernan et al. 1985). In another study, soil core samples taken from a roadside in Times Beach, MO, contained levels of 2,3,7,8-TCDD ranging from 0.8 to 274 ppb. Many roadways in Times Beach had been sprayed with waste oil containing Dioxins for dust control (Freeman et al. 1986). In a third study conducted by Hoffman et al. (1986), 2,3,7,8-TCDD was measured in soil samples from the Quail Run Mobile Home Park in Gray Summit, MO. A maximum soil concentration of 2,200 ppb (single non-composited sample) was detected at one site; however, concentrations typically ranged from 39 to 1,100 ppb in composite soil samples.

In conclusion, soil concentrations of Dioxins are typically higher in urban areas than in rural areas. Soil concentrations associated with industrial sites are clearly the highest, with Dioxin levels ranging from the hundreds to thousands of ppt. In general, as the degree of chlorination increases, the concentrations increase. HpCDD and OCDD congeners are generally found at higher concentrations in soil and sediments than the TCDD, PeCDD, and HxCDD congeners.

3.6.5 Levels in Food

From 1979 to 1984 under the FDA's Total Diet Program, the FDA conducted limited analyses for the higher chlorinated Dioxins (HxCDD, HpCDD, and OCDD) in market-basket samples collected (Firestone et al. 1986). Food samples which were found to contain PCP residues > 0.05 μg/g were analyzed for 1,2,3,4,6,7,8-HpCDD and OCDD. In addition, selected samples of ground beef, chicken, pork, and eggs from the market-basket survey were analyzed for these Dioxin congeners (wet weight basis), regardless of the results of the PCP analysis. HxCDD was not found in any of the foods sampled; however, the detection limit (10–40 pg/g) was very high. Generally low concentrations (< 300 pg/g) of HpCDD and OCDD were found in bacon, chicken, pork chops, and beef liver. Several beef livers had higher concentrations of OCDD residues (614–3,830 pg/g), and one beef liver contained 428 pg/g (ppt) of HpCDD. HxCDD, HpCDD, and OCDD were not detected in milk, ground beef, or seafood samples, but the detection limits (10–40 ppt) were very high. No Dioxins were found in 17 egg samples collected in various parts of the US. Of the 18 pork samples, OCDD was detected only in 2 samples (27 ppt and 53 ppt) and in 2 of the 16 chicken samples (29 ppt and 76 ppt). One chicken sample with PCP residues (> 0.05 μg/g) contained concentrations of 1,2,3,4,6,7,8-HpCDD (28 ppt) and OCDD (252 ppt). The CDD residues (21–1,610 pg/g) in eggs from Houston, Texas, and Mena, Arkansas, with PCP residues > 0.05 μg/g collected in 1982 and 1983–84, respectively, contained 1,2,3,4,6,7,8-HpCDD concentrations ranging from 21 to 588 ppt, and OCDD concentrations ranging from 80 to 1,610 ppt. These residues were attributed to local PCP contamination problems in these areas (Firestone et al. 1986). Milk samples contaminated with PCP at levels ranging from 0.01 μg/g to 0.05 μg/g PCP contained no detectable Dioxins. It should be noted that the reported LOD (10–40 ppt) for the FDA analyses from these older samples, are higher than concentrations of CDDs observed in foods from more recent studies. Samples of beef liver, pork chops, chicken, ground beef, and eggs collected in the US and analyzed for HpCDD and OCDD contained average concentrations of

HpCDD (2.2–9.6 ppt) and OCDD (6.3–47.6 ppt) (Jasinski 1989). Eggs contained the lowest levels of HpCDD and OCDD, and beef liver contained the highest levels.

LaFleur et al. (1990) analyzed the concentration of 2,3,7,8-TCDD and 2,3,7,8-TCDF (wet weight basis) in a variety of food products collected randomly from grocery stores located in the southern, Midwestern, and northwestern regions of the United States. Concentrations of 2,3,7,8-TCDD ranged from 17 to 62 pg/kg for ground beef, were not detectable in ground pork, ranged from 12 to 37 pg/kg for beef hot dogs, and ranged from 7.2 to 9.4 pg/kg for canned corned beef hash on a whole-weight basis. Concentrations of 2,3,7,8-TCDF were generally much less than concentrations of 2,3,7,8-TCDD, with the exception of ground pork and corned beef hash. For ground pork, TCDD concentrations were not detectable and 2,3,7,8-TCDF concentrations ranged from 13 to 20 pg/kg and for corned beef hash, concentrations of TCDD ranged from 7.2 to 9.4 pg/kg, while concentrations of TCDF ranged from 9.8 to 10 pg/kg.

In a study carried out by Schecter et al. (1994d), congener-specific analyses for Dioxins and Furans were performed on 18 dairy, meat, and fish products obtained from a supermarket in upstate New York. Total Dioxin concentrations (on a wet weight basis) ranged from 0.35 to 2.91 ppt in fish, 0.6–59.3 ppt for meats, and 0.6–14 ppt in dairy products. The TEQ for both the CDDs and CDFs on a wet weight basis for these food samples ranged from 0.02 to 1.5 ppt, 0.02–0.13 ppt for fish products, 0.03–1.5 ppt for meat products, and 0.04–0.7 ppt for dairy products, with ground beef having the highest TEQ.

The EPA and U.S. Department of Agriculture (USDA) conducted the statistically designed surveys of the occurrence and concentrations of Dioxins/Furans in beef fat (Ferrario et al. 1996; Winters et al. 1996), pork fat (Lorber et al. 1997), poultry fat (Ferrario et al. 1997), and the U.S. milk supply (Lorber et al. 1998). It was clear from the results, that 1,2,3,4,6,7,8-HpCDD and 1,2,3,4,5,6,7,8-OCDD were typically found at the highest concentrations in all food samples. Concentrations of 2,3,7,8-TCDD were highest in heavy fowl (0.43 ppt) and young turkeys (0.24 ppt); much lower concentrations were found in beef (0.05 ppt), pork (0.10 ppt), young chickens (0.16 ppt), light fowl (0.03 ppt) and milk (0.07 ppt). The total concentrations of Dioxins/Furans were highest in pork fat (75.67 ppt) and milk (15.43 ppt), and ranged from 5.68 to 14.09 ppt for all other types of foods tested. The TEQ value for Dioxins/Furans combined was highest for pork fat (1.30 ppt), heavy fowl (0.98 ppt), young turkeys (0.93 ppt), and beef fat (0.89 ppt), with lower TEQ values of 0.40–0.82 ppt for young chickens, light fowl, and milk. Dioxins were detected in 8 samples of cow's milk in Germany at concentrations ranging from 0.2 ppt for 2,3,7,8-TCDD (LOD 0.2 ppt) to < 10 ppt of OCDD (LOD less than blanks) (Beck et al. 1987). In a Swedish study, only 1 of 10 samples of milk held in either glass bottles or paper cartons contained a detectable level of 2,3,7,8-TCDD (0.46 pg/g milk fat; paper carton; LOD 0.4 pg/g).

3.7 CONTAMINATION TRENDS OF FURANS

Furans are present in the atmosphere both in the vapor and particulate phase (Hites 1990). The ratio of the vapor to particulate phase Furans in air increases with increasing temperature.

3.7.1 Levels in Air

The levels of Furans determined in the ambient air in North America are presented in Tables 3.9–3.11. Based on the sources of emissions, the concentrations of Furans in air show geographical variation. Generally, the levels show the following trend: industriavauto tunnel > urban > suburban > rural (Eitzer and Hites 1989a). Even in a particular area, the level shows daily and seasonal variability. For example, the concentrations of CDFs are generally higher on rainy days with high humidity and on less windy days (Nakano et al. 1990). The levels are also higher in winter than in summer, due to increases in the contribution from combustion sources (heating) (Hunt et al. 1990). Tables 3.9–3.11 indicates that the concentrations of total tetra-, penta-, hexa-, hepta-, and octaCDFs in ambient urban/suburban air can vary within the ranges of 0.13–7.34, 0.09–5.10, < 0.09–12.55, 0.08–2.71, and 0.13–3.78 pg/m^3, respectively. In rural areas, the concentrations of total tetra-, penta- hexa-, hepta-, and octaCDFs were below their detection limits. It has also been determined that the vapor/particulate phase ratio of the Furans in ambient air depends on the season of the year and the number of chlorine substituents. Generally, the terra- and pentaCDFs are present at higher ratios in the vapor phase, while hepta- and octaCDF are present predominantly in the particulate phase in the atmosphere. This ratio of vapor/particulate phase increases during summer, compared to winter (Eitzer and Hites 1989a; Hunt et al. 1990; Nakano et al. 1990). The congener profile in the atmosphere followed the congener profile of their sources, that is, if the major source of Furans in the atmosphere is a municipal incinerator, the congener pattern in the air follows the congener pattern in flue gas from that municipal incinerator (Edgerton et al. 1989; Eitzer and Hites 1989a).

The majority of Furans found in the air are non-2,3,7,8-substituted congeners, which are much less toxic than 2,3,7,8-substituted congeners. Among the 2,3,7,8-substituted isomers in the air, the 1,2,3,4,6,7,8-heptaCDF congener dominates, followed by 2,3,7,8-tetraCDF. It has been shown that 2,3,7,8-tetraCDF constitutes ≈ 9 percent of total tetraCDFs; 1,2,3,7,8-penta- and 2,3,4,7,8-pentaCDF constitute ≈ 9 percent and 10.4 percent, respectively, of total pentaCDFs; 1,2,3,4,7,8-hexa-, and 1,2,3,6,7,8-hexaCDF constitute ≈ 9.4 percent and 18.1 percent, respectively, of the total hexaCDFs; and 1,2,3,4,6,7,8-hepta and 1,2,3,4,7,8,9-heptaCDF constitute ≈ 64.7 percent and 4.4 percent, respectively, of the total heptaCDFs present in the air near a municipal solid waste incinerator in Dayton, Ohio (Tiernan et al. 1989b). Considerably higher concentrations of Furans have been detected in the indoor air and wipe samples of buildings after accidental fires involving PCB capacitors/transformers. For example, the concentrations of total Furans and 2,3,7,8-tetraCDF (plus co-eluting isomers) in wipe samples from the transformer vault after the 1983 transformer fire in Chicago were 12,210 and 410 ng/100 cm^2, respectively (Hryhorczuk et al. 1986). The concentrations of total tetraCDFs in air and wipe samples inside the vault 4 months after the 1983 San Francisco transformer fire were 1,000–3,000 pg/m^3 and 1,000–23,000 ng/100 cm^2, respectively (Stephens 1986). Seven months following the fire, the maximum concentration of 2,3,7,8-substituted Furans in air of the building that contained the transformer vault was 19.5 pg/m^3. The concentrations of total tetraCDFs, 2,3,7,8-tetraCDF (plus co-eluting isomers) and total pentaCDFs of indoor air in a Binghamton, New York, office building 1.5–2 years after cleanup following a 1981 electric fire were ≤ 23, 195, and 60 pg/m^3, respectively (Smith et al.

Table 3.9: Overall national average of Dioxins and Furans in fat of meat and milk on a lipid basis (ppt, or pg/g).

Dioxin/Furan Congener	Beef (n = 63)	Pork Fat (n = 78)	Young Chickens (n = 39)	Light Fowl (n = 12)	Heavy Fowl (n = 12)	Young Turkeys (n = 15)	Milk (Composites) (n = 8)
2,3,7,8-TCDD	0.05 (0.03)	0.10 (0.01)	0.16 (0.15)	0.05 (0.03)	0.43 (0.42)	0.24 (0.24)	0.07 (0.07)
1,2,3,7,8-PeCDD	0.35 (0.04)	0.45 (0.01)	0.24 (0.12)	0.15 (0.00)	0.32 (0.22)	0.32 (0.23)	0.32 (0.32)
1,2,3,4,7,8-HxCDD	0.64 (0.18)	0.52 (0.10)	0.18 (0.05)	0.15 (0.00)	0.24 (0.13)	0.16 (0.03)	0.39 (0.39)
1,2,3,6,7,8-HxCDD	1.42 (1.21)	1.10 (0.80)	0.39 (0.33)	.34 (0.29)	0.71 (0.70)	0.79 (0.77)	1.87 (1.87)
1,2,3,7,8,9-HxCDD	0.53 (0.26)	0.47 (0.04)	0.39 (0.29)	0.15 (0.01)	0.60 (0.51)	0.17 (0.06)	0.55 (0.55)
1,2,3,4,6,7,8-HpCDD	4.48 (4.39)	10.15 (9.93)	1.53 (1.53)	0.93 (0.93)	2.04 (2.02)	0.54 (0.52)	5.03 (5.03)
OCDD	4.78 (3.26)	52.77 (52.40)	5.31 (5.31)	2.07 (2.07)	7.67 (7.67)	0.75 (0.68)	4.89 (4.89)
2,3,7,8-TCDF	0.03 (0.00)	0.09 (0.004)	0.28 (0.28)	0.25 (0.25)	0.48 (0.47)	0.57 (0.57)	0.08 (0.08)
1,2,3,7,8-PeCDF	0.31 (0.00)	0.45 (0.00)	0.21 (0.06)	0.18 (0.05)	0.14 (0.02)	0.36 (0.25)	0.05 (0.00)
2,3,4,7,8-PeCDF	0.36 (0.06)	0.56 (0.14)	0.25 (0.12)	0.22 (0.11)	0.18 (0.09)	0.53 (0.47)	0.28 (0.28)
1,2,3,4,7,8-HxCDF	0.55 (0.27)	0.98 (0.60)	0.23 (0.10)	0.16 (0.04)	0.17 (0.06)	0.20 (0.13)	0.39 (0.39)
1,2,3,6,7,8-HxCDF	0.40 (0.12)	0.58 (0.58)	0.20 (0.07)	0.15 (0.03)	0.15 (0.01)	0.17 (0.03)	0.25 (0.25)
1,2,3,7,8,9-HxCDF	0.31 (0.00)	0.45 (0.00)	0.15 (0.00)	0.15 (0.00)	0.15 (0.00)	0.15 (0.00)	0.05 (0.00)
2,3,4,6,7,8-HxCDF	0.39 (0.10)	0.57 (0.16)	0.21 (0.08)	0.14 (0.02)	0.15 (0.02)	0.15 (0.03)	0.28 (0.28)
1,2,3,4,6,7,8-HpCDF	1.00 (0.75)	3.56 (3.35)	0.27 (0.20)	0.15 (0.05)	0.20 (0.10)	0.15 (0.02)	0.83 (0.83)
1,2,3,4,7,8,9-HpCDF	0.31 (0.00)	0.57 (0.17)	0.17 (0.17	0.15 (0.00)	0.15 (0.00)	0.15 (0.00)	0.05 (0.00)
OCDF	1.88 (0.00)	2.30 (1.85)	0.34 (0.07)	0.29 (0.00)	0.31 (0.04)	0.29 (0.00)	0.05 (0.00)
Total Dioxins/Furans, pg/g	17.79 (10.67)	75.67 (70.14)	10.51 (8.82)	5.68 (3.88)	14.09 (12.48)	5.69 (1.03)	15.43 (15.23)
Dioxins/Furans I-TEQ, pg/g	0.89 (0.35)	1.30 (0.46)	0.64 (0.41)	0.40 (0.16)	0.98 (0.80)	0.93 (0.76)	0.82 (NR)

Concentrations calculated at no-detects (ND) equal the detection limit (result for ND = 0 are in parentheses)

Source: Ferrario et al. 1996, 1997; Lorber et al. 1997; Winters et al. 1996

Table 3.10: Concentration of Furans in ambient indoor and outdoor air in North America.

Site	Sampling Year	Furans	Conc. (pg/m^3)	Reference
Bridgeport, CT (outdoor)	1987–1988	2,3,7,8-TCDF	0.078	Hunt and Maisel 1990
		Total tetra CDF	0.856	
		1,2,3,7,8-PeCDF	0.031	
		2,3,4,7,8-PeCDF	0.047	
		Total PeCDF	0.547	
		1,2,3,4,7,8-HxCDF	0.106	
		1,2,3,6,7,8-HxCDF	0.039	
		2,3,4,6,7,8-HxCDF	0.087	
		1,2,3,7,8,9-HxCDF	0.007	
		Total HxCDF	0.580	
		1,2,3,4,6,7,8-HpCDF	0.212	
		1,2,3,4,6,7,8-HpCDF	0.033	
		Total HpCDF	0.369	
		OCDF	0.211	
Toronto Island, Canada (outdoor)	1988–1989	Total TetraCDF	0.404	Steer et al. 1990
		Total PeCDF	0.118	
		Total HxCDF	0.204	
		Total HpCDF	0.240	
		OCDF	0.142	
Dorset, Canada (outdoor)	1988–1989	Total TetraCDF	0.164	Steer et al. 1990
		Total PeCDF	0.200	
		Total HxCDF	0.074	
		Total HpCDF	0.52	
		OCDF	0.194	
Windsor, Canada (outdoor)	1988–1989	Total TetraCDF	0.733	Steer et al. 1990
		Total PeCDF	0.383	
		Total HxCDF	0.333	
		Total HpCDF	0.55	
		OCDF	0.182	
Boston, MA office Building (indoor)	No data	2,3,7,8-TCDF	(0.37)–1.4	Kominsky and Kwoka 1989
		Total TetraCDF	(0.64)–6.2	
		Total PeCDF	(0.12)–1.9	
		Total HxCDF	(0.39)–(1.5)	
		OCDF	(0.54)–(1.8)	
Albany, NY (outdoor)	1987–1988	Total TetraCDF	3.86	Smith et al. 1990
		2,3,7,8-TCDF/ unknown isomer	0.89	
		Total PeCDF	2.00	
		Total HxCDF	0.28	
		Total HpCDF	< 0.34	
		OCDF	< 0.50	

Table 3.10 contd. ...

...Table 3.10 contd.

Ste	Sampling Year	Furans	Conc. (pg/m³)	Reference
Binghamton, NY (outdoor)	1988	Total TetraCDF	0.94	Smith et al. 1990
		2,3,7,8-TCDF/ unknown isomer	0.18	
		Total PeCDF	0.25	
		Total HxCDF	< 0.09	
		Total HpCDF	< 0.14	
		OCDF	< 0.30	
Los Angeles, CA (outdoor)	1987	2,3,7,8-TCDF	0.021	Maisel and Hunt 1990
		other tetra CDF	0.30	
		1,2,3,7,8-PeCDF	0.077	
		2,3,4,7,8-PeCDF	0.077	
		other PeCDF	0.41	
		1,2,3,4,7,8-HxCDF	0.151	
		1,2,3,6,7,8-HxCDF	0.25	
		2,3,4,6,7,8-HxCDF	< 0.069	
		1,2,3,7,8,9-HxCDF	< 0.083	
		other HxCDF	< 0.080	
		1,2,3,4,6,7,8-HpCDF	< 0.190	
		1,2,3,4,6,7,8-HpCDF	< 0.018	
		other HpCDF	0.26	
		OCDF	0.056	
Dayton, OH (outdoor-suburb/ roadside)	1988	Total TetraCDF	0.13	Tieman et al. 1989
		Total PeCDF	0.24	
		Total HxCDF	0.14	
		Total HpCDF	0.11	
		OCDF	< 0.07	
Dayton, OH (outdoor-municipal solid waste incinerator)	1988	Total TetraCDF	1.23	Tieman et al. 1989
		2,3,7,8-TetraCDF	0.11	
		Total PeCDF	5.10	
		1,2,3,7,8-PeCDF/ unknown isomer	0.46	
		2,3,4,7,8-PeCDF	0.53	
		Total HxCDF	12.55	
		1,2,3,4,7,8-HxCDF/ unknown isomer	1.18	
		1,2,3,6,7,8-HxCDF	2.27	
		2,3,4,6,7,8-HxCDF	< 0.06	
		1,2,3,7,8,9-HxCDF	< 0.41	
		Total HpCDF	12.71	
		1,2,3,4,6,7,8-HpCDF	8.22	
		1,2,3,4,6,7,8-HpCDF	0.56	
		OCDF	3.78	

Table 3.10 contd. ...

...Table 3.10 contd.

Ste	Sampling Year	Furans	Conc. (pg/m³)	Reference
Dayton, OH (outdoor-rural area)		Total TetraCDF	< 0.02	Tieman et al. 1989
		Total PeCDF	< 0.02	
		Total HxCDF	< 0.05	
		Total HpCDF	< 0.07	
		OCDF	< 0.17	
Windsor Island, Canada (outdoor)		Total TetraCDF	0.21	Bobet et al. 1990
		Total PeCDF	0.09	
		Total HxCDF	0.10	
		Total HpCDF	0.08	
		OCDF	0.13	
Walpole Island, Canada (outdoor)		Total TetraCDF	< 0.05	Bobet et al. 1990
		Total PeCDF	< 0.07	
		Total HxCDF	< 0.10	
		Total HpCDF	< 0.07	
		OCDF	< 0.14	
Lake Trout, WI (outdoor)	1987	Total tetra CDF	0.083	Edgarton et al. 1989
		Total PeCDF	0.067	
		Total HxCDF	0.031	
		Total HpCDF	0.012	
		OCDF	0.006	
Akron OH (outdoor)	1987	2,3,7,8-TetraCDF	0.200	Edgarton et al. 1989
		Total tetra CDF	1.23	
		1,2,3,7,8 PeCDF	0.029	
		2,3,4,7,8 PeCDF	0.036	
		Total PeCDF	0.590	
		1,2,3,4,7,8-HxCDF	0.083	
		1,2,3,6,7,8-HxCDF	0.065	
		2,3,4,6,7,8-HxCDF	< 0.021	
		1,2,3,7,8,9-HxCDF	0.032	
		Total HxCDF	0.620	
		1,2,3,4,6,7,8-HpCDF	0.237	
		1,2,3,4,6,7,8-HpCDF	< 0.029	
		Total HpCDF	0.383	
		OCDF	0.180	

Table 3.10 contd. ...

...Table 3.10 contd.

Ste	Sampling Year	Furans	Conc. (pg/m^3)	Reference
Columbus, OH (outdoor)	1987	2,3,7,8-TetraCDF	0.405	Edgarton et al. 1989
		Total tetra CDF	2.85	
		1,2,3,7,8 PeCDF	0.045	
		2,3,4,7,8 PeCDF	< 0.056	
		Total PeCDF	0.995	
		1,2,3,4,7,8-HxCDF	0.165	
		1,2,3,6,7,8-HxCDF	0.141	
		2,3,4,6,7,8-HxCDF	< 0.02	
		1,2,3,7,8,9-HxCDF	0.079	
		Total HxCDF	0.785	
		1,2,3,4,6,7,8-HpCDF	0.335	
		1,2,3,4,6,7,8-HpCDF	< 0.021	
		Total HpCDF	0.450	
		OCDF	< 0.260	
Waldo, OH (outdoor)	1987	2,3,7,8-TetraCDF	0.130	Edgarton et al. 1989
		Total tetra CDF	0.890	
		1,2,3,7,8 PeCDF	0.021	
		2,3,4,7,8 PeCDF	< 0.033	
		Total PeCDF	0.500	
		1,2,3,4,7,8-HxCDF	0.098	
		1,2,3,6,7,8-HxCDF	0.014	
		2,3,4,6,7,8-HxCDF	< 0.008	
		1,2,3,7,8,9-HxCDF	0.097	
		Total HxCDF	0.510	
		1,2,3,4,6,7,8-HpCDF	0.220	
		1,2,3,4,6,7,8-HpCDF	0.019	
		Total HpCDF	0.290	
		OCDF	0.077	

1986a). Similarly, concentrations of tetraCDF, pentaCDF, hexaCDF, heptaCDF and octaCDF $\leq$ 0.4, 0.6, 2.2, 4.4, and 4.8 ng/100 cm^2, respectively, were present in the wipe samples of a building used for the improper incineration of PCBs over 12 years ago (Thompson et al. 1986).

3.7.2 Levels in Water

The concentrations of Furans in most waters are so low that it is difficult to determine the levels in drinking water and surface water, unless the surface water is sampled close to points of effluent discharge containing CDFs. Because of their low water solubilities and high K_{oc} values, the Furans partition from the water to sediment in environmental water or in sludge during the treatment of waste waters. Therefore, more monitoring data are available for Furans levels in the latter two media.

Table 3.11: Levels of Furans in human adipose tissue.

Congener	Sample Source and Mean Conc. (ppt on Fat Basis)				
	Japan[a]	Sweden[a]	Germany[a]	Canada[b]	United States[c]
2,3,7,8-TetraCDF	9	3.9	0.9	3.3	9.1[d]
2,3,4,7,8-PeCDF	25	54	44	33.3	40.0[e]
1,2,3,4,7,8-HxCDF	15	6	10	37[f]	9.3
1,2,3,6,7,8-HxCDF	14	5	6.7	37[f]	5.4
2,3,4,6,7,8-HxCDF	8	2	3.8	52	1.8
1,2,3,4,6,7,8-HpCDF	No data	11	19.5	37.1	21.0[e]
OCDF	No data	4	< 1	12.0	60.0[d]

[a] Rappe et al. 1987
[b] Le Bel et al. 1990
[c] Derieved from Rappe 1989
[d] Stanley et al. 1986
[e] EPA 1989
[f] These were not separated

Table 3.12: Levels of Furans in human milk.

Congener	Levels of Furans in Human Milk			
	Sweden[a]	West Germany[b]	United States[c]	Japan[d]
2,3,7,8-TetraCDF	4.2	1.7	2.85	2.9
1,2,3, 7,8-PeCDF	< 1.0	0.5	0.45	1.8
2,3,4,7,8-PeCDF	21.3	26.7	7.3	23.0
1,2,3,4,7,8-HxCDF	4.7	7.8	5.55	3.9
1,2,3,6,7,8-HxCDF	3.4	6.5	3.2	2.5
2,3,4,6,7,8-HxCDF	1.4	3.4	1.85	1.9
1,2,3,4,6,7,8-HxCDF	7.4	5.5	4.1	3.3
OCDF	3.2	1.4		< 2.0

[a] Rappe et al. 1987; [b] Furst et al. 1992; [c] Schecter et al. 1991; [d] Rappe et al. 1992

A drinking water sample in Sweden contained 2,3,4,7,8-pentaCDF at a concentration of 0.002 ppq (Rappe 1991). The levels of Furans in drinking water from 20 communities in New York state were measured (Meyer et al. 1989). Total tetraCDFs at a concentration of 2.6 ppq (pg/L) and octaCDF at a concentration 0.8 ppq are the only two congener groups detected in 1 of 20 water supplies (Lockport, New York). The concentration of 2,3,7,8-tetraCDF in water from Lockport was 1.2 ppq. The raw water that served as the source of this drinking water contained several CDFs at the following concentrations (ppq): total tetraCDF, 18.0; 2,3,7,8-tetraCDF, not detected (detection limit 0.7); 1,2,3,7,8-pentaCDF, 2.0; total pentaCDF, 27.0; 1,2,3,4,7,8-hexaCDF, 39.0; 1,2,3,6,7,8-hexaCDF, 9.2; total hexaCDF, 85.0; 1,2,3,4,6,7,8-heptaCDF, 210; total heptaCDF, 210; and octaCDF, 230. Since the finished drinking water contained 2,3,7,8-tetraCDF, and the raw water did not contain any detectable level of this compound, the source of 2,3,7,8-tetraCDF in the drinking water must be the chlorination process. Considerably higher concentrations of Furans were detected in the sediment of the raw water. This provides more indirect evidence that chlorination may be partially responsible for the *in situ* production of Furans.

Effluents from bleached kraft and sulfite mill pulp in the United States, Canada, and Europe contained total tetraCDFs in the concentration range of < 0.01–4,100 ppt, whereas the concentrations of 2,3,7,8-tetraCDF varied from < 0.002 to 8.4 ppt. The octaCDF levels in these effluents ranged from < 0.05 to 0.5 ppt. The sludge from the treated effluents from paper mills contained much higher concentrations of Furans. In one case, the sludge from a chloralkali process contained ≤ 52,000 ppt of 2,3,7,8-tetraCDF and 81,000 ppt of octaCDF (Clement et al. 1989a, 1989b; Waddell et al. 1990; Whitemore et al. 1990).

3.7.3 Levels in Soil and Sediment

The maximum 2,3,7,8-tetraCDF and 2,3,7,8-substituted Furan concentrations of 0.3 ppt (ng/kg) and 11.0 ppt, respectively, were determined for sediments from an uncontaminated river (Elk River) in Minnesota (Reed et al. 1990). The maximum concentrations of total pentaCDFs, hexaCDFs, heptaCDFs, and octaCDF in sediment samples from the same river were 25.0, 12.0, 30.0, and 23.0 ppt, respectively. In all cases, the analyte was not detected in some samples. The concentrations of 2,3,7,8-tetraCDF in sediment from the lower Hudson River (New York), Cuyahoga River (Ohio), Menominee River (Wisconsin), Fox River (Wisconsin), Raisin River (Michigan), and Saginaw River (Wisconsin) ranged from 5 to 97 ppt (O'Keefe et al. 1984; Smith et al. 1990b). The concentration of 2,3,7,8-tetraCDF in sediment from an uncontaminated lake (Lake Pepin) in Wisconsin was < l ppt, while its concentration in sediment from Lake Michigan in Green Bay (Wisconsin) was 24 ppt (Smith et al. 1990a). The concentrations of 2,3,7,8-tetraCDF in estuarine sediment varied from 15.0 ppt for an uncontaminated sediment in Long Island Sound (New York) to 4,500 ppt in sediment from an estuary adjacent to a 2,4,5-production facility in Newark, New Jersey (Bopp et al. 1991; Norwood et al. 1989). A concentration ≤ 1,400 ppt was also detected in sediment from New Bedford Harbor (Massachusetts) near a Superfund site (Norwood et al. 1989). The concentrations of 2,3,7,8-tetraCDF and other 2,3,7,8-substituted congeners of pentaCDF were higher in contaminated sediments than uncontaminated sediments (Norwood et al. 1989). In a survey of harbor sediment near a wood treatment facility at Thunder Bay (Ontario), the concentration of tetraCDFs and pentaCDFs were below the detection limit, while the levels of the higher congeners increased with the degree of chlorination (maximum of 6.5 ng/g for HexaCDF to 400 ng/g for OCDF) (McKee et al. 1990). The concentrations (ppt) of Furans in uncontaminated soils from the vicinity of Elk River, Minnesota were as follows (detection limit in parentheses): 2,3,7,8-tetraCDF, not detected (0.8); total tetraCDF, not detected (0.8) to 1.2; total hexaCDFs, 6.7–150; 1,2,3,4,6,7,8-heptaCDF, 26–72; total heptaCDFs, 30–260; and octaCDF, not detected (3) to 270 (Reed et al. 1990). The concentrations (ppt) of CDFs in soils adjacent to a refuse incineration facility in Hamilton, Ontario, were as follows (detection limit in parenthesis): total tetraCDFs, not detected (0.3) to 71; total pentaCDFs, not detected (1.3) to 6.0; total hexaCDFs, not detected (1.3); total heptaCDFs, not detected (1.3) to 180; and octaCDF, not detected (0.8) to 811 (McLaughlin et al. 1989). These levels were not elevated compared to urban control samples. Similarly, the levels of Furans in soils adjacent to a municipal incinerator in England were indistinguishable from background levels (Mundy et al. 1989). On the other hand, much higher levels of Furans were detected in soils from PCP-containing waste landfill in Germany. For example, the concentrations (ppt) of CDFs in the landfill soil were as follows: 1,2,3,7,8/1,2,3,4,8-pentaCDF, 17,000; 2,3,4,7,8-pentaCDF, 7,000; 1,2,3,4,7,8/1,2,3,4,7,9-hexaCDF, 152,000; 1,2,3,6,7,8-hexaCDF, 48,000; 1,2,3,7,8,9-hexaCDF, 3,000; and 2,3,4,6,7,8-hexaCDF, 24,000 (Hagenmaier and Berchtold 1986).

3.7.4 Levels in Humans

The general population is exposed to Furans by inhalation, ingestion of drinking water, consumption of food, and through the use of certain consumer products. Furans by inhalation and ingestion of drinking water would be low. It has been shown that inhalation exposure was not a major pathway of human exposure to Furans (Travis and Hattermer-Frey 1989). The estimate that inhalation exposure contributes 2 percent of the total average human intake of Dioxins/Furans (Hattermer-Frey and Travis 1989) has been questioned as too low by other investigators (Goldfarb and Harrad 1991). The concentrations of Dioxins/Furans in foods consumed by a typical German were determined, and the intake of total Dioxins/Furans from food expressed as TE to 2,3,7,8-TCDD was estimated to be 1.2 pg TE/kg body weight/day (International Dioxin toxic equivalent) (Fürst et al. 1990). The estimated intake of CDD/CDF from typical Canadian food was 1.5 pg TE/kg body weight/day (Birmingham et al. 1989). From detailed determinations of the levels of TCDD/TCDF in air, water, soil, food, and consumer products in Canada, the estimated intakes of CDD/CDF were 0.07 pg TE/kg body weight/day from air, 0.002 pg TE/kg body weight/day from water, 0.02 pg TE/kg body weight/day from ingestion of soil, 2.328 pg TE/kg body weight/day from food, and 0.005 pg TE/kg body weight/day from consumer products (Birmingham et al. 1989b).

Numerous data are available regarding the levels of Furans in body tissue and fluids of exposed and background (no obvious source of exposure) population (Nagayama et al. 1977; Ryan 1986; Schecter et al. 1987; Tiernan et al. 1984; Young 1984). Furans are lipophilic and tend to concentrate in fatty tissues. A positive correlation between 2,3,4,7,8-pentaCDF, 1,2,3,4,7,8-hexaCDF, 2,3,4,6,7,8-hexaCDF in adipose tissue and age of donor (higher concentrations at older age) was found (Le Bel et al. 1990). A similar correlation between 1,2,3,4,7,8-/1,2,3,6,7,8-hexaCDF and age of donor was also reported among the urban population in California (Stanley et al. 1989). No significant correlation between either the level of 2,3,7,8-tetraCDF, 1,2,3,4,6,7,8-heptaCDF, and octaCDF in adipose tissue and age of donor or between any Furans and sex was discernable (Le Bel et al. 1990). The latter findings are different from the case of 2,3,7,8-tetraCDD where higher concentrations of 2,3,7,8-tetraCDD were detected in female donors than male donors and a positive correlation between 2,3,7,8-tetraCDD levels and age of donors was found (Patterson et al. 1986a). The average levels of 2,3,7,8-substituted Furans in human fat of exposed and background populations of different countries have been reviewed (Jensen 1987). More recent data for the background levels of 2,3,7,8-substituted Furans in human adipose tissues from different countries are given in Table 3.11 A comparative study of Furan contents in liver and adipose tissue of control humans (Germany) showed that on a fat basis, the concentrations of Furans were higher in the liver than in adipose tissue (Beck et al. 1990; Thoma et al. 1990).

Several studies indicate that the levels of Furans in the adipose tissue of exposed populations exceeds the levels detected in background or control populations. For example, adipose tissue levels of Furans in an exposed patient of the Binghamton State Office Building (Schecter et al. 1985a, 1985c, 1986; Schecter and Ryan 1989), Yusho victims in Japan (Miyata et al. 1989; Ryan et al. 1987a), and three patients with fatal PCP poisoning (Ryan et al. 1987b) are all higher than control populations. However, no conclusive evidence of higher Furan exposure was found in seven people exposed during the Missouri Dioxin episode and in Vietnam veterans (Kang et al. 1991; Needham et al. 1987). Certain municipal incinerator workers, such as those engaged in ash cleaning are exposed to higher levels of Furans. The whole blood level of total Furans in pooled blood of 56 such workers was 102.8 ppt (on lipid basis) compared to 47.0 ppt in pooled blood of 14 control subjects (Schecter et al. 1991c). The concentrations of 2,3,7,8-tetraCDF, 1,2,3,7,8-pentaCDF, 1,2,3,4,7,8-hexaCDF, 1,2,3,7,8,9-hexaCDF, 2,3,4,6,7,8-hexaCDF, 1,2,3,4,7,8,9-heptaCDF, and octaCDF were also higher in the pooled blood of workers compared to pooled blood of control subjects. No information on Furan levels in the tissues of sport fishermen or subsistence fishermen in the United States is available (Kimbrough 1991), although the levels of 1,2,3,4,7,8-hexaCDF and 1,2,3,4,6,7,8-heptaCDF in the serum lipids of people in Baltic regions who eat fish regularly was higher than those of a control population (Svensson et al. (1991). The estimated bioconcentration factor for 2,3,7,8-tetraCDF in human fat (on lipid basis) was 591 and was higher than other chlorinated aromatics including PCBs, octachlorostyrene, OCDD, and octaCDF (Geyer et al. 1987).

A large number of data is available on the levels of Furans in human milk from different countries (Dewailly et al. 1991; Schecter and Gasiewicz 1987a, 1987b; Schecter et al. 1989b). The levels of Furans in human milk derived from different countries are shown in Table 3.13. Levels of Furans in human milk from other countries including South and North Vietnam and the former Soviet Union are also available (Schecter et al. 1989d, 1990c). From these data, it appears that the most prevalent congener in human milk is 2,3,4,7,8-pentaCDF, followed by 1,2,3,4,7,8-hexaCDF. In one study, no correlation was found between consumption of contaminated fish and accumulation of Furans in the milk from nursing mothers (Hayward et al. 1989). During the breast feeding period, the level of Furans in milk lipid-is highest in the first week and slowly decreases thereafter (Beck et al. 1992; Fürst et al. 1989b). The level of Furans in breast milk is highest for women having their first child and distinctly lower for women having their second and third child (Beck et al. 1992).

The levels of Furans in human whole blood from various countries are listed in Table 3.13. Plasma levels of Furans in people from different countries have been measured and the individual congener concentrations on a fat basis in control populations (not exposed to obvious sources of Furans) vary from a minimum of < 0.1 ppt for 2,3,7,8-tetraCDF to a maximum of 80 ppt for 2,3,4,7,8-pentaCDF (Chang et al. 1990; Nygren et al. 1988; Rappe 1991; Schecter 1991). The highest 2,3,4,7,8-pentaCDF concentration was found in a high fish-consuming population around the Baltic Sea (Svensson et al. 1991). The most prevalent congener in human plasma lipids in the United States was 1,2,3,4,6,7,8-heptaCDF, followed by 1,2,3,7,8- and 2,3,4,7,8-pentaCDF. This pattern was reversed in the plasma lipids of Swedish people where 2,3,4,7,8-pentaCDF was the prevalent congener followed by 1,2,3,4,6,7,8-heptaCDF (Chang et al. 1990). A similar pattern of high 2,3,4,7,8-pentaCDF level in blood was observed in human blood from Germany (Schecter et al. 1991b). Using a multivariate analysis, the concentration of Furans in

Table 3.13: Mean levels of Furans in human whole blood (ppt lipid) from various countries[a].

	Germany		USA	Vietnam		
Congeners	**N = 85**	**Standard Deviation**	**n = 100[b]**	**Ho CHi Minh City N = 50[b]**	**Dong Nai N = 33[b]**	**Hanoi N = 32[b]**
2,3,7,8-TetraCDF	2.5	1.8	3.1	4.6	3.9	2.6
1,2,3,7,8-PeCDF	ND		2.8	3.2	2.9	< 1.1
2,3,4,7,8-PeCDF	36.8	16.8	13.0	21	22	8.6
Total PeCDF	36.8		15.8	24.2	24.9	9.2
1,2,3,4,7,8-HxCDF	17.5[c]		15.0	14.0	27.0	6.5
1,2,3,6,7,8-HxCDF	13.7[c]		14.0	11.0	27.0	6.4
1,2,3, 7,8,9-HxCDF	ND[c]		ND (1.2)[d]	ND (1.4)[d]	ND (1.2)[d]	ND (1.4)[d]
2,3,4,6,7,8-HxCDF	ND[c]		3.6	3.3	5	1.8
Total HxCDF	32.1[c]	20.8	32.6	28.3	59	14.7
1,2,3,4,6,7,8-HpCDF	23.8[c]		36.0	22	31	12
1,2,3,4,7,8,9-HpCDF	ND		ND (1.8)[d]	2.6	2.7	< 1.2
Total HpCF	24.1[c]	12.0	36.0	24.6	33.7	12.6
OCDF	5.5	3.5	4.2	ND (5.5)[d]	11.0	< 3.0

[a] Schecter 1991; [b] These samples were pooled into one; [c] these values are derieved from Papke et al. 1989; [d] the values in parentheses are limit of Detection

the plasma of exposed Vietnam veterans from the United States were determined to be slightly higher than matched controls (Nygren et al. 1988). It was also determined that higher chlorinated Furans do not appear to partition according to the lipid content of whole blood. As the degree of chlorination increases, the percent associated with the protein fraction also increases. Therefore, it was concluded that partitioning of higher chlorinated CDFs is not dependent on lipid content, but specific binding to the protein fraction of serum and whole blood (Patterson et al. 1989; Schecter et al. 1991a).

3.7.5 Levels in Food

The analysis of meat, fish, and dairy products purchased from a supermarket in upstate New York for Furans showed of Furans concentrations of 0.14–7.0, 0.07–1.14, and 0.3–5 ppt (wet weight), respectively (Schecter et al. 1993). The concentrations of 2,3,7,8-TCDF in these meat, fish, and dairy products were 0.01–0.1, 0.02–0.73, and 0.02–0.15 ppt (wet weight), respectively (Schecter et al. 1993). A large number of data concerning the levels of Furans in fish collected from different waters have been obtained (De Vault et al. 1989; Gardner and White 1990; O'Keefe et al. 1984; Petty et al. 1983; Smith et al. 1990b; Zacharewski et al. 1989). 2,3,7,8-tetraCDF was the prevalent Furan congener present in fish, followed by 2,3,4,7,8-pentaCDF. The mean level of total 2,3,7,8-substituted Furans in gutted whole fish from the St. Maurice River, Quebec, caught immediately downstream of a kraft mill was 260 pg/g, but the level declined to 112 ppt at 95 km downstream (Hodson et al. 1993). Data on 2,3,7,8-substituted Furan congeners in aquatic fauna were analyzed by principal component analysis. In this method, the congener profile in aquatic fauna can be used to predict the principal source of contamination such as pulp mill effluent, deposition from combustion source, and effluent from magnesium production (Zitko 1992).

The percent migration of 2,3,7,8-tetraCDF from commercial articles of food contact products (e.g., milk packaged in cartons, coffee filters, paper cups and plates, popcorn bags) to foods may range from 0.1 to 35 percent under normal use conditions (Cramer et al. 1991). Therefore, the concentration of Furans in packaged whole milk depends on the packaging material. Usually, commercial milk packaged in glass contains less Furans than milk packaged in cartons (Rappe et al. 1990c). Hayward et al. (1991) found that the mean concentration of 2,3,7,8-tetraCDF in whole milk packaged in cartons from California was 0.45 pg/g wet weight. All other 2,3,7,8-substituted Furans were either not detected or detected at very low levels (Hayward et al. 1991). Commercial milk from Sweden contained significant levels of other 2,3,7,8-substituted Furans (Rappe et al. 1990c). The intake of Dioxins/Furans from all bleached paper food-contact articles was estimated to be 8.8 pg toxic equivalent (TE)/person/day (Cramer et al. 1991). However, with the reduction of Dioxins/Furans levels in paper pulp available at the present time, the exposure may be considerably less than this estimate (Cramer et al. 1991).

CHAPTER 4

Health Effects of Dioxins and Furans

4.1 HEALTH EFFECTS

The seven 2,3,7,8-chlorine substituted congeners are the most toxic Dioxin congeners, with 2,3,7,8-tetrachlorodibenzo-p-dioxin (2,3,7,8-TCDD) being one of the most toxic. Polychlorinated dibenzofurans are structurally and toxicologically related chemicals as are certain "Dioxin-like" PCBs. TCDD is a known cancer-causing agent, and other Dioxin like compounds are known to cause cancer in laboratory animals. Additionally, Dioxin exposure has been linked to a number of other diseases, including type 2 diabetes and ischemic heart disease, to name a few. Short-term exposure of humans to high levels of Dioxins may result in skin lesions, such as chloracne and patchy darkening of the skin, and altered liver function. Long-term exposure is linked to impairment of the immune system, the developing nervous system, the endocrine system, and reproductive functions. Chronic exposure of animals to Dioxins has resulted in several types of cancer. TCDD was evaluated by the WHO's International Agency for Research on Cancer (IARC) in 1997 and 2012. Based on animal data and on human epidemiology data, TCDD was classified by IARC as a "known human carcinogen". However, TCDD does not affect genetic material and there is a level of exposure below which cancer risk would be negligible. In humans, excess risks were observed for all cancers, without any specific cancer predominating. In specific cohorts, excess risks were observed for reproductive cancers (breast female, endometrium, breast male, testis) but, overall, the pattern is inconsistent. In animals, endocrine, reproductive and developmental effects are among the most sensitive to Dioxin exposure. Decreased sperm counts in rats and endometriosis in rhesus monkeys occur at concentrations 10 times higher than current human exposure. In humans, results are inconsistent regarding changes in concentrations of reproductive hormones. A modification of the sex ratio at birth was described in Seveso. There exist no data on effects such as endometriosis or time-to-pregnancy. Small alterations in thyroid function have occasionally been found. Increased risk for diabetes was seen in Seveso and a herbicide applicators cohort but, overall, results were inconsistent. Experimental data indicate that endocrine and reproductive effects should be among the most sensitive effects in both animals and humans. Epidemiological studies have evaluated only a few of these effects.

Due to the omnipresence of Dioxins, all people have background exposure and a certain level of Dioxins in the body, leading to the so-called body burden. Current normal background exposure is not expected to affect human health on average. However, due to the high toxic potential of this class of compounds, efforts need to be undertaken to reduce current background exposure.

4.2 EXPOSURE TO DIOXINS AND FURANS

Human exposure to background contamination with PCDD/PCDF is possible via several routes:

- Inhalation of air and intake of particles from air,
- Ingestion of contaminated soil,
- Dermal absorption,
- Food consumption.

Dioxins being lipophilic are slowly metabolised and tend to bioaccumulate. The half-life of these compounds varies, but TEQ of the mixtures to which humans are exposed are usually determined by just few compounds. The halflife of TCDD has been estimated to be 7–8 yrs in humans. This half-life may vary with dose, age, sex, and body composition. Most of the effects are believed to be mediated through the aryl hydrocarbon receptor. Various Dioxin effects, including enzyme induction, immunotoxicity, developmental effects, tend to be similar irrespective of whether the exposure is acute or chronic. Ninety percent of the daily Dioxin intake (from background contamination) results from ingestion (WHO 1999). Foodstuffs of animal origin are especially responsible for the daily intake of approximately 2 pg TEQ/(kg bw·d). Other non-fatty foodstuffs are of minor importance in terms of Dioxins/Furans intake. Due to many measures to reduce emissions of Dioxins/Furans in the environment, reduction of Dioxins/Furans contamination in food was observed. As a consequence, the daily intake via food decreased. Whereas in Germany in 1991, the average daily intake was 127.3 pg TEQ/d, the present daily intake for an average German adult is estimated to 69.6 pg TEQ/d. The strongest decline was observed for fish. In 1991, fish contributed for ca. 30 percent of the daily intake (same percentage as for dairy and meat products): today only 10 percent of the daily intake is due to fish.

In 2001, the EU conducted a scientific assessment of PCDD/PCDF in food (SCOOP 2000). Data on health effects in humans following exposure to CDDs have come from studies on accidental, occupational, and residential exposure and from studies on the use of 2,3,7,8-TCDD-contaminated pesticides.

4.2.1 Occupational Exposure

Exposures to 2,3,7,8-TCDD, one of the most potent of the Dioxin congeners, have occurred occupationally in workers involved in the manufacture and application of trichlorophenols and the chlorophenoxy acid herbicides 2,4-dichlorophenoxyacetic acid (2,4-D) and 2,4,5-trichlorophenoxyacetic acid (2,4,5-T). Holmstedt (1980) has reviewed the history of industrial exposures that have occurred between 1949 and 1976, and Kogevinas et al. (1997) summarized recent data on these cohorts. The first reported cases of industrial poisoning were in 1949 at a 2,4,5-T producing factory in Nitro, West Virginia. 2,3,7,8-TCDD formation resulted from uncontrolled conditions in the reactor producing 2,4,5-TCP from tetrachlorobenzene in methanol and sodium hydroxide. Approximately 228 workers (including production workers, laboratory personnel, and medical personnel) were affected. Between 1949 and 1968, three other explosive releases were reported: the first involved 254 workers at the BASF AG facility in Ludwigshafen, Germany, in 1953 (Goldman 1973; Thiess et al. 1982; Zober et al. 1990, 1993); a second similar accident in 1963 involving 106 workers at the Philips-Duphar facility in Amsterdam, Netherlands was a problem since the seriousness of the 2,3,7,8-TCDD exposure was not anticipated and cleanup workers were exposed (Holmstedt 1980); and the third was an explosion in a 2,4,5-TCP manufacturing facility in Coalite, England, involving 90 workers (May 1973). Holmstedt (1980) cited papers describing occupational exposure in 24 additional factories producing TCPs or 2,4,5-T during the same period of time. Exposure data on most of these incidents were limited; various numbers of workers were affected, and many of the published reports are anecdotal. Ott et al. (1994) measured serum 2,3,7,8-TCDD levels in 138 of the 254 exposed workers several decades after the explosion at the BASF facility. More than 35 years after the explosion, serum 2,3,7,8-TCDD levels of < 1–553 pg/g lipid were found; these correspond to serum levels of 3.3–12,000 pg/g lipid (calculated using a 7-year half-life) at the time of the accident.

Some of the most comprehensive studies on occupational exposure were conducted by the National Institute for Occupational Safety and Health (NIOSH). They are cross-sectional studies of workers at U.S. chemical facilities involved in the manufacture of 2,3,7,8-TCDD-contaminated products between 1942 and 1984 (Calvert et al. 1991, 1992; Egeland et al. 1994; Fingerhut et al. 1991; Sweeney et al. 1993). Serum 2,3,7,8-TCDD levels were measured in the workers at two of the plants. The mean 2,3,7,8-TCDD serum lipid level in 281 production workers in the Newark, New Jersey, and Verona, Missouri, plants was 220 ppt (range, 2–3,390 ppt) 18–33 years after exposure termination; the referent group of 260 people who had no self-reported occupational exposure and were matched by neighborhood, age, race, and sex had a mean serum 2,3,7,8-TCDD level of 7 ppt (Calvert et al. 1992; Sweeney et al. 1993). Sweeney et al. (1990) estimated current mean lipid-adjusted 2,3,7,8-TCDD levels of 293.4 ppt (range, 2–3,390 ppt) in 103 production workers at the New Jersey facility and 177.3 ppt (range, 3–1,290 ppt) in 32 workers at the Missouri facility; the mean half-life extrapolated levels (using a half-life of 7 years) were 2,664.7 ppt (range, 2–30,900 ppt) and 872.3 ppt (range, 3–6,100 ppt) in the two facilities, respectively. It should be noted that serum 2,3,7,8-TCDD levels were only measured in workers at these two facilities, and it is not known if the levels in these workers are reflective of serum 2,3,7,8-TCDD levels in workers at the other ten facilities.

There are also a number of studies of chlorophenol and phenoxy herbicide applicators. Some of these studies used job histories, questionnaires, and interviews to determine which phenoxy herbicides the workers had used. Many of the studies did not measure exposure levels or internal doses; rather, 2,3,7,8-TCDD exposure was assumed if the worker was exposed to a phenoxy herbicide known to be contaminated with 2,3,7,8-TCDD, such as 2,4,5-T. However, the level of exposure to these 2,3,7,8-TCDD contaminated products was generally not determined.

A number of studies have examined the possible association between Agent Orange exposure and adverse health effects in Vietnam veterans and Vietnamese residents living in the area of spraying. The results of a study comparing blood 2,3,7,8-TCDD levels in Vietnam veterans and the general U.S. population found that on average there was no significant difference between blood 2,3,7,8-TCDD levels between Vietnam veterans and comparison populations (CDC 1987). Thus, "service in Vietnam" or self-reported exposure to Agent Orange is not a reliable index of 2,4,5-T or 2,3,7,8-TCDD exposure. Studies of Air Force personnel participating in Operation Ranch Hand have found increased serum 2,3,7,8-TCDD levels in some of the persons (CDC 1987; USAF 1991). The median level in serum lipids for 888 Ranch Hand personnel was 12.4 ppt (range, 0 to 617.7 ppt) in contrast to 4.2 ppt (0–54.8 ppt) in a comparison group of 856 matched Air Force personnel (Wolfe et al. 1995). The median and high serum 2,3,7,8-TCDD levels would extrapolate to original serum levels of 43 and 3,135 ppt, respectively, based on 20 years of elapsed time, and a half-life of 8.5 years. Since the tour of duty in Vietnam for the majority of U.S. veterans was generally less than 1 year, the military exposure was considered to be of intermediate duration if not stated otherwise in the original study.

4.2.1.1 Death

Several epidemiology studies have investigated mortality in populations occupationally or environmentally exposed to 2,3,7,8-TCDD or chemicals contaminated with 2,3,7,8-TCDD or other CDD congeners. However, none of the studies examining humans acutely exposed to high concentrations of 2,3,7,8-TCDD or other CDD congeners (as contrasted with long-term studies) reported acute instances of death. No significant increases in the number of deaths were observed in workers at phenoxy herbicide or chlorophenol manufacturing facilities (Cook et al. 1986, 1987b; Fingerhut et al. 1991; Ott et al. 1980, 1987; Zack and Suskind 1980) or in workers exposed to 2,3,7,8-TCDD as a result of the accident at the BASF AG facility in Germany (Ott and Zober 1996; Thiess et al. 1982; Zober et al. 1990). Additionally no increases in mortality were observed in the 10-year period after the Seveso accident (Bertazzi et al. 1989b) or in Vietnam veterans involved in Operation Ranch Hand (Wolfe et al. 1985). Although none of these studies found significant increases in the overall mortality rate, several studies found statistically significant increases in cause-specific mortality. For example, Flesch-Janys et al. (1995) found a significant risk of cardiovascular disease and ischemic heart disease mortality in workers exposed to 2,3,7,8-TCDD and other congeners during the BASF AG accident, and Fingerhut et al. (1991) found a significantly increased risk of cancer mortality in phenoxy herbicide and chlorophenol production workers.

4.2.1.2 Systemic Effects

The effects of 2,3,7,8-TCDD exposure in humans exposed in occupational or environmental settings have been described in several studies. Few studies provided precise exposure levels. However, for some cohorts, blood lipid 2,3,7,8-TCDD levels in samples collected shortly after exposure and stored frozen for several years have been analyzed.

4.2.1.3 Respiratory Effects

No respiratory effects were associated with exposure to 2,3,7,8-TCDD-contaminated herbicides in a group of Vietnam Air Force veterans involved in Operation Ranch

Hand examined more than 10 years after the war (Wolfe et al. 1985). In the 1987 follow-up (USAF 1991), no association was found between the initial or current serum level of 2,3,7,8-TCDD and incidences of asthma, bronchitis, pleurisy, pneumonia, or tuberculosis; abnormal spirometric measurements were often associated with CDD blood levels, but according to the authors (USAF 1991), the differences in the mean level between high- and low-exposure subjects were not clinically important. The authors suggested that these findings may have been related to the association between 2,3,7,8-TCDD and body fat because obesity is known to cause a reduction in vital capacity. Calvert et al. (1991) found no significant differences in ventilatory function between a group of 281 workers employed 15 years earlier in the production of NaTCP, 2,4,5-T ester, or hexachlorophene and 260 referents. At the time of the examination, the lipid-adjusted mean serum 2,3,7,8-TCDD concentration was 220 ppt in the exposed workers compared to 7 ppt for the referents.

Cardiovascular examination did not reveal any changes in 17 individuals who were treated for dermal lesions following acute exposure to 2,3,7,8-TCDD in the Seveso industrial accident (Reggiani 1980) or in a group of Missouri residents living in 2,3,7,8-TCDD-contaminated areas for a chronic period of time (mean 2.8 years in one area, 4.9 years in others) (Hoffman et al. 1986). In the 10-year period following the Seveso accident, there was a significant increase in the relative risk (RR) of death from chronic ischemic heart disease in men (RR = 1.56; 95% CI = 1.2–2.1). Flesch-Janys et al. (1995) found significant increases in mortality from heart and circulatory diseases in workers exposed to 2,3,7,8-TCDD and other CDD congeners during the accident at BASF AG. Relative risks for cardiovascular disease and ischemic heart disease mortality were 1.96 (95% CI = 1.15–3.34) and 2.48 (95% CI = 1.32–4.66), respectively, for workers with extrapolated serum lipid 2,3,7,8-TCDD levels of 348 pg/g (ppt) (current 2,3,7,8-TCDD levels were used to estimate 2,3,7,8-TCDD levels at the end of exposure. An international study comprising 36 cohorts from 12 countries and a total of 21,863 workers exposed to phenoxyacid herbicides and chlorophenols followed from 1939 to 1992 detected an increased risk for death from cardiovascular disease, especially ischemic heart disease (RR = 1.67; 95% CI = 1.23–2.26) among the exposed workers (Vena et al. 1998).

Gastrointestinal Effects. Earlier studies of individuals with exposure to substances contaminated with 2,3,7,8-TCDD found significant elevations in self-reported ulcers (Bond et al. 1983; Suskind and Hertzberg 1984), but a study of Vietnam veterans (USAF 1991) failed to find such effects. A more recent study evaluated the gastrointestinal effects of exposure to substances contaminated with 2,3,7,8-TCDD in an occupational cohort (Calvert et al. 1992). The mean serum 2,3,7,8-TCDD level (on a lipid basis) for the workers was 220 ppt and was found to be highly correlated with years of exposure to 2,3,7,8-TCDD-contaminated substances; controls had a mean serum 2,3,7,8-TCDD concentration of 7 ppt. The only significant finding from the physical examination was a statistically significant association between decreased anal sphincter tone and 2,3,7,8-TCDD exposure.

4.2.1.4 Hematological Effects

Contact with 2,3,7,8-TCDD-contaminated soil in Missouri by physical or recreational activities for 6 months at 100 ppb or for 2 years at 20–100 ppb resulted in a slight but statistically significant increase in total white blood cell (WBC) counts using a prevalence test (5.3% were increased above 10,000 WBC/mm^3 compared to 0.7 percent for controls, but the increase was slight) (Hoffman et al. 1986). A follow-up study of the same population found no differences in the number of red blood cells, white blood cells, or platelets between exposed and non-exposed individuals (Evans et al. 1988). A health study of Vietnam veterans involved in Operation Ranch Hand indicated an association between high initial and current serum 2,3,7,8-TCDD levels and increased erythrocyte sedimentation (Wolfe et al. 1995), and an earlier study by Wolfe et al. (1985) indicated an increase in mean corpuscular volume; however, these changes were minor and were not observed in the 1991 follow-up (USAF 1991). Higher serum 2,3,7,8-TCDD levels were also associated with positive dose-response trends for increases in white blood cell and platelet levels.

4.2.1.5 Dermal Effects

The most commonly observed effect of 2,3,7,8-TCDD exposure in humans is chloracne (Jirasek et al. 1976; Kimbrough et al. 1977; May 1973; Oliver 1975; Reggiani 1980). Chloracne is characterized by follicular hyperkeratosis (comedones) occurring with or without cysts and pustules (Crow 1978). Unlike adolescent acne, chloracne may involve almost every follicle in an involved area and may be more disfiguring than adolescent acne (Worobec and DiBeneditto 1984). Chloracne usually occurs on the face and neck, but may extend to the upper arms, back, chest, abdomen, outer thighs, and genitalia. In mild cases, the lesions may clear several months after exposure ceases, but in severe cases they may still be present 30 years after initial onset (Crow 1978; Moses and Prioleau 1985). In some cases, lesions may resolve temporarily and reappear later. Scarring may result from the healing process. Other chlorinated organic chemicals can also cause chloracne. Skin lesions from environmental exposures to 2,3,7,8-TCDD have been most thoroughly studied in the population exposed in Seveso, Italy. Reggiani (1980) described dermal lesions for 17 persons (primarily children) hospitalized shortly after the accidental release in Seveso. The incidence of chloracne was examined in a group of 3 men and 4 women who were among 231 workers exposed to Dioxins at a chemical factory in Ufa, Russia, approximately 25 years prior to blood collection in 1991 and 1992 (Schecter et al. 1993). Five of the seven (three males and two females) were diagnosed with chloracne after working in the manufacture of 2,4,5-T contaminated with 2,3,7,8-TCDD between 1965 and 1967. Other effects manifested as dermal changes have also been noted to accompany chloracne. In addition to chloracne, hyperpigmentation and hirsutism (also known as hypertrichosis or abnormal distribution of hair) were also reported in 2,3,7,8-TCDD-exposed workers (Jirasek et al. 1976; Oliver 1975; Poland et al. 1971; Suskind and Hetzberg 1984). In the cohort examined by Suskind and Hetzberg (1984), hypertrichosis was observed 25 years after exposure, particularly among workers with persistent chloracne upon clinical examination.

4.2.1.6 Neurological Effects

Symptoms of intoxication including lassitude, weakness of the lower limbs, muscular pains, sleepiness or sleeplessness, increased perspiration, loss of appetite, headaches, and mental and sexual disorders were reported in several of the 117 workers with severe chloracne who had been exposed to 2,3,7,8-TCDD in an occupational setting

(Moses et al. 1984; Suskind 1985). Abnormal neurological effects that included peripheral neuropathy, sensory impairment, tendency to orthostatic collapse, and reading difficulties were reported in workers exposed to 2,3,7,8-TCDD in an industrial accident in Germany (Goldman 1973). Abnormal neurological symptoms were observed in a group of 41 Missouri residents with measured 2,3,7,8-TCDD serum lipid levels (Webb et al. 1989). Psychological effects have been associated with 2,3,7,8-TCDD exposure in some human studies. Personality changes were reported following acute exposure (Oliver 1975). Depression (Levy 1988; Wolfe et al. 1985), hypochondria, hysteria, and schizophrenia (Wolfe et al. 1985) were found more often in Vietnam veterans exposed to 2,3,7,8-TCDD-contaminated herbicides than in the control group of veterans. The overall evidence from case reports and epidemiological studies showed that exposure to CDDs is associated with signs and symptoms of both central and peripheral nervous system shortly after exposure. In some cases, the effects lasted several years. However, evaluation of individuals 5 to 37 years after the last exposure has not revealed any long-lasting abnormalities.

4.2.1.7 Reproductive Effects

Several studies have reported alterations in the sex ratio of children of men and women exposed to high levels of CDDs. Mocarelli et al. (1996) observed decreases in the sex ratio of children born to parents living in area A at the time of the accident in Seveso, Italy. More females than males (48 females versus 26 males; normal ratio is 100 females to 106 males) were born between April 1977 (nine months after the accident) and December 1984. The effects of 2,3,7,8-TCDD exposure on gonadal function (production of germ cells and secretion of sex hormones) has not been extensively investigated. A health study in Vietnam veterans involved in Operation Ranch Hand found a significant association between decreased testicular size and serum 2,3,7,8-TCDD levels, but no association was found for low serum testosterone levels (USAF 1991). No alterations in sperm count or the percentage of abnormal sperm were observed in Vietnam veterans involved in Operation Ranch Hand (Wolfe et al. 1985). Basharova (1996) reported an alteration in sex ratio (more females than males) in children of workers exposed to 2,3,7,8-TCDD contaminated 2,4,5-T at a production facility in Ufa, Russia. No additional information on the percentage of male and female children or statistical analysis of data was provided. Similarly, more females than males (51.4% *vs* 48.6%; 19,675 births) were born to 9,512 male workers exposed to chlorophenate wood preservatives contaminated with CDDs (Dimich-Ward et al. 1996). James (1997) statistically analyzed the results of this study and found that the sex ratio was statistically significant, as compared to the expected Caucasian live birth sex ratio of 0.514.

4.2.1.8 Developmental Effects

In residents of Seveso, Italy, a significant rise in the incidence of birth defects, as compared to pre-accident levels, was observed the year after the accident (Bisanti et al. 1980). A variety of birth defects were observed, but the incidence for any particular defect was not elevated. The authors suggest that the rise in birth defects may not be related to 2,3,7,8-TCDD exposure. Prior to 1976, birth defects in Italy were usually under reported; the authors note that the reported incidences of birth defects after the accident (23 per 1,000 births) were similar to incidences reported in other western countries.

In a study of residents of Northland, New Zealand exposed to 2,4,5-T during aerial spraying, no significant alterations in the total number of birth defects were observed in children born between 1973 and 1976, as compared to the incidence in children born between 1959 and 1960 (before the aerial 2,4,5-T spraying began) (Hanify et al. 1981). Stockbauer et al. (1988) studied the Missouri cohort and found no statistically significant excess risk of birth defects among infants from exposed mothers (n = 410) compared to an unexposed referent group (n = 820). Several studies investigated the outcome of pregnancies fathered by Vietnam veterans potentially exposed to 2,3,7,8-TCDD-contaminated herbicides. Two case-control studies (Aschengrau and Monson 1990; Erickson et al. 1984) have examined the risk of Vietnam veterans having a child with birth defects. The Erickson et al. (1984) study used a cohort of 7,133 infants with birth defects registered by the Metropolitan Atlanta Congenital Defects Program and 4,246 control infants; information on military service and possible exposure to Agent Orange was obtained during interviews with the mother and father. The overall risk of having a child with birth defects was not significantly increased in the Vietnam veterans (OR of 0.97, 95% CI = 0.83–1.14). However, Vietnam veterans fathered a higher proportion of the children with some birth defects (spina bifida, cleft lips, and congenital tumors including dermoid cysts, teratomas, hepatoblastomas, central nervous system tumors, and Wilm's tumors) (Erickson 1984).

Michalek et al. (1998) examined birth records of children born between 1959 and 1992 to Operation Ranch Hand veterans. A slight increase in the incidence of preterm births (not statistically significant) was observed in the low (current CDD level of 10 ppt) and high (extrapolated initial CDD level of > 79 ppt) exposure groups but not in the medium (extrapolated initial CDD level of 79 ppt) exposure group. An increase in the relative risk of infant deaths was observed in all three groups.

4.2.1.9 Cancer

The carcinogenicity of 2,3,7,8-TCDD in humans has been assessed in numerous case-control and mortality cohort studies of chemical manufacturing and processing workers and phenoxy herbicide and chlorophenols applicators, Vietnam veterans exposed to Agent Orange, and residents of Seveso, Italy. A major weakness in many of these studies is the lack of adequate exposure data. Exposure levels or 2,3,7,8-TCDD body burdens were not measured, rather surrogates of exposure such as exposure to chemicals contaminated with 2,3,7,8-TCDD or chloracne were used to identify subjects likely exposed to 2,3,7,8-TCDD. Increases in the overall cancer risk were observed in a number of large cohort mortality studies of chemical manufacturing workers and phenoxy herbicide applicators (Becher et al. 1996; Fingerhut et al. 1991; Hooiveld et al. 1998; Kogevinas et al. 1993, 1997; Manz et al. 1991; Ott and Zober 1996; Zober et al. 1990). Most of the subjects in these studies were males working in chlorophenoxy herbicide, or trichlorophenol manufacturing facilities. In one of the few studies assessing the carcinogenicity of 2,3,7,8-TCDD in women, Kogevinas et al. (1993) found a significantly elevated risk for cancer in women probably exposed to 2,3,7,8-TCDD during the production or application of chlorophenoxy herbicides and/or chlorophenols. The Zober et al. (1990), Fingerhut et al. (1991), Manz et al. (1991),

Ott and Zober (1996), and Hooiveld et al. (1998) studies used current serum 2,3,7,8-TCDD levels in surviving workers to estimate exposure. In the Flesch-Janys et al. (1995) study, risk ratios for the cohort were estimated with year-of-birth stratified Cox regression using seven exposure levels (the reference cohort, the first four quintiles and the ninth and tenth deciles of the estimated 2,3,7,8-TCDD levels and total TEQ); an external cohort of gas supply workers (n = 2,158) served as an unexposed control group. The estimated mean 2,3,7,8-TCDD level for the entire cohort at the end of employment in the plant was 141.4 ng/kg blood fat (median of 38.2 ng/kg). The estimated mean total Dioxin equivalents, calculated as the weighted sum of combined PCDD/F congeners, was 296.5 ng/kg (median of 118.3 ng/kg). There was a dose-dependent increase in cancer mortality with increasing levels of 2,3,7,8-TCDD (p = 0.01 for trend), predominantly due to increased risk ratio (3.30; 95% CI = 2.05–5.31) in the highest-dose group (344.7–3,890.2 ng/kg).

Cohort mortality studies have also found increases in the incidences of soft-tissue sarcomas. Fingerhut et al. (1991) found significant increase in deaths from soft-tissue sarcomas (SMR = 922; 95% CI = 190–2,695) in the high-exposure cohort, although this was only based on three deaths. In the Saracci et al. (1991) multinational cohort, an increase in deaths from soft-tissue sarcoma was observed in phenoxy herbicide sprayers (three deaths, SMR = 297; 95% CI = 61–868) and in workers dying 10–19 years after first exposure (four deaths, SMR = 606; 95% CI = 165–1,552). Similarly, the Kogevinas et al. (1997) study of the IARC cohort found three cases of soft-tissue sarcoma in workers exposed for 10–19 years (SMR = 6.52; 95% CI = 1.35–19.06); shorter exposure durations did not result in significantly elevated SMRs. Additionally, when workers were divided into latency groups, the SMRs were not elevated. Duration of probable exposure to 2,3,7,8-TCDD did not appear to influence soft tissue sarcoma cancer risks. Cohort mortality studies by Fingerhut et al. (1991), Zober et al. (1990), Manz et al. (1991) (including the Flesch-Janys et al. [1998] follow-up data), and Kogevinas et al. (1997) found significant increases in the risk of respiratory tract cancer.

Increases in the risk of several types of cancer have been observed in residents of Seveso, Italy. In the residents with the highest exposure (zone A), no increases in the risk ratio of all malignancies were observed (Bertazzi et al. 1993). However, the small number of zone A residents (724) limits the statistical power of the analysis. Among residents living in zone B (4,824 people), significant increases were observed for the risk of hepatobiliary cancer (risk ratio of 3.3; 95% CI = 1.3–8.1) and multiple myeloma (risk ratio of 5.3; 95% CI = 1.2–22.6) in women and lymphoreticulosarcoma in men (RR of 5.7; 95% CI = 1.7–19).

The available epidemiology data suggest that 2,3,7,8-TCDD may be a human carcinogen. Statistically significant increases in risks for all cancers were found in highly exposed workers with longer latency periods. Although the estimated SMRs are low, they are consistent across studies with the highest exposures. The evidence for site-specific cancers is weaker, with some data suggesting a possible relationship between soft-tissue sarcoma, non-Hodgkin's lymphoma, or respiratory cancer with 2,3,7,8-TCDD exposure. It should be emphasized that some of the human studies do not provide adequate exposure data and were confounded by concomitant exposure to other chemicals.

4.3 DIETARY INTAKE OF DIOXINS AND FURANS

Evaluation of dietary intake is important in the exposure assessment of Dioxins/Furans as more than 90 percent of exposure occurs through food intake. Table 4.1 shows the mean or median dietary intake of Dioxins/Furans expressed as TEQ/kg/day or pg/TEQ/day among adults, children, and breast fed infants in various countries. In most Western countries Dioxin intake originated from various foods, such as meat, meat products, milk, dairy products and fish, whereas in Japan, Finland, and Norway, the Dioxin intake was derived from fish and fish products (Tsutsumi et al. 2001; Becher et al. 1998). Breast-fed babies have an extremely high Dioxin intake. Based on Dioxin content in breast milk and the volume of milk consumed daily, the Dioxin intake among breast-fed babies was estimated at 26–170 pg TEQ/kg/day (Beck et al. 1992; Wearne et al. 1996) which is much higher than the TDI. Several researchers reported that after 1970s–1980s the content of Dioxins/Furans in breast milk sharply declined in developed countries.

4.3.1 Toxicokinetics

Humans can absorb Dioxins by the inhalation, oral, and dermal routes of exposure. Dioxins, when administered orally, are well absorbed by experimental animals, but they are absorbed less efficiently when administered by the dermal route. In a human volunteer, > 86 percent of the administered single oral dose appeared to have been absorbed. In general, absorption is vehicle-dependent and congener-specific. Passage across the intestinal wall is predominantly limited by molecular size and solubility.

The predominant Dioxins carriers in human plasma are serum lipids and lipoproteins, but chlorine substitution plays a role in the distribution in these fractions. For most mammalian species, the liver and adipose tissue are the major storage sites of Dioxins; in some species, skin and adrenals also can act as primary deposition sites. 2,3,7,8-Substituted Dioxins are the predominant congeners retained in tissues and body fluids.

Table 4.1: Estimated mean dietary intake of Dioxins/Furans compounds in adults, children, and breast fed infants in various countries.

Country	pgTEQ/kg/day	pgTEQ/day	Year of Survey	References
US	0.52–2.57		1995	Scecter et al. 1996
Germany	1.3	93.5		Beck et al. 1989
Germany	2			Beck et al. 1992a
Germany	1.2	85	1984–88	Furst et al. 1990
Germany	0.88		1993–1996	Rainer et al. 1998
Germany	1.60		1998	Wittsiepe et al. 2000
New Zealand	0.18–0.44	14.5–30.6		Buckland et al. 1988
Venetia		42	1994–1996	Zanotto et al. 1999
Japan	0.89	44.69	1999–2000	Tsutsumi et al. 2001
Japan	0.47		2002	Ministry of Environment 2003
Belgium	1.00	65.3	2000–2001	Focant et al. 2002
Spain		210	1996	Domingo et al. 1999
Spain		63.8	2002	Bocio et al. 2005

4.3.2 Absorption

No quantitative data were located regarding absorption of CDDs in humans following inhalation exposure, dermal exposure. The absorption of 2,3,7,8-TCDD was estimated to be > 87 percent in a human volunteer following ingestion of a single radioactively labelled dose of 0.00114 μg 2,3,7,8-TCDD/kg in corn oil (Poiger and Schlatter 1986). Furthermore, data on levels of CDDs in blood from populations with above-background exposures (i.e., occupational, accidental) also suggest that dermal absorption occurs in humans. Due to the relatively low vapor pressure and high lipid solubility, dermal uptake of 2,3,7,8-TCDD in the workplace may be a significant source of occupational exposure (Kerger et al. 1995).

4.3.3 Distribution

When human hepatic and adipose tissues were examined for the presence of 2,3,7,8-TCDD, the concentration detected in the liver was about 1/10 of that in the adipose tissue on a whole-tissue-weight basis. However, on the basis of the total tissue lipid, the concentration in adipose tissue lipid was one-half that in the liver lipid (Thoma et al. 1990). Increased adipose tissue levels of Dioxins were reported in populations with known high residential or occupational exposure (Beck et al. 1989c; Fingerhut et al. 1989; Patterson et al. 1989b; Schecter et al. 1994). High levels of 2,3,7,8-TCDD were found in fat (42–750 ppt) and serum lipid (61–1,090 ppt) of Missouri chemical workers (Patterson et al. 1989b). Measurable Dioxins and Furans level were reported in the liver tissue of human stillborn neonates suggesting that the transplacental intrauterine transfer of these persistent chemicals resulted from environmentally exposed mothers (Schecter et al. 1990). In addition, Dioxins are distributed to human milk (Fürst et al. 1994; Schecter et al. 1987a, 1987b, 1989e) and numerous studies have published concentrations of various congeners in human milk samples. Levels of Dioxins in human milk have been found to be significantly and positively associated with of proximity of residence to waste sites and to dietary fat intake per week.

4.3.4 Metabolism

No data were reported regarding metabolic pathways of Dioxins in humans. However, there is some evidence that 2,3,7,8-TCDD is partially excreted in the faeces in the form of metabolites (Wendling et al. 1990). Most *in vivo* studies and epidemiological studies on humans have been concerned with rates of elimination of these compounds; *in vitro* work has shown that a wide variety of metabolites can be found, either in hydroxylated format (typically in faeces) or as conjugates (in urine). Metabolism of this group of compounds seems always to represent a detoxification process rather than one of bioactivation (Hu and Bunce 1999).

4.3.5 Elimination and Excretion

A median half-life of 7.1 years was estimated for 2,3,7,8-TCDD in a group of 36 Vietnam veterans (CDC 1987; Pirkle et al. 1989). In a study of 48 German workers at a pesticide facility who were exposed to a mixture of Dioxins/Furans, a median half-life of 7.2 years was estimated for 2,3,7,8-TCDD (Flesch-Janys et al. 1996). Needham et al. (1994) estimated a half-life of 8.2 years in 27 Seveso residents with initial serum 2,3,7,8-TCDD levels of 130 to 3,830 ppt. Using data from a human subject ingesting a single dose of 1.14 ng/kg 2,3,7,8-TCDD, Poiger and Schlatter (1986) calculated a half-life of 2,120 days (5.8 years). There are limited data available on the elimination of other Dioxin congeners in humans. In the Flesch-Janys et al. (1996) study of 48 workers at a German pesticide facility, elimination half times were estimated for several CDD congeners. The estimated half-lives were 15.7 years for 1,2,3,7,8-PeCDD, 8.4 years for 1,2,3,4,7,8-HxCDD, 13.1 years for 1,2,3,6,7,8-HxCDD, 4.9 years for 1,2,3,7,8,9-HxCDD, 3.7 years for 1,2,3,4,6,7,8-HpCDD, and 6.7 years for OCDD. In a study of six German workers with high CDD/CDF body burdens, elimination half-lives corrected for alterations in body weight ranged from 3.5 years for 1,2,3,4,6,7,8-HpCDF to 7.9 years for 2,3,7,8-TCDD and 15 years for 1,2,3,4,7,8-HxCDD (Rohde et al. 1997).

Dioxins are lipophilic compounds which can concentrate in maternal milk. Therefore, lactation provides an efficient mechanism for decreasing the body burden of these compounds (Schecter and Gasiewicz 1987a). Dioxins levels in breast milk samples from 193 German women ranged from 2.5 to 47 ng TEQ/kg milk fat (mean of 13 ng/kg) (Fürst et al. 1989b). More than 50 percent of the total Dioxins detected in samples was represented by OCDD, which was detected at a mean concentration of 195 ng/kg milk fat (range, 13–664 ng/kg). The amounts of other congeners in human milk decreased with decreasing chlorination; the mean concentration of 2,3,7,8-TCDD in milk fat was 2.9 ng/kg (range, < 1–7.9 ng/kg). Fürst et al. (1989b) also found that the levels of Dioxins found in the milk of mothers breast-feeding their second child were about 20–30 percent lower than in those breast-feeding their first child. It was further noted that the highest excretion of Dioxins was during the first few weeks after delivery.

4.3.5.1 Toxicity

Acute exposure of humans to 2,3,7,8-TCDD can cause chloracne and hepatic effects. Specifying the route of exposure in these human cases is difficult because the individuals were probably exposed by a combination of routes. Furthermore, human data did not provide any information regarding exposure levels and co-exposure to other chemicals confounds the results. Also, in most cases, the exposed subjects were examined long after exposure occurred. Acute oral exposure to 2,3,7,8-TCDD caused delayed type of death in all animal species tested. No deaths were observed with other congeners (2,7-DCDD, 1,2,3,4,6,7,8-HpCDD, 1,2,3,4,6,7,8,9-OCDD) tested, and 2,3,7,8-TCDD proved to be the most toxic Dioxin. However, interspecies and interstrain differences were found in the susceptibility to PCDDs. Systemic effects observed in animals after acute oral exposure to 2,3,7,8-TCDD included cardiovascular, gastrointestinal, hematological, hepatic, renal, endocrine, dermal effects, and body weight loss. Hepatic and body weight effects were the main signs of 2,3,7,8-TCDD toxicity and occurred also after exposure to a mixture of 1,2,3,6,7,8-HxCDD and 1,2,3,7,8,9-HxCDD.

No information was located regarding health effects of other congeners in humans, and limited data exist about effects caused by an acute exposure to these congeners in animals. Intermediate duration exposure of humans to PCDDs has occurred after industrial accidents or in population groups exposed to herbicides contaminated by Dioxins. Hepatic and dermal changes were the main effects noted, and an association between incidence of diabetes and exposure to 2,3,7,8-TCDD has been reported. The main adverse effects in animals following intermediate duration oral and dermal exposure to 2,3,7,8-TCDD included chloracne, wasting syndrome, and liver effects. A number of epidemiology studies have examined the toxicity of Dioxins following chronic exposure to phenoxy herbicides and chlorophenols contaminated with 2,3,7,8-TCDD. Although a number of effects have been observed, interpretation of the results is confounded by a number of factors including lack of adequate exposure information, long post exposure periods, concomitant exposure to other chemicals, and small cohorts.

4.3.5.2 Carcinogenicity

Properties-Several epidemiological studies of phenoxy herbicide and chlorophenol producers found increases in cancer mortality in populations exposed to 2,3,7,8-TCDD. Exposure to this compound has been especially associated with the development of soft tissue sarcoma after a prolonged latency period. The human data suggest that 2,3,7,8-TCDD may be a human carcinogen, however, the interpretation of many of these studies is limited by confounding factors (e.g., small cohorts, short latency periods, coexposure to other chemicals, inadequate exposure data). There are no reliable human studies on the carcinogenicity of other Dioxins. Animal studies provided sufficient evidence that 2,3,7,8-TCDD is a carcinogen after oral and dermal exposure. Furthermore, 2,3,7,8-TCDD has promoting ability on tumours initiated by diethylnitrosourea. Similarly, chronic oral exposure of rodents to a mixture of 1,2,3,6,7,8-HxCDD and 1,2,3,7,8,9-HxCDD or to 2,7-DCDD resulted in carcinogenic effects. No studies were located regarding cancer effects in animals following inhalation exposure to Dioxins.

4.3.5.3 Effects

Studies in humans and animals indicated that 2,3,7,8-TCDD can cross the placenta and is excreted in milk. Studies on the developmental toxicity of 2,3,7,8-TCDD in humans are inconclusive. Some studies have found significant increases in the risk of certain birth defects, while other found no significant alterations. However, a number of limitations (e.g., lack of exposure data, small sample sizes, and the lack of reliable data for birth defects prior to 2,3,7,8-TCDD exposure) limits the interpretation of the results of these studies. Developmental toxicity has been observed in animals orally exposed to 2,3,7,8-TCDD and 2,7-DCDD. The most common effects were cleft palate, hydronephrosis, impaired development of the reproductive system, immunotoxicity, and death. No studies were located regarding developmental effects in animals after inhalation and dermal exposure.

Data from studies on reproductive effects in humans are inconclusive. Reproductive effects have been observed in oral animal studies. Increased incidences of pre and post implantation losses were observed in 2,3,7,8-TCDD-exposed rodents, and monkeys. Adverse effects have also been observed in the reproductive organs (decreased weight), hormone levels, and gametes of male rats. None of the acute duration exposure studies assessed the potential of PCDDs to impair fertility. Reduced fertility, increased incidence of abortions, altered estrus cycle, and endometriosis were observed in animals exposed for intermediate or chronic durations. Reproductive effects have also been observed in animals exposed to mixed HxCDD, but not following exposure to 2MCDD, 2,3-DCDD, 2,7-DCDD, 1,2,3,4-TCDD, or OCDD.

Data on the reproductive toxicity of Dioxins following dermal exposure is limited to a single animal study which found no adverse effects on reproductive organs of mice chronically exposed to 2,3,7,8-TCDD. No animal inhalation reproductive toxicity studies were located.

Inconclusive results were obtained regarding genotoxicity of PCDDs in human as well as in animal studies. Structural chromosomal changes were found in some groups of exposed individuals. Positive and negative results at the chromosomal level as well as at the gene level were reported in animal studies. Furthermore, negative results were obtained in dominant lethal tests and sex linked recessive lethal tests in rats and Drosophila, respectively. In addition, mostly negative results were obtained in prokaryotic organisms. Some studies indicated that the covalent binding of 2,3,7,8-TCDD to DNA is low, and that this mechanism does not operate in Dioxins genotoxicity.

4.3.6 Toxicity and Mode of Action

2,3,7,8-substituted tetra-and pentachlorinated congeners are well absorbed. In contrast, OCDD was poorly absorbed from the gastrointestinal tract of rats (Birnbaum and Couture 1988), but absorbed more on chronic exposure (Birnbaum et al. 1989a). Absorption is also vehicle-dependent (Poiger and Schlatter 1980). The mechanism of absorption of CDDs by the inhalation and dermal routes of exposure is not known. Transfer of CDDs from the aqueous environment of the intestine across cell membranes is predominantly limited by molecular size and lipid solubility. Absorption is also vehicle-dependent (Poiger and Schlatter 1980). Highly chlorinated congeners, although absorbed in small amounts, can accumulate in the liver. Results from studies in thoracic duct-cannulated rats showed that 2,3,7,8-TCDD was transported primarily via the lymphatic route and was predominantly associated with chylomicrons (Lakshmanan et al. 1986). Several studies have examined the distribution of CDDs between blood and adipose tissue.

Experiments of *in vivo* binding of CDD congeners to various serum fractions revealed that as chlorine content increased, binding to lipoproteins gradually decreased, 75 percent of 2,3,7,8-TCDD was found bound to lipoprotein compared to 45 percent for OCDD (Patterson et al. 1989b). However, binding to other proteins increased with chlorine content (20% for 2,3,7,8-TCDD *vs* 50 percent for OCDD). Also, fewer CDDs (< 10%) were bound to the chylomicrons in serum. In studies regarding *in vitro* with human whole blood, 80 percent of the applied amount of 2,3,7,8-TCDD was associated with lipoproteins, 15 percent with proteins, and 5 percent with cellular components (Henderson et al. 1988). Also, there is some evidence that 2,3,7,8-TCDD and related stereoisomers may be associated with plasma prealbumin (McKinney et al. 1985a; Pedersen et al. 1986). Within the lipoprotein fraction and per mole of lipoprotein, 2,3,7,8-TCDD has highest affinity for very low density lipoprotein (VLDL), followed by LDL and HDL (Marinovich et al. 1983). A study using cultured human fibroblasts presented some evidence that specific binding to LDL and the LDL receptor pathway may explain in part the rapid early uptake of 2,3,7,8-TCDD with LDL entry (Weisiger et al. 1981).

The extraordinary potency of 2,3,7,8-TCDD in evoking a dose-related induction of cytochrome P-450 associated AHH activity, the stereospecificity among related halogenated aromatic compounds to evoke this response, and the tissue specificity of enzyme induction, led Poland and Glover (1973b) to postulate the existence of an induction receptor. This receptor, the Ah receptor (Ah for aromatic hydrocarbon), was later identified in the cytosol of mouse liver cells (Poland et al. 1976) and, subsequently, in hepatic and extrahepatic tissues of a variety of laboratory animals, mammalian cell cultures, human organs and cell cultures, and also non-mammalian species (Okey et al. 1994). Results from structure-binding relationships for a series of CDD congeners using mouse hepatic cytosol showed that not all the congeners had the same affinity for the Ah receptor; affinity was found to be determined by the chlorine-substitution pattern (Mason et al. 1986; Poland et al. 1976, 1979). 2,3,7,8-TCDD and structurally related compounds induce a wide range of biological responses, including alterations in metabolic pathways, body weight loss, thymic atrophy, impaired immune responses, hepatotoxicity, chloracne and related skin lesions, developmental and reproductive effects, and neoplasia. The expression of these responses is thought to be initiated by the binding of individual congeners (or ligands) with the Ah receptor.

4.4 EXPOSURE TO FURANS

Furans are found at very low levels in the environment of industrial countries and at even lower levels in non industrial countries. People are exposed to very small levels of Furans by breathing air, drinking water, and eating food, but most human exposure comes from food containing Furans. Table 4.2 gives the route of exposure, symptoms, and remedy of Dioxins and Furans. The levels of Furans in air are usually higher in city and suburb areas than in rural areas. Furans are not found in soils that have not been polluted. Furans have been detected in the stack emissions and ash from certain industries and processes that are sources of these compounds in air at levels that are thousands of times higher than the levels in the air that people usually breathe. Once emitted in the air from stacks, Furans are dispersed by the cleaner air and the level of Furans drops substantially. Similarly, the concentrations of Furans in waste waters from certain industries and in soil at dumpsites can be thousands to millions times higher than the levels found in clean water and soil. Since Furans tend to concentrate in the fat, and milk contains fat, mother's milk can be a source of Furans for babies. But considering the small amounts of Furans in milk and the other beneficial effects of human milk to a baby and the length of time a baby uses mother's milk, scientists

Table 4.2: Dioxins and Furans-route of exposure, symptoms, and remedy.

Route of Exposure	Symptoms	First Aid
Inhalation	Chloracne. Symptoms may be delayed.	Fresh air rest. Refer for medical attention.
Skin	May be absorbed! Redness. Pain. (See Inhalation).	Remove contaminated clothes. Rinse and then wash skin with water and soap. Refer for medical attention.
Eyes	Redness. Pain.	First rinse with plenty of water for several minutes and then take to a doctor.
Ingestion	(See Inhalation).	Give slurry of activated charcoal in water to drink. Induce vomiting (only in conscious persons!). Refer for medical attention.

believe that mother's milk, on balance, is still beneficial to babies. Cow's milk and formula usually contain lower amounts of Furans than human milk.

Children playing in dumpsites may come in contact with Furans through their skin and by eating dirt. It has been estimated that over 90 percent of the total daily intake of Furans for the general adult population occurs from eating food containing them. The rest comes from air, consumer products, and drinking water. Meat and meat products, fish and fish products, and milk and milk products contribute equally to intake of Furans from food, while intake from vegetable products contributes much less. Eating large amounts of fatty fish from water contaminated by Furans may increase one's daily intake from food.

People in certain occupations may be exposed to higher levels of Furans than the general population. Exposure in the workplace occurs mostly by breathing air and touching substances that contain Furans. Workers involved with cleaning up after transformer fires, workers in the pulp and paper mill industry, workers in municipal incinerators, and workers in sawmills may be exposed to higher levels of Furans than the general population. Contact with Furans at hazardous waste sites can happen when workers breathe air or touch soil containing Furans.

4.4.1 Inhalation Exposure

No studies were located regarding the death, systemic effects, immunologic effects, neurologic effects, and reproductive effects in humans or animals after inhalation exposure to CDFs.

4.4.2 Oral Exposure

Much of the information that pertains to human health effects of CDFs comes from large numbers of people who consumed rice oil contaminated with PCBs heat exchange fluid in Japan in 1968 (Yusho incident) and Taiwan in 1979 (Yu-Cheng incident) (Chen and Hsu 1986; Kuratsune 1989; Kashimoto and Miyata 1986; Okumura 1984; Rogan 1989). Twenty-four deaths were observed in 2,061 cases of Yu-Cheng poisoning identified by the end of 1983 (Hsu et al. 1985). Clinical observations strongly suggest that Yusho and Yu-Cheng patients experienced frequent or more severe respiratory infections (Kuratsune 1989; Rogan 1989). Chronic bronchitis accompanied by persistent cough and sputum production was observed in 40–50 percent of some examined patients, with symptoms gradually improving during 5–10 years following onset (Nakanishi et al. 1985; Shigematsu et al. 1971, 1977). Early symptoms in 89 male and 100 female Yusho patients included vomiting (23.6% and 28% frequencies) and diarrhea (19.1% and 17%) (Kuratsune 1989). Various changes in immune status have been reported in Yusho and Yu-Cheng patients, including decreased serum IgA and IgM levels and lymphocyte subpopulations, diminished phagocyte complement and IgG receptors, and diminished delayed-type skin hypersensitive response (Chang et al. 1981, 1982a, 1982b; Lu and Wu 1985; Nakanishi et al. 1985; Shigematsu et al. 1971). Various neurological symptoms, including numbness, weakness and neuralgia of limbs, hypesthesia and headaches are common in Yusho and Yu-Cheng victims (Chia and Chu 1984, 1985; Kuratsune 1989; Rogan 1989). Irregular menstrual cycles and abnormal basal body temperature patterns were observed in $\approx$ 60 percent and 85 percent of female Yusho patients, respectively (Kusuda 1971). These alterations were accompanied by decreased urinary excretion of estrogens, pregnanediol, and pregnanetriol, and possibly suggest corpus luteum insufficiency and retarded follicular maturation. Fertility, fecundity, and rates of spontaneous abortion have not been studied in Yusho and Yu-Cheng patients (Kuratsune 1989; Rogan 1989). Approximately half of 24 deaths observed in 2,061 Yu-Cheng victims by the end of 1983 were attributed to hepatoma, cirrhosis, or unspecified liver diseases with hepatomegaly (Hsu et al. 1985).

4.4.3 Dermal Exposure

No studies were located regarding death in humans after dermal exposure to Furans.

4.4.4 Toxicokinetics

Humans can absorb CDFs by the inhalation, oral, and dermal routes of exposure. CDFs, when administered orally, are well absorbed by experimental animals, but are absorbed less efficiently when administered by the dermal route. Due to their lipophilic nature, CDFs tend to accumulate in lipid-rich tissues. High amounts of CDFs are usually found in the liver, adipose, skin, and muscle. Accumulation in tissues is strongly dependent on the chlorine substitution pattern, which in turn, determines the rate of metabolism. CDFs are metabolized predominantly by cytochrome P-450 to polar metabolites that undergo glucuronidation. Substitutions in positions 4 and 6 impair biotransformation, so that CDFs with these positions substituted, in addition to lateral substitutions (2,3,7,8), are preferentially retained or excreted unchanged. This is true for humans and animals.

4.4.4.1 Absorption

Absorption of CDFs can be inferred from the fact that CDFs have been detected in tissues and blood of subjects after accidental or occupational exposure to airborne CDFs (Schecter and Ryan 1989; Schecter et al. 1991). A recent study examined absorption of CDFs from maternal milk in a 3-month-old infant (Jodicke et al. 1992). Analysis of the milk and the infant's stools showed that some highly chlorinated CDF congeners were more concentrated in the stools than in the milk suggesting that these congeners were less well absorbed or more resistant to entero-hepatic circulation than those with lower chlorine content. Results from a balance sheet analysis showed that over 95 percent of the total CDFs in the milk were removed from the intestinal tract of the infant (Jodicke et al. 1992).

4.4.4.2 Distribution

Data from autopsy reports from two adult individuals, revealed the presence of Furans in four tissues: abdominal and subcutaneous fat, liver, muscle, and kidney (Ryan et al. 1985a). Furans were reported in the liver and adipose tissue of a breast-fed infant born to a mother with Yu-Cheng (Masuda et al. 1985). Beck et al. (1990) detected Furans in the brain, adipose tissue, thymus, spleen, and liver of three infants who died of sudden infant death before reaching 1 year of age. Maternal exposures were not reported. Of the three infants, only one had been breast fed for a significant period of time ($\approx$ 6 months). Analyses of tissues of a Yu-Cheng patient who died two years after poisoning revealed that the liver had the highest concentration of Furans ($\approx$ 35 ppb);

the concentration in other tissues was one or more than one order of magnitude lower than in the liver (Chen et al. 1985b). The major Furan congeners retained in the liver were 1,2,4,7,8-pentaCDF, 2,3,4,7,8-pentaCDF, and 1,2,3,4,7,8-hexa-CDF.

4.4.4.3 Metabolism

No data were located regarding metabolism of Furans in humans. However, some information can be derived from Yusho and Yu-Cheng patients. These subjects ingested contaminated rice oil in which $\approx$ 40 different Furan congeners were identified. Analysis of hepatic adipose tissues of some patients revealed the presence of highly chlorinated congeners and congeners Furans that lacked adjacent unsubstituted carbon atoms (Chen et al. 1985b; Masuda et al. 1985). This indicates that the presence of unsubstituted adjacent carbon atoms favors metabolism, which is consistent with data for other chlorinated aromatic hydrocarbons. Highly chlorinated congeners have slower metabolic rates, possibly due to steric hindrance.

4.4.4.4 Elimination and Excretion

Since Furans have been found in human milk samples from a number of countries (Schecter 1991; Van den Berg et al. 1986), breast feeding is a potential source of excretion (and exposure for the infant) for these compounds. Furans were reported in the liver and adipose tissue of a breast-fed infant born to a mother with Yu-Cheng (Masuda et al. 1985). Analysis of the stools from Yusho patients 22 years after the contamination episode showed a high concentration of penta and hexa-CDFs relative to control subjects (Iida et al. 1992). For example, 2,3,4,7,8-penta-CDF was on the average 20 times more concentrated in the stools of Yusho patients than in controls. Results from a recent study in which Furans were monitored in an infant's feces after breast feeding suggested that the feces may be the preferred route of elimination of highly chlorinated Furan congeners (Jodicke et al. 1992).

4.4.4.5 Toxicity

Essentially all of the information pertaining to health effects of Furans in humans is from the Yusho and Yu-Cheng rice oil poisoning incidents. No definite information is available on human health effects of acute oral exposure to Furans because exposure during these incidents predominately involved intermediate-duration exposure. Information on humans exposed to PCB fires, particularly PCB mixtures not containing chlorinated benzenes (which can form Dioxins), could possibly help characterize health effects of Furans following acute dermal and/or inhalation exposure. Health effects associated with the Binghamton State Office Building electrical transformer fire cannot be attributed solely to Furans or any of the other components of the soot due to the mixture of chemicals, which included chlorinated benzenes and Dioxins, and other confounding factors.

Relatively little information is available on systemic effects of acute duration oral exposure to Furans in animals. Several effects have been observed, including histopathologic and possible functional changes in the kidneys and gastrointestinal tract and evidence of wasting and anaemia. Many of these effects occurred at lethal or other high doses, although effects in the guinea pig, which is the most sensitive species tested in acute oral studies, are relatively well characterized. Since acute toxicity of Furans may depend more on total dose, rather than frequency of dosing, and is characteristically delayed in expression, some information from intermediate duration oral studies is relevant to acute exposure. Data on effects in animals following acute dermal or inhalation exposure to Furans are not available, although mobilization of Furans from adipose tissue to target organs is likely to be similar, regardless of the route of exposure.

Acute dermal studies are relevant because skin is a route of concern for exposure at or near hazardous waste sites, particularly due to possibilities for brief contact. Acute inhalation studies are unlikely to be relevant, due to the low potential for inhalation exposure in the vicinity of hazardous waste sites and ambient air.

Most of the existing toxicity information for Furans is available from intermediate duration studies of orally-exposed humans following Yusho and Yu-Cheng poisoning and animals. Dermal and ocular effects; mild anaemia; mild and transient hepatic alterations, including increased serum levels of triglycerides and liver enzymes and related ultrastructural changes; and bronchitis and other respiratory effects secondary to infection, were most consistently observed in the exposed humans. Although some estimates of doses associated with some effects of Yusho and Yu-Cheng exposure are available, these probably do not reflect the most sensitive toxic end points, as indicated by studies in rats, guinea pigs, and monkeys. Some systemic effects of intermediate duration oral Furans exposure in animals are consistent with the effects observed in humans, but the animal studies better characterize progression of certain effects (e.g., liver toxicity) and have identified other systemic effects (e.g., wasting syndrome, stomach mucosal lesions). Hepatic effects in rats were used as a basis for an intermediate-duration oral MRL. Because of limitations in the database, it is unclear whether different species should be used for studying effects on different target organs. The only information available on systemic toxicity of intermediate duration dermal exposure is from a study in mice, which found effects in the stomach and liver and on body weight, however, these data are suggestive of similar effects by both dermal and oral routes. No data were located regarding effects in animals after intermediate-duration inhalation exposure, but inhalation is a minor route of concern for humans. No information is available on effects in humans or animals following chronic exposure to Furans by any route.

Various toxic effects have been observed in children born to mothers exposed during the Yusho and Yu-Cheng incidents, including dermal lesions, decreased birth weights, neurobehavioral deficits, and some prenatal deaths. Although no exposure-related congenital malformations have been reported in these children, oral studies in mice and rats have documented induction of hydronephrosis and/or cleft palate by 2,3,7,8-substituted tetra-, penta-, and hexaCDF congeners. Tissues other than kidney and palate were examined only in the rat studies, which provide some evidence indicating that rats are more susceptible to Furans than mice and that neonatal thymic toxicity is a more sensitive developmental end point than fatal mortality or cleft palate in rats. Irregular menstrual cycles, abnormal basal body temperature patterns, and decreased urinary excretion of estrogens and pregnanediol were observed in female Yu-Cheng patients. Some intermediate duration oral studies showed no histological alterations in the ovaries, uterus, or testes of rats treated with various Furans, although there is some evidence from other oral studies (intermediate duration in rats and acute duration in guinea pigs) that the testes are a target. No information is available on reproductive effects of Furans in animals or humans following dermal or inhalation exposure, but limited available toxicokinetic data suggest that the potential for reproductive toxicity is likely to be qualitatively similar across routes. Limited information is available regarding genotoxic effects of Furans in humans.

Examination of lymphocytes of Yu-Cheng individuals revealed an increased frequency of sister chromatid exchanges. This effect could be attributed to PCBs that were found in the serum of these subjects at a concentration level 1,000 times higher than Furans, because genotoxic effects of halogenated aromatic hydrocarbons are not known to be Ah receptor-mediated. Only 2,3,7,8-tetra-CDF has been tested for genotoxicity in eukaryotic organisms (S. cerevisiae yeast), and only mono-CDFs, octa-CDF, and 2,3,7,8-tetra-CDF have been tested in prokaryotes (S. typhimurium bacteria). The results of these studies showed that 2,3,7,8-CDF was not mutagenic in S. cerevisiae and that Furans were generally not mutagenic in various strains of S. typhimurium, with only 2- and 3-mono-CDF inducing some activity.

It has been suggested that in the gastrointestinal tract, ingested Furans are incorporated into chylomicra particles that enter the blood stream (Patterson et al. 1989a). In the blood stream, Furans are bound to different, very low density lipoproteins, low-density lipoproteins, high-density lipoproteins, and also to protein, most likely pre-albumin. In human blood, it has been shown that as the degree of chlorination of the Furan congener increases (from four chlorines up) the percentage of Furan associated with the protein fraction increases, suggesting that higher chlorinated Furans do not partition according to the lipid content of the fraction (Patterson et al. 1989a). This could indicate the presence of specific interactions between the Furan congeners and the carrier proteins or other proteins.

4.4.4.6 Carcinogenicity

A retrospective mortality study of Yusho victims and an informal survey of Yu-Cheng deaths provide inconclusive evidence of liver cancer. An intermediate duration study in mice showed no skin neoplastic activity following dermal application of 2,3,4,7,8-pentaCDF or 1,2,3,4,7,8-hexaCDF alone, although these congeners as well as 2,3,7,8-tetraCDF promoted development of mouse skin neoplasms. These congeners also promoted development of liver tumors in rats following subcutaneous injection, providing further evidence of tumor promotion by Furans. Results of a 2-year carcinogenicity study in which 2,3,4,7,8-pentaCDF or 1,2,3,4,7,8-hexaCDF were administered to rats by 1 single or 4 weekly subcutaneous injections are inconclusive due to small numbers of tested animals.

4.4.5 Toxicity and Mode of Action

The mechanism of action of Furans is based on structure receptor binding relationships, structure-induction relationships, and structure-toxicity relationships (Goldstein and Safe 1989; Safe 1990b, 1991). Most of the studies providing this information investigated compounds other than Furans, particularly 2,3,7,8-TCDD and other Dioxins, and used parenteral routes of exposure and/or *in vitro* test systems. It is beyond the scope of this profile to discuss these studies in detail. The concept of a common mechanism explains why all of these compounds, including Furans, elicit the same responses and differ only in their relative potency. The expression of the toxic response, which is species and strain dependent, is initiated by the binding of individual congeners with the Ah receptor. The responsiveness of a particular organ or cell depends on the presence of a functional Ah receptor. Initial binding of a Furan congener to the Ah receptor is followed by an activation or transcription step and subsequent accumulation of occupied nuclear receptor complexes. These complexes interact with a specific deoxyribonucleic acid (DNA) sequence in the CYPlAl gene (which regulates the expression of cytochrome P-4501Al isozymes), changing its secondary and supersecondary structure (Elferinck and Whitlock 1990), which leads to enhancement of the CYPlAl gene expression. A specific nucleotide sequence present in multiple copies to which the nuclear complex binds has been identified (Denison et al. 1989). Ultimately, newly synthesized enzymes and macromolecules resulting from the pleiotropic response to the CDF-receptor complex are responsible for many of the effects caused by CDFs and other halogenated aromatic hydrocarbons.

4.5 TOLERABLE INTAKES

Different international expert groups have performed health risk assessments of Dioxins and related compounds. A Nordic expert group (for five Scandinavian countries) proposed a tolerable daily intake (TDI) for 2,3,7,8-Cl4DD and structurally similar chlorinated PCDD and PCDF of 5 pg/kg body weight (bw), based on experimental studies on cancer, reproduction, and immunotoxicity. Germany used a tried approach and recommended 1 pg I-TEQ/kg bw·d as a desirable target to be achieved in the long-term and that actions should be taken if the daily exposure exceeds 10 pg I-TEQ/kg bw. day. In the USA, ATSDR (Agency for Toxic Substances and Disease Register) has set a minimal risk level (MRL) for Dioxins of 1 picogram per kilogram body weight (ATSDR). The U.S. federal government has made the following recommendations to protect human health (ATSDR 1998):

- The EPA has set a limit of 0.00003 micrograms of 2,3,7,8 Cl_4DD per liter of drinking water (0.00003 mg/L).
- Discharges, spills, or accidental releases of 1 pound or more of 2,3,7,8-Cl_4DD must be reported to EPA.
- The Food and Drug Administration (FDA) recommends against eating fish and shellfish with levels of 2,3,7,8-Cl4DD greater than 50 parts per trillion (50 ppt).

A first World Health Organization (WHO) meeting, in 1990, established a TDI of 10 pg/kg bw for 2,3,7,8-Cl4DD, based on liver toxicity, reproductive effects and immunotoxicity, and making use of kinetic data in humans and experimental animals. Since then new epidemiological and toxicological data have emerged, in particular with respect to neuro-developmental and endocrinological effects. In May 1998, a joint WHO-ECEH (World Health Organization-European Centre for Environmental Health) and International Programme on Chemical Safety (IPCS) expert group re-evaluated the old TDI and came out with a new TDI (which is a range) of 1–4 pg TEQ/kg bw, which includes all 2,3,7,8-substituted PCDD and PCDF as well as Dioxin-like PCB. The TDI is based on the most sensitive adverse effects, especially hormonal, reproductive and developmental effects, which occur at low doses in animal studies; for example, in rats and monkeys at body burdens in the range of 10–50 ng/kg bw. Human daily intakes corresponding with body burdens similar to those associated with adverse effects in animals were estimated to be in the range of 10–40 pg/kg bw·d. The 1998 WHO-TDI does not apply an uncertainty factor to account for interspecies differences in toxicokinetics since body burdens have been used to scale doses across species. However, the estimated human intake was based on Lowest Observed Adverse Effect Levels (LOAELs) and not on No Observed Adverse Effect Levels (NOAELs). For many endpoints humans might be less sensitive than animals, uncertainty still remains regarding animal to human extrapolations. Further, differences between animals and humans exist in the half-lives for the different PCDD/PCDF congeners. To account for all these uncertainties, a composite uncertainty factor of 10 was recommended. As subtle effects might already be occurring in the general population in developed countries at current background levels of exposure to Dioxins and related compounds, the WHO expert group recommended that every effort should be made to reduce exposure to below 1 pg TEQ/kg bw·d (WHO 1998; van Leeuwen and Younes 1998). Since the WHO expert consultation has established the new TDI of 1–4 pg WHO-TEQ/kg bw·d, countries started to move towards this recommendation; for example, Japan established a TDI of 4 pg WHO-TEQ/kg bw·d as its environmental standard.

Most recently, on 4–14 June 2001, the Joint FAO/WHO Expert Committee on Food Additives (JECFA) held its 57th meeting in Rome (JECFA 2001). The Committee decided to express tolerable intakes as monthly values due to the long half-lives of PCDD, PCDF, and Dioxin-like PCB. Thus, a monthly-based period would be a much more appropriate period to better reflect the average intakes as daily ingestion has a small or even negligible effect on overall exposure. A provisional tolerable monthly intake (PTMI) of 70 pg/kg·bw·month was finally chosen as midpoint of two studies: data by Ohsako et al. (2001) would results in a TMI of 100 pg/kg·bw·month, whereas the data by Faqi et al. (1998) would result in a TMI of 40 pg/kg·bw·month. Similar to the other evaluation, the TEQ includes PCDD, PCDF, and Dioxin-like PCB (JECFA 2000).

4.6 LATEST HEALTH EFFECTS STUDIES

A new study, funded in part by NIEHS, found that Dioxin affects not only the health of an exposed rat, but also unexposed descendants through a mechanism of epigenetic trans-generational inheritance. The study was conducted in the laboratory of Michael Skinner, Ph.D., a professor in the Center for Reproductive Biology in the Department of Biological Sciences at Washington State University (WSU) who designed the study. Although not designed for risk assessment, these results have implications for the human populations that are exposed to Dioxin and are experiencing declines in fertility and increases in adult onset disease, with a potential to transmit them to later generations.

4.6.1 Dangers of Dioxin

In the Skinner group's experiments, exposure to Dioxin caused changes in the DNA methylation patterns of sperm that were transmitted across generations, in an imprinted-like manner, to affect the health of multiple generations of descendents. The grandchildren of exposed rats showed Dioxin-induced effects ranging from polycystic ovarian disease to kidney disease. The work raises the serious concern that even if toxic chemicals, such as Dioxin, were completely removed from the environment, they could continue to cause disease for multiple generations.

4.6.2 Health Effects of Dioxin—Early onset of Puberty in Females

Skinner's group (2012) used low *in vivo* doses of Dioxin, so that toxic effects were not expected. Female rats were exposed while pregnant, and both their direct progeny and descendants two generations removed were examined. Although the most prominent phenotypes were kidney disease in males and polycystic ovarian disease in females, a number of other effects including abscesses, colon impaction, lung abnormalities, and missing testes were also observed in animals from the Dioxin-treated lineage. Additionally, females from the Dioxin-exposed lineage experienced the early onset of puberty. Conversely, males showed delayed puberty, suggesting sex-specific effects of exposure. Early puberty in humans has increased over recent decades and is believed to have an environmental link.

4.6.3 Alteration of Methylation Patterns in Germ Line DNA across Generations

The researchers were able to identify 50 specific regions of DNA that were differentially methylated in the Dioxin-treated animals. These regions were permanently reprogrammed and protected from DNA methylation, in a manner that allowed them to be passed down across generations. In the future, these regions may serve as biomarkers that would allow early detection of exposure and risk for disease.

Other chemical compounds, including bisphenol A, phthalates, the insecticide DEET, and the jet fuel JP8 have all been shown to promote disease across generations, through a similar mechanism of epigenetic trans-generational inheritance. This pathway of disease propagation exists not only in rats, but also in humans, mice, worms, flies, and even plants. Thus, future research will be needed to see if other environmental compounds may also lead to health effects across generations.

CHAPTER 5

Analytical Methods for Dioxins and Furans

5.1 INTRODUCTION

The process of method development and validation has a direct on the quality of the data. For any matrix to be analysed for organics, it is necessary to extract the desired analyte in such an organic solvent in which it has maximum solubility or affinity. Extraction involves the bringing into contact a solution with another immiscible solvent. The solvent is also soluble with a specific analyte contained in the solution. In this step, the sample is homogenized and extracted with a suitable solvent or solvent mixture to remove the bulk of the sample matrix and the analyte residue is transferred into the solvent. Both the selection of the proper solvent and the method of extraction are critical in obtaining a satisfactory recovery of any analyte from the sample matrix. In the present research, the selected polychlorinated aromatic hydrocarbons viz. 2,3,7,8-tetrachlorodibenzo-p-dioxin (2,3,7,8-TCDD), 2,3,7,8-tetrachlorodibenzofuran (2,3,7,8-TCDF), octachlorodibenzo-p-dioxin (OCDD), and octachlorodibenzofuran (OCDF) have been separated out and analysed by GC-ECD and GC-MS following the same principal. For liquid–liquid and soxhlet/rotary shaker extraction techniques such a solvent is selected that has a high affinity for the compound of interest but not miscible with water and which doesn't reacts with soil matrix. The use of highly sophisticated GC with specific and sensitive detector is required for the identification and analysis of Dioxins and Furans, which are present in very low concentration (nanogram to picogram levels). The retention time is indicative of the particular PCDD/PCDF congener and peak height/area represents the concentrations. The identification of 2,3,7,8-TCDD, 2,3,7,8-TCDF OCDD and OCDF congeners is based on their elution at their exact retention time (within 0.005 retention time units measured in the routine calibration) and the simultaneous detection of the three most abundant ions in the molecular ion region. The quantization of the individual congeners is achieved in conjunction with the establishment of a five point calibration curve for each homologue, during which each calibration solution is analyzed once.

5.2 ANALYTICAL METHODS FOR DIOXINS AND FURANS

Methods used to prepare environmental samples are similar to those used for biological samples: organic solvent extraction of Dioxins and Furans from the sample and concentration, clean up, and fractionation of the Dioxins using evaporative and column chromatography techniques. The following section describes the methods available for the different types of environmental samples.

- US EPA Method 8280A: The Analysis of Polychlorinated Dibenzo-p-dioxins and Polychlorinated Dibenzofurans by High Resolution Gas Chromatography/Low Resolution Mass Spectrometry (HRGC/LRMS).
- US EPA Method 8290: Polychlorinated Dibenzodioxins (PCDDs) and Polychlorinated Dibenzofurans (PCDFs) by High Resolution Gas Chromatography/High Resolution Mass Spectrometry (HRGC/HRMS).

5.3 METHOD 1613 B -TETRA- THROUGH OCTA-CHLORINATED DIOXINS AND FURANS BY ISOTOPE DILUTION HRGC/HRMS

Scope and Application

This method is for determination of tetra- through octa-chlorinated dibenzo-p-dioxins (CDDs) and dibenzofurans (CDFs) in water, soil, sediment, sludge, tissue, and other sample matrices by high resolution gas chromatography/high resolution mass spectrometry (HRGC/HRMS). The seventeen 2,3,7,8-substituted CDDs/CDFs listed in below table may be determined by this method. Specifications are also provided for separate determination of 2,3,7,8-tetrachloro-dibenzo-p-dioxin (2,3,7,8-TCDD) and 2,3,7,8-tetrachloro-dibenzofuran (2,3,7,8-TCDF).

Compounds	
2,3,7,8-TCDD	1,2,3,4,7,8-HxCDF
Total TCDD	1,2,3,6,7,8-HxCDF
2,3,7,8-TCDF	1,2,3,7,8,9-HxCDF
Total TCDF	2,3,4,6,7,8-HxCDF
1,2,3,7,8-PeCDD	Total-HxCDF
Total-PeCDD	1,2,3,4,6,7,8-HpCDD
1,2,3,7,8-PeCDF	Total-HpCDD
2,3,4,7,8-PeCDF	1,2,3,4,6,7,8-HpCDF
Total-PeCDF	1,2,3,4,7,8,9-HpCDF
1,2,3,4,7,8-HxCDD	Total-HpCDF
1,2,3,6,7,8-HxCDD	OCDD
1,2,3,7,8,9-HxCDD	OCDF
Total-HxCDD	

The detection limits and quantitation levels in this method are usually dependent on the level of interferences rather than instrumental limitations. The Method Detection Limit (MDL) for 2,3,7,8-TCDD has been determined as 4.4 pg/L (parts-per-quadrillion) using this method.

Safety

The 2,3,7,8-TCDD isomer has been found to be acnegenic, carcinogenic, and teratogenic in laboratory animal studies. It is soluble in water to approximately 200 ppt and in organic solvents to 0.14%. On the basis of the available toxicological and physical properties of 2,3,7,8-TCDD, all of the PCDD/PCDFs should be handled only by highly trained personnel thoroughly familiar with handling and cautionary procedures and the associated risks.

Protective equipment-Disposable plastic gloves, apron or lab coat, safety glasses or mask, and a glove box or fume hood adequate for radioactive work should be used. During analytical operations that may give rise to aerosols or dusts, personnel should wear respirators equipped with activated carbon filters. Eye protection equipment (preferably full face shields) must be worn while working with exposed samples or pure analytical standards. Latex gloves are commonly used to reduce exposure of the hands. When handling samples suspected or known to contain high concentrations of the PCDD/PCDFs, an additional set of gloves can also be worn beneath the latex gloves.

Equipments

- Gas Chromatograph—Shall have splitless or on-column injection port for capillary column, temperature program with isothermal hold.
- GC column for CDDs/CDFs and for isomer specificity for 2,3,7,8-TCDD—60 ± 5 m long x 0.32 ± 0.02 mm ID; 0.25 µm 5% phenyl, 94% methyl, 1% vinyl silicone bonded-phase fused-silica capillary column (J&W DB-5, or equivalent).
- GC column for isomer specificity for 2,3,7,8-TCDF—30 ± 5 m long x 0.32 ± 0.02 mm ID; 0.25 µm bonded-phase fused-silica capillary column (J&W DB-225, or equivalent).
- Mass Spectrometer—28–40 eV electron impact ionization, shall be capable of repetitively selectively monitoring 12 exact m/z's minimum at high resolution (≥ 10,000) during a period of approximately one second.
- GC/MS Interface—The mass spectrometer (MS) shall be interfaced to the GC such that the end of the capillary column terminates within 1 cm of the ion source but does not intercept the electron or ion beams.
- Data System—Capable of collecting, recording, and storing MS data.

Reagents and Standards

Extraction

Solvents—Acetone, toluene, cyclohexane, hexane, methanol, methylene chloride, and nonane; distilled in glass, pesticide quality, and lot-certified to be free of interferences are used as extraction solvents.

Adsorbents for Sample Cleanup

Silica gel

- *Activated silica gel*: 100–200 mesh, rinsed with methylene chloride, baked at 180°C for a minimum of one hour, cooled in a dessicator, and stored in a precleaned glass bottle with screwcap that prevents moisture from entering.
- *Acid silica gel (30% w/w)*: Thoroughly mix 44.0 g of concentrated sulfuric acid with 100.0 g of activated silica gel in a clean container. Break up aggregates with a stirring rod until a uniform mixture is obtained. Store in a bottle with a fluoropolymer-lined screw-cap.
- *Basic silica gel*: Thoroughly mix 30 g of 1N sodium hydroxide with 100 g of activated silica gel in a clean container. Break up aggregates with a stirring rod until a uniform mixture is obtained. Store in a bottle with a fluoropolymer-lined screw-cap.

Alumina

Either one of two types of alumina, acid or basic, may be used in the cleanup of sample extracts. The same type of alumina must be used for all samples, including those used to demonstrate initial precision and recovery and ongoing precision and recovery.

- Acid alumina: Activate by heating to 130°C for a minimum of 12 hours.
- Basic alumina: Activate by heating to 600°C for a minimum of 24 hours. Alternatively, activate by heating in a tube furnace at 650–700°C under an air flow rate of approximately 400 cc/minute. Do not heat over 700°C, as this can lead to reduced capacity for retaining the analytes. Store at 130°C in a covered flask. Use within five days of baking.

Carbon

- Carbopak C
- Celite 545

  Thoroughly mix 9.0 g Carbopak C and 41.0 g Celite 545 to produce an 18% w/w mixture. Activate the mixture at 130°C for a minimum of six hours. Store in a dessicator.

Standard Solutions: Purchased or prepared from materials having 98% or greater chemical purity. Standards are stored in the dark at room temperature in screw-capped vials with fluoropolymer-lined caps. A mark is placed on the vial at the level of the solution so that solvent loss by evaporation can be detected. If solvent loss has occurred, the solution should be replaced.

Stock Solutions: Dissolve an appropriate amount of assayed reference material in solvent. For example, weigh 1–2 mg of 2,3,7,8-TCDD to three significant figures in a 10 mL ground-glass-stoppered volumetric flask and fill to the mark with nonane. After the TCDD is completely dissolved, transfer the solution to a clean 15 mL vial with fluoropolymer-lined cap. Stock standard solutions should be checked for signs of degradation prior to the preparation of calibration or performance test standards.

PAR Stock Solution of all CDDs/CDFs: Using the stock solutions as prepared above, prepare the PAR stock solution to contain the CDDs/CDFs at the required concentrations by dilution with nonane.

Precision and Recovery (PAR) Standard: Used for determination of initial and ongoing precision and recovery. Dilute 10 μL of the precision and recovery standard to 2.0 mL with acetone for each sample matrix for each sample batch.

Labeled-Compound Spiking Solution

All CDDs/CDFs: From stock solutions, or from purchased mixtures, prepare Labeled-Compound Spiking solution to contain the labeled compounds in nonane at the concentrations 2 ng/L except $^{13}C_{12}$ at 4 ng/L. This solution is diluted with acetone prior to use. Dilute a sufficient volume of the labeled compound solution by a factor of 50 with acetone to prepare a diluted spiking solution. Use the solution within the same day.

Cleanup Standard: Prepare ^{37}Cl-2,3,7,8-TCDD in nonane at 0.8 ng/L concentration. The cleanup standard is added to all extracts prior to cleanup to measure the efficiency of the cleanup process.

Internal Standard(s)-All CDDs/CDFs: Prepare the internal standard solution to contain ^{13}C 12-1,2,3,4-TCDD and ^{13}C 12-1,2,3,7,8,9-HxCDD in nonane at the concentration at 200 ng/L (labelled compound stock solution).

Calibration Standards—Prepare five calibration solutions in nonane.

Sample Collection, Preservation, Storage, and Holding Times

(i) Collect samples in amber glass containers following conventional sampling. Aqueous samples that flow freely are collected in refrigerated bottles. Solid samples are collected as grab samples using wide-mouth jars.

(ii) Maintain aqueous samples in the dark at 0–4°C from the time of collection until receipt at the laboratory. If residual chlorine is present in aqueous samples, add 80 mg sodium thiosulfate per liter of water. If sample pH is greater than 9, adjust to pH 7–9 with sulfuric acid. Maintain solid, semi-solid, oily, and mixed-phase samples in the dark at < -10°C.

Holding Times

(i) There are no demonstrated maximum holding times associated with CDDs/CDFs in aqueous, solid, semi-solid, tissues, or other sample matrices. If stored in the dark at 0–4°C and preserved as given above (if required), aqueous samples may be stored for up to one year. Similarly,

(ii) Store sample extracts in the dark at < -10°C until analyzed. If stored in the dark at < -10°C, sample extracts may be stored for up to one year.

Sample Preparation

Sample preparation involves modifying the physical form of the sample so that the CDDs/CDFs can be extracted efficiently. In general, the samples must be in a liquid form or in the form of finely divided solids in order for efficient extraction to take place. Suggested sample quantities to be extracted for various matrices are—Aqueous sample: 1000 mL and Solid sample: 10 g.

Extraction and Concentration

Extraction procedures include separatory funnel and solid phase for aqueous liquids; Soxhlet/Dean-Stark for solids, filters, and SPE disks; and Soxhlet extraction and HCl digestion for tissues. Acid/base back-extraction is used for initial cleanup of extracts. Macro-concentration procedures include rotary evaporation, heating mantle, and Kuderna-Danish (K-D) evaporation. Micro-concentration uses nitrogen blow down.

Separatory funnel extraction of filtrates and of aqueous samples with visibly absent particles

Pour the spiked sample or filtrate (with labelled-compound spiking solution) into a 2 L separatory funnel. Rinse the bottle or flask twice with 5 mL of reagent water and add these rinses to the separatory funnel. Add 60 mL methylene chloride to separatory funnel, stopper, and, extract the sample by shaking the funnel for two minutes with periodic venting. Allow the organic layer to separate from the aqueous phase for a minimum of 10 minutes. If an emulsion forms and is more than one third the volume of the solvent layer, employ mechanical techniques to complete the phase separation. Drain the methylene chloride extract through a solvent-rinsed glass funnel approximately one-half full of granular anhydrous sodium sulfate into a solvent-rinsed concentration device. Extract the water sample two more times with 60 mL portions of methylene chloride. Drain each portion through the sodium sulfate into the concentrator. After the third extraction, rinse the separatory funnel with at least 20 mL of methylene chloride, and drain this rinse through the sodium sulfate into the concentrator. Concentrate the extract by macro-extraction using rotary evaporator or heating mantle or Kuderna-Danish. Proceed with micro-concentration and solvent exchange. If the extract is to be cleaned up by column chromatography (alumina, silica gel, Carbopak/Celite, or Florisil), bring the final volume to 1.0 mL with hexane. Proceed with column cleanups. If the extract is to be concentrated for injection into the GC/MS, quantitatively transfer the extract to a 0.3 mL conical vial for final concentration, rinsing the larger vial with hexane and adding the rinse to the conical vial. Reduce the volume to approximately 100 μL. Add 10 μL of nonane to the vial, and evaporate the solvent to the level of the nonane. Solid Phase extraction can be used for samples containing less than 1% solids.

Extract Cleanup

Cleanup may not be necessary for relatively clean samples (e.g., treated effluents, groundwater, drinking water).

Gel permeation chromatography: It removes high molecular weight interferences that cause GC column performance to degrade. It should be used for all soil and sediment extracts and may be used for water extracts that are expected to contain high molecular weight organic compounds (e.g., polymeric materials, humic acids).

Silica Gel Cleanup—used to remove nonpolar and polar interferences. Place a glass-wool plug in a 15 mm ID chromatography column. Pack the column bottom to top with: 1 g silica gel, 4 g basic silica gel, 1 g silica gel, 8 g acid silica gel, 2 g silica gel, and 4 g granular anhydrous sodium sulfate. Tap the column to settle the adsorbents. Pre-elute the column with 50–100 mL of hexane. Close the stopcock when the hexane is within 1 mm of the sodium sulfate. Discard the eluate. Check the column for channeling. If channeling is present, discard the column and prepare another. Apply the concentrated extract to the column. Open the stopcock until the extract is within 1 mm of the sodium sulfate. Rinse the receiver twice with 1 mL portions of hexane, and apply separately to the column. Elute the CDDs/CDFs with 100 mL hexane, and collect the eluate. Concentrate the eluate for further cleanup or injection into the HPLC or GC/MS.

Alumina Cleanup—used to remove nonpolar and polar interferences. Place a glass-wool plug in a 15 mm ID chromatography column. If using acid alumina, pack the column by adding 6 g acid alumina. If using basic alumina, substitute 6 g basic alumina. Tap the column to settle the adsorbents. Pre-elute the column with 50–100 mL of hexane. Close the stopcock when the hexane is within 1 mm of the alumina. Discard the eluate. Check the column for channeling. If channeling is present, discard the column and prepare another. Apply the concentrated extract to the column. Open the stopcock until the extract is within 1 mm of the alumina. Elute the interfering compounds with 100 mL hexane and discard the eluate. If using acid alumina, elute the CDDs/CDFs from the column with 20 mL methylene chloride:hexane (20:80 v/v). Collect the eluate. If using basic alumina, elute the CDDs/CDFs from the column with 20 mL methylene chloride:hexane (50:50 v/v). Collect the eluate. Concentrate the eluate for further cleanup or injection into the HPLC or GC/MS.

Carbon Column—used to remove nonpolar interferences. Cut both ends from a 10 mL disposable serological pipette to produce a 10 cm column. Fire-polish both ends and flare both ends if desired. Insert a glass-wool plug at one end, and pack the column with 0.55 g of Carbopak/Celite to form an adsorbent bed approximately 2 cm long. Insert a glass-wool plug on top of the bed to hold the adsorbent in place. Pre-elute the column with 5 mL of toluene followed by 2 mL of methylene chloride: methanol:toluene (15:4:1 v/v), 1 mL of methylene chloride:cyclohexane (1:1 v/v), and 5 mL of hexane. If the flow rate of eluate exceeds 0.5 mL/minute, discard the column. When the solvent is within 1 mm of the column packing, apply the sample extract to the column. Rinse the sample container twice with 1 mL portions of hexane and apply separately to the column. Apply 2 mL of hexane to complete the transfer. Elute the interfering compounds with two 3 mL portions of hexane, 2 mL of methylene chloride:cyclohexane (1:1 v/v), and 2 mL of methylene chloride:methanol:toluene (15:4:1 v/v). Discard the eluate. Invert the column, and elute the CDDs/CDFs with 20 mL of toluene. If carbon particles are present in the eluate, filter through glass-fiber filter paper. Concentrate the eluate for further cleanup or injection into the HPLC or GC/MS.

Florisil Column: Pre-elute the activated Florisil column with 10 mL of methylene chloride followed by 10 mL of hexane:methylene chloride (98:2 v/v) and discard the solvents. When the solvent is within 1 mm of the packing, apply the sample extract (in hexane) to the column. Rinse the sample container twice with 1 mL portions of hexane and apply to the column. Elute the interfering compounds with 20 mL of hexane:methylene chloride (98:2) and discard the eluate. Elute the CDDs/CDFs with 35 mL of methylene chloride and collect the eluate. Concentrate the eluate for further cleanup or for injection into the HPLC or GC/MS.

HRGC/HRMS Analysis

Inject 1.0 μL or 2.0 μL of the concentrated extract containing the internal standard solution, using on-column or splitless injection. The volume injected must be identical to the volume used for calibration. Start the GC column initial isothermal hold upon injection. Start MS data collection after the solvent peak elutes. Stop data collection after the OCDD and OCDF have eluted. If only 2,3,7,8-TCDD and 2,3,7,8-TCDF are to be determined, stop data collection after elution of these compounds. Return the column to the initial temperature for analysis of the next extract or standard (Below table).

Suggested GC Operating Conditions

Parameters	Conditions
Injector temperature	270°C
Interface temperature	290°C
Initial temperature	200°C
Initial time	Two minutes
Temperature program	200–220°C, at 5°C/minute: 220°C for 16 minutes 220–235°C, at 5°C/minute 235°C for 7 minutes 235–330°C, at 5°C/minute

Quantitative Determination

Isotope Dilution Quantitation—By adding a known amount of a labelled compound to every sample prior to extraction, correction for recovery of the CDD/CDF can be made because the CDD/CDF and its labelled analog exhibit similar effects upon extraction, concentration, and gas chromatography. Relative response (RR) values are used in conjunction with the initial calibration data to determine concentrations directly, so long as labelled compound spiking levels are constant, using the following equation:

$$C_{ex}\ (\text{ng/nL}) = \frac{(A1_n + A2_n)\, C_l}{(A1_l + A2_l)\, RR}$$

where,

C_{ex} = The concentration of the CDD/CDF in the extract,

$A1_n$ and $A2_n$ = The areas of the primary and secondary m/z's for the CDD/CDF.

$A1_l$ and $A2_l$ = The areas of the primary and secondary m/z's for the labelled compound.

C_l = The concentration of the labelled compound in the calibration standard.

C_n = The concentration of the native compound in the calibration standard.

Using the concentration in the extract determined above, compute the percent recovery of the ^{13}C-labeled compounds using the following equation:

$$\text{Recovery (\%)} = \frac{\text{Concentration found (ug/nL)}}{\text{Concentration spiked (ug/nL)}} \times 100$$

The concentration of a CDD/CDF in the solid phase of the sample is computed using the concentration of the compound in the extract and the weight of the solids.

$$\text{Concentration in solid (ng/kg)} = \frac{(C_{ex} \times V_{ex})}{W_s}$$

where,
C_{ex} = the concentration of the compound in the extract
V_{ex} = the extract volume in mL
W_s = the sample weight (dry weight) in kg

The concentration of a CDD/CDF in the aqueous phase of the sample is computed using the concentration of the compound in the extract and the volume of the water extracted.

$$\text{Concentration in aqueous phase (pg/L)} = \frac{(C_{ex} \times V_{ex})}{V_s}$$

where,
C_{ex} = the concentration of the compound in the extract
V_{ex} = the extract volume in mL
V_s = the sample volume in Litres

5.4 ANALYTICAL METHOD DEVELOPMENT

A laboratory method was developed for Identification and Measurement of 2,3,7,8-TCDD, 2,3,7,8-TCDF, OCDD and OCDF in water, flyash and soil samples. The details of the method are given below.

Apparatus and Reagents

Equipments and glassware

- Gas Chromatograph/Mass Spectrometer with Data System (GC/MS/MS-DS).
- The Varian CP 3800 GC equipped with temperature programming, and all required accessories, including syringes, gases, and a capillary column was used.
- Gas Chromatograph—Mass Spectrometer Interface: The gas chromatograph coupled directly to the mass spectrometer source was used.
- Mass Spectrometer: The mass spectrometer was operated in a full scan Elcctron Impact (EI) mode with a total cycle time (including voltage reset time) of one second or less.
- Data System: A dedicated computer data system was employed to control the rapid multiple ion monitoring process and to acquire the data. Quantification of data (peak areas) and EI traces (displays of intensities of each m/z being monitored as a function of time) were acquired during the analyses. Quantifications were reported based upon the computer-generated peak areas (chart recording).
- GC Column: A fused silica column (25 m × 0.32 mm I.D.) coated with DB-5, 1.0 μm film thickness was used to isolate 2,3,7,8-TCDD, 2,3,7,8-TCDF, OCDD, and OCDF.

Maintenance of equipment and glassware

All the equipments, glassware used for sampling, storage, extraction, and analysis must be free of all the interferences/impurities. The GC columns used for separation of analyte needs proper conditioning to remove lodged impurities and moisture. Conditioning of the column is carried out by heating the column at temperature 50ºC higher than the working temperature but not exceeding the maximum temperature with one end of the column attached to the injector port and other end left open. n-Hexane is flushed through the column so as to remove the impurities if present.

All glassware required for analyzing Dioxins/Furans are scrupulously cleaned before initial use as well as after each analysis. The glassware is cleaned as soon as possible after analysis by first rinsing with water or with solvent that was last used. This is followed by washing with soap or detergent, rinsing with tap water and distilled water. Volumetric flasks used for storing standards are immersed overnight in chromic acid for removal of trace organics. Finally the glassware is rinsed with double distilled water. Contaminated glassware other than volumetric flask are oven dried at 300ºC for removal of trace organics.

Microsyringes are properly cleaned after every 12–15 injections with acetone and solvent used for injection. Deposition if any is removed with cleaning wire or by immersing the syringe in acetone till the particles are dislodged. Syringes are stored in their respective cases.

5.5 INSTRUMENTATION

Optimization of GC-MS Conditions

An ion trap GC-MS equipped with Electron Impact (EI) mode was used for 2,3,7,8-TCDD, 2,3,7,8-TCDF, OCDD and OCDF analysis. The various conditions set for the GC-MS and ion-trap for the GC resolution and mass spectra of 2,3,7,8-TCDD, 2,3,7,8-TCDF, OCDD and OCDF are given in Table 5.1 to Table 5.3. The ion trap was held at 200°C. The manifold temperature was 80°C. The ion trap was connected by a heated (300°C) transfer line to GC. DB-5 minibore capillary column has been used and connected to on-column injector directly into ion source of quadrapole mass spectrometer. The EI mode was operated at electron energy of 70 eV. The gas flow of 1.5 mL/min has been optimized. A 59-min temperature programme has been used to separate 2,3,7,8-TCDD, 2,3,7,8-TCDF, OCDD and OCDF. The samples (1–2 μL) were injected and the column temperature (Table 5.3) has been programmed as follows 70°C isothermal for 1.5 min, with rise of 10°C/min to 235°C for 10 min, 5°C/min to 275°C for 3 min, 5°C/min to 325°C for 10 min. The filament of the ion source was switched off during elution of the solvent for mass specific detection (mass fragmentography) of 2,3,7,8-TCDD, 2,3,7,8-TCDF, OCDD, and OCDF.

GC-MS System Calibration

Initial calibration of the GC–MS System is required before any samples are analysed for 2,3,7,8-TCDD. Solutions of 10, 50, 100, 200, 300, and 500 pg μL^{-1} of 2,3,7,8-TCDD were prepared for the initial calibration. The instrument was tuned with a reference compound perfluorotributylamine (PFTBA). A six point standard calibration

Table 5.1: GC-MS conditions optimized for PCDDs and PCDFs.

Parameters	**Conditions**
Model	Saturn WS
Make	Varian
Carrier gas	Helium
Flow rate (mL/min)	1.5
Split ratio	1:20
Trap temperature (°C)	200
Manifold temperature (°C)	80
Transfer line temperature (°C)	310
Column	DB-5 capillary column Length: 30 m Diameter: 0.25 mm Film thickness: 0.25 μm

Table 5.2: GC-MS Ion trap conditions optimized for PCDDs and PCDFs.

Parameters	**Conditions**
Temperature (°C)	200
Emission Current (μA)	90
RF Storage	48
Scan rate (Sec/Scan)	0.75
Filament delay (min)	10
Threshold (count)	2
Background mass (amu)	99
Mass range (amu)	90–500
PFTBA* tuning	Target

*: Perfluorotributylamine

Table 5.3: GC-MS column oven programme for PCDDs and PCDFs.

Temperature (°C)	**Rate (°C/min)**	**Hold (min)**	**Total (min)**
70	0.0	1.50	1.50
235	10.0	10.00	28.00
275	5.0	3.0	39.00
325	5.0	10.00	59.00

curve was determined using six different concentrations each of 2,3,7,8-TCDD, 2,3,7,8-TCDF, OCDD, and OCDF solution (Figs. 5.1 to 5.4).

The identification of any unknown compound was made by matching the retention time with that of the standard obtained under same conditions. Confirmation of the compound was carried out by monitoring the three major abundant ions and the signal-to-noise ratio (S/N) for the GC peak at each exact m/z greater than or equal to 2.5 for each specified congener detected in a sample extract, and greater than or equal to 10 for each specified congener in the calibration standard.

The quantity of the compound present is proportional to the area of the peak and can be used to determine the concentration of the compound in the sample. A

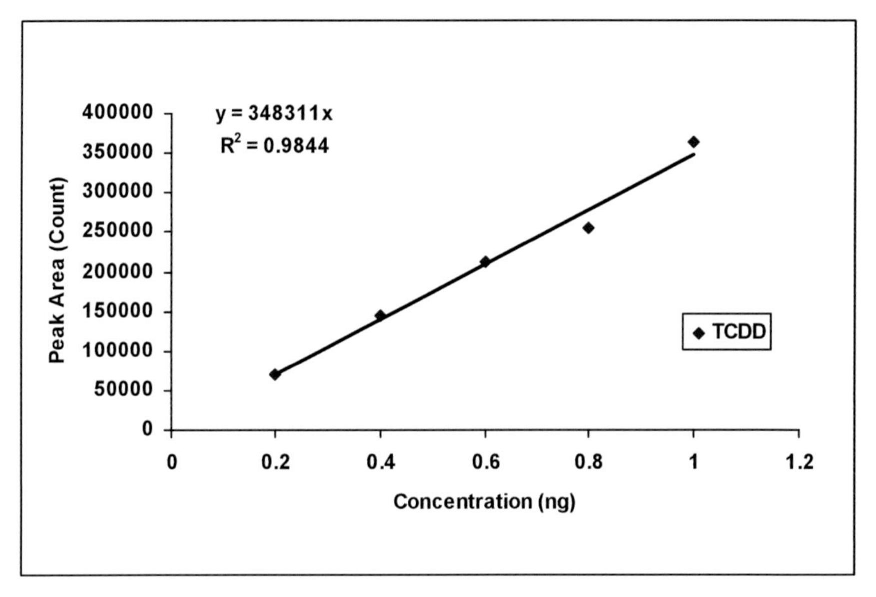

Fig. 5.1: Calibration plot for 2,3,7,8-TCDD.

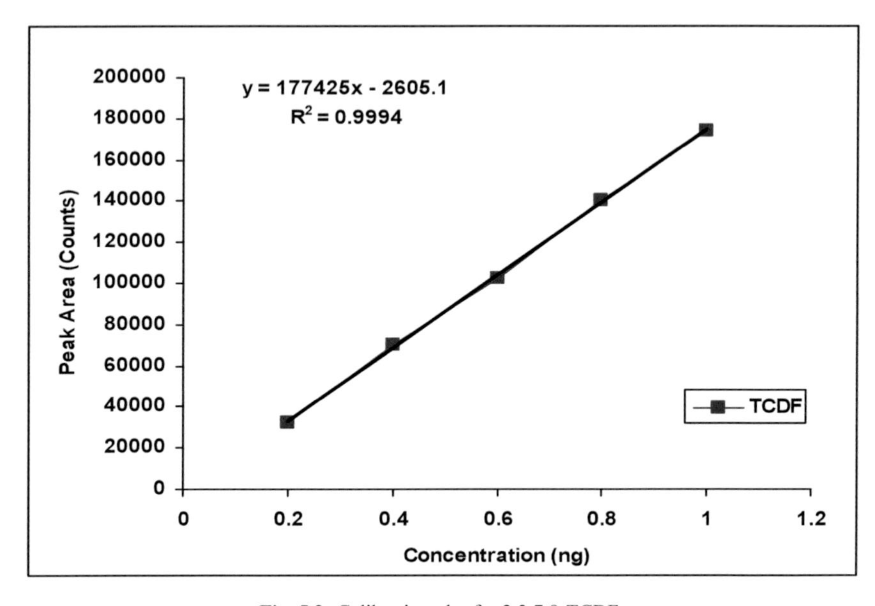

Fig. 5.2: Calibration plot for 2,3,7,8-TCDF.

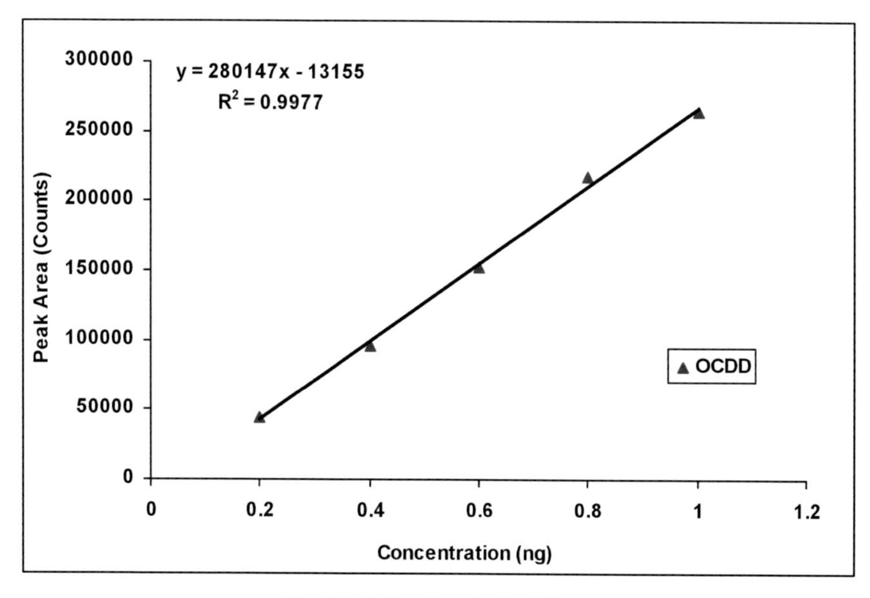

Fig. 5.3: Calibration plot for OCDD.

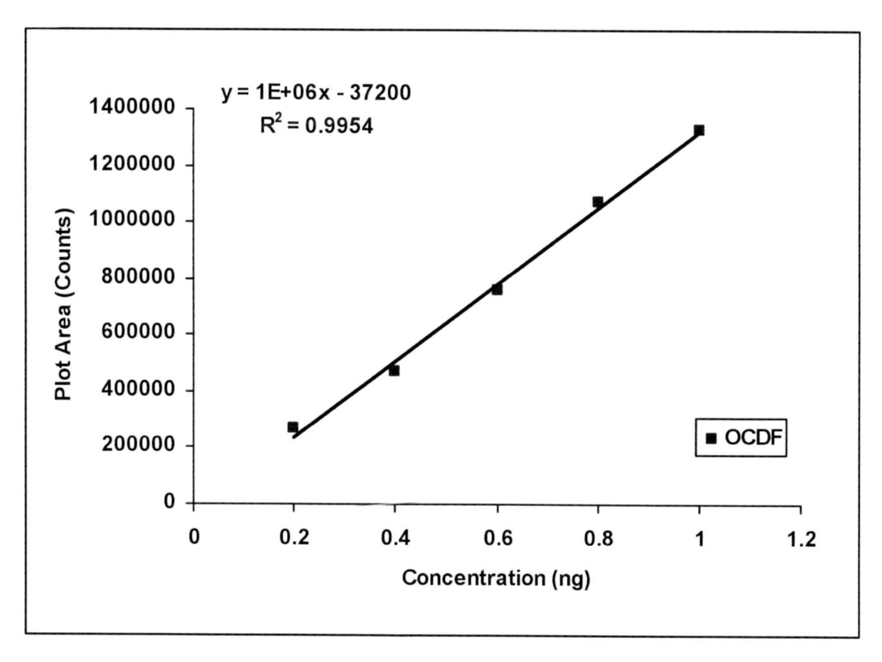

Fig. 5.4: Calibration plot for OCDF.

calibration plot is drawn using different concentrations of standard compound but injecting the same volume. This calibration plot can be used to determine the quantity of the unknown compound.

Gas Chromatographic Operations

The GC-MS/MS and GC-ECD parameters were optimized for resolution of different selected PCDD and PCDF congeners on the selected DB-5 capillary column. The optimized GC-MS and GC-ECD conditions are depicted in Tables 5.1 to 5.5. Calibration curves were plotted by preparing individual standards of different concentrations of 2,3,7,8-TCDD, 2,3,7,8-TCDF, OCDD and OCDF (Figs. 5.1 to 5.4). The gas chromatograph of 2,3,7,8-TCDD, 2,3,7,8-TCDF, OCDD, and OCDF is given in Fig. 5.5. During calibration exactly the same volume of the different concentration of standard solution was injected in the GC. To test the repeatability of the injection volumes, replicates of a single standard was injected and the standard deviation was determined. The gas chromatograms and mass fragmentograms of 2,37,8-TCDD, 2,37,8-TCDF and OCDF standard are given in Figs. 5.6 to 5.11.

Table 5.4: GC-ECD conditions optimized for PCDDs and PCDFs.

Parameters	**Conditions**
Model	Perkin Elmer
Make	Clarus-500
Carrier gas	Nitrogen
Flow rate (mL/min)	1.0
Split ratio	1:20
Column	DB-5 capillary column Length: 30 m Diameter: 0.25 mm Film thickness: 0.25 μm

Table 5.5: GC-ECD column oven programme for PCDDs and PCDFs.

Temperature (°C)	**Rate (°C/min)**	**Hold (min)**	**Total (min)**
150	0.0	1.00	1.00
200	20.0	1.00	3.50
300	3.0	20.00	53.33
325	5.0	10.00	57.83

5.6 METHOD DEVELOPMENT

Endeavors were made to develop an easy and economic method for the analysis of identified 2,3,7,8-TCDD, 2,3,7,8-TCDF, OCDD, and OCDF congeners in soil and water samples. Soxhlet and rotary shaker were used to extract 2,3,7,8-TCDD, 2,3,7,8-TCDF, OCDD, and OCDF from soil samples. Different extracting solvents viz. hexane+acetone (1:1) mixture and dichloromethane were also used and efficiency of extraction was calculated. An alternative rotary shaker extraction method was developed for analysis of 2,3,7,8-TCDD, 2,3,7,8-TCDF, OCDD, and OCDF congeners in soil samples. The liquid-liquid method has been optimized for the analysis of

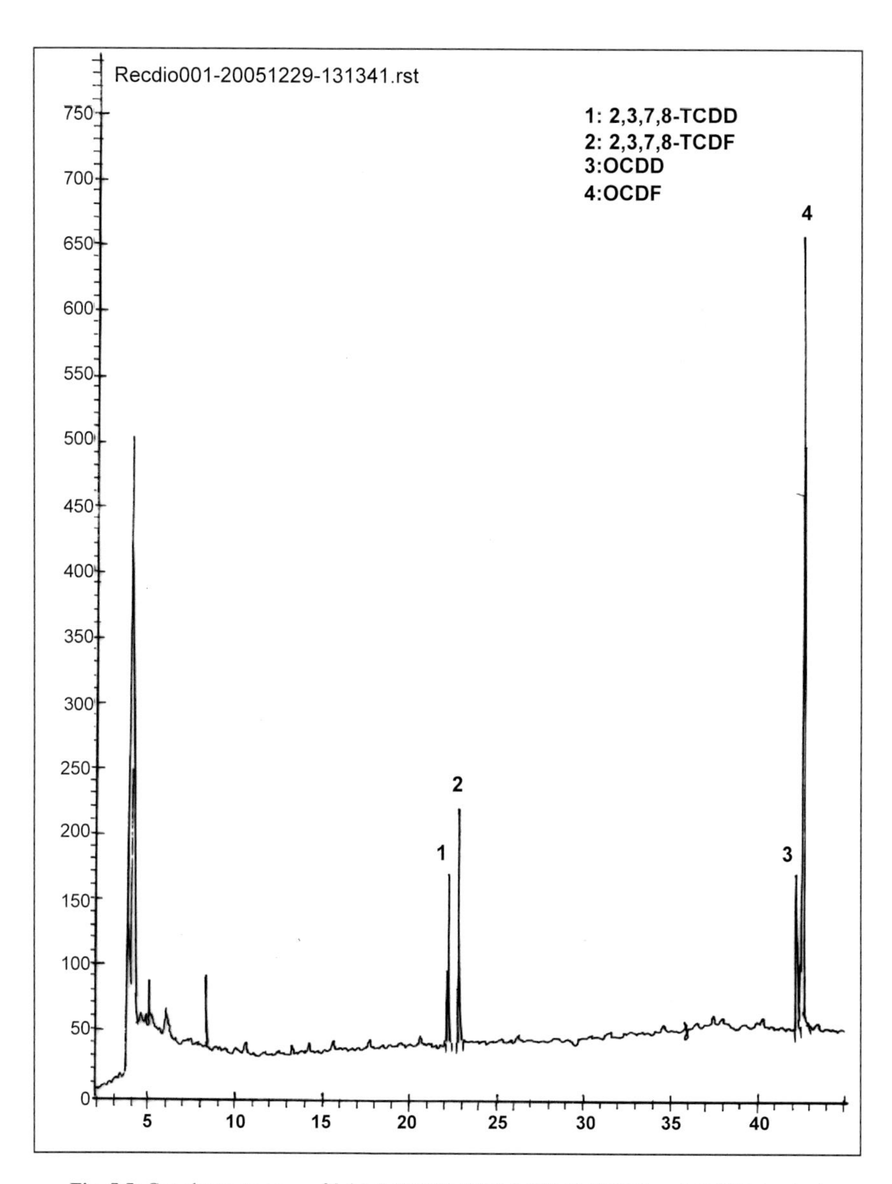

Fig. 5.5: Gas chromatogram of 2,3,7,8-TCDD, 2,3,7,8-TCDF, OCDD, and OCDF standard on DB-5 capillary column.

2,3,7,8-TCDD, 2,3,7,8-TCDF, OCDD, and OCDF in water samples. After extraction, the sample extracts were subjected to sequential clean-up process followed by GC-MS/GC-ECD analysis. The analyte specific column analysis was performed using DB-5 column. After proper clean up and concentration, 1 μL of sample extract was injected in the GC-MS and the major fragment ions of the analyte were compared with that of the reference fragmentogram of the library. Sample blanks were also processed in conjunction with samples using same reagents, materials, and equipments.

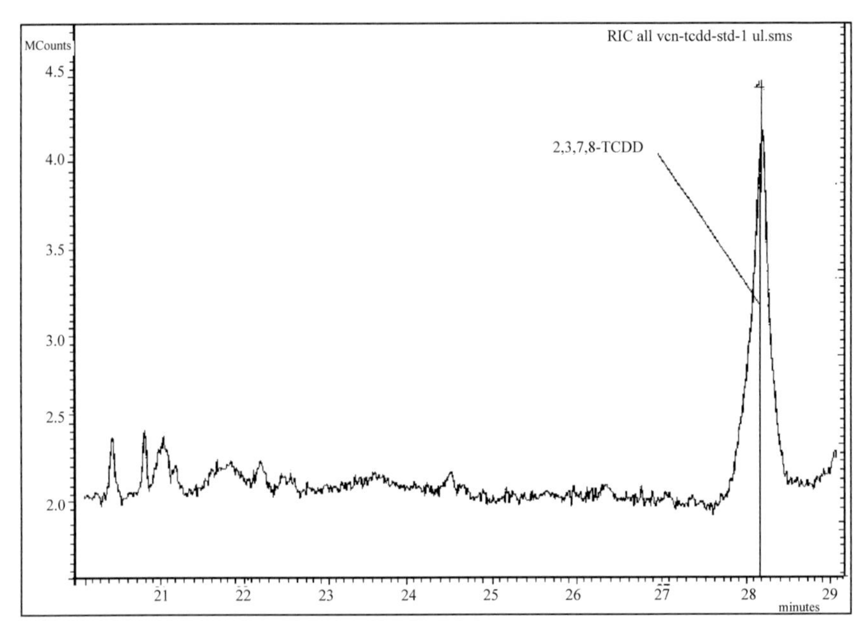

Fig. 5.6: Gas chromatogram of standard 2,3,7,8-TCDD on DB-5 capillary column.

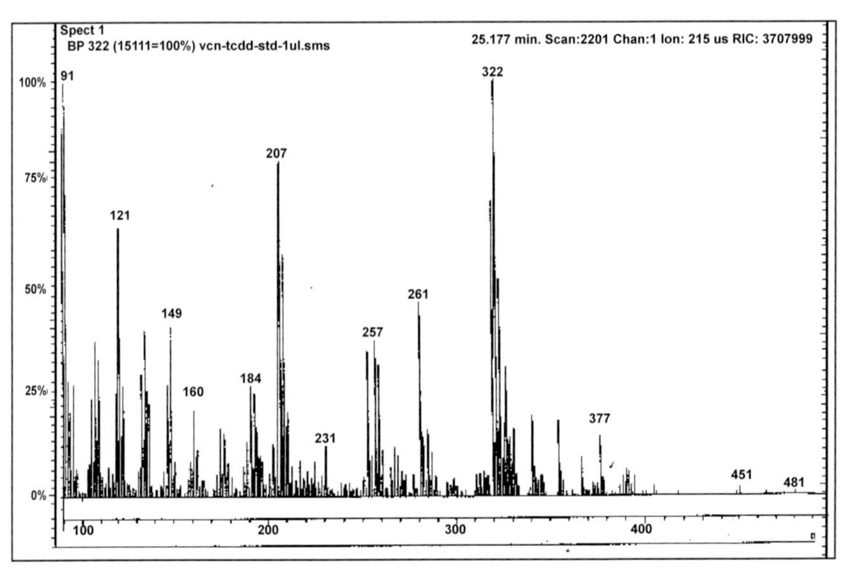

Fig. 5.7: Mass spectra of standard 2,3,7,8-TCDD on DB-5 capillary column.

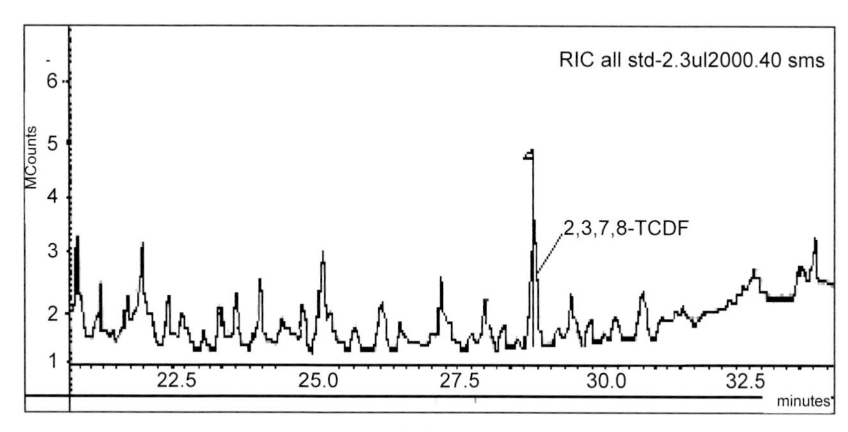

Fig. 5.8: Gas chromatogram of 2,3,7,8-TCDF standard on DB-5 capillary column.

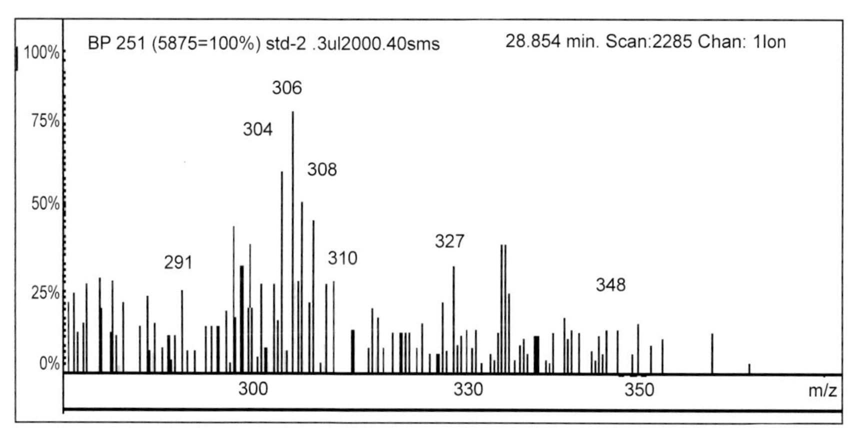

Fig. 5.9: Mass spectra of 2,3,7,8-TCDF standard on DB-5 capillary column.

5.6.1 Soil

The simulated soil samples of selected polychlorinated aromatic hydrocarbons, 2,3,7,8-TCDD, 2,3,7,8-TCDF, OCDD, and OCDF at ppt levels were analysed using two extraction techniques to establish the most efficient extraction method. Separate simulated soil samples spiked with 2,3,7,8-TCDD, 2,3,7,8-TCDF, OCDD, and OCDF were prepared. Individual 2,3,7,8-TCDD, 2,3,7,8-TCDF, OCDD, and OCDF were extracted from homogenized soil samples by rotary shaker and Soxhlet extraction methods using three different organic solvents viz. dichloromethane, acetone-hexane (1:1) mixture, and toluene for extraction. After the extraction, the bulk of the solvent was removed by evaporation in especially fabricated K-D equipment. Interfering materials were removed from the extracts by successive clean-up processes using chromatographic columns of silica gel, alumina, and carbopack-C/celite columns. Extracts were adjusted to the appropriate concentration for determinative analysis by GC-MS. The GC-MS chromatogram and mass spectra of 2,3,7,8-TCDD in soil

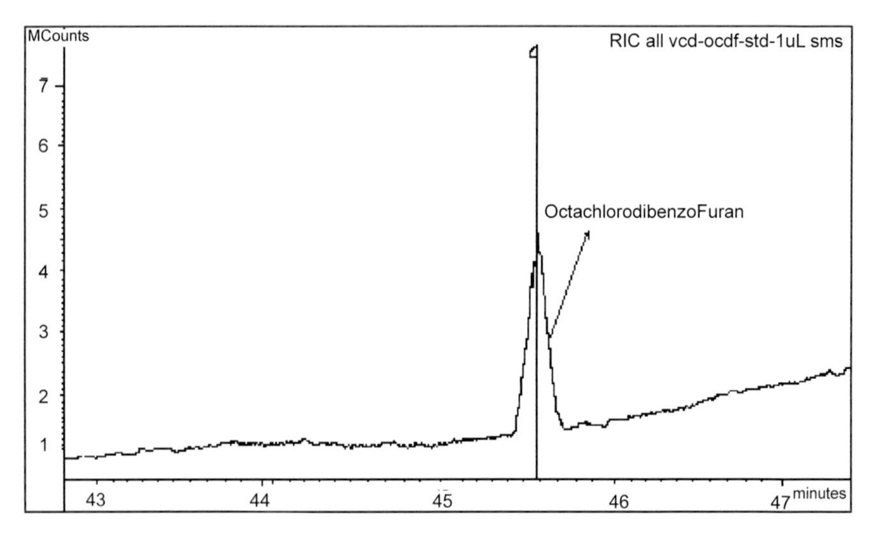

Fig. 5.10: Gas chromatogram of OCDF standard on DB-5 capillary column.

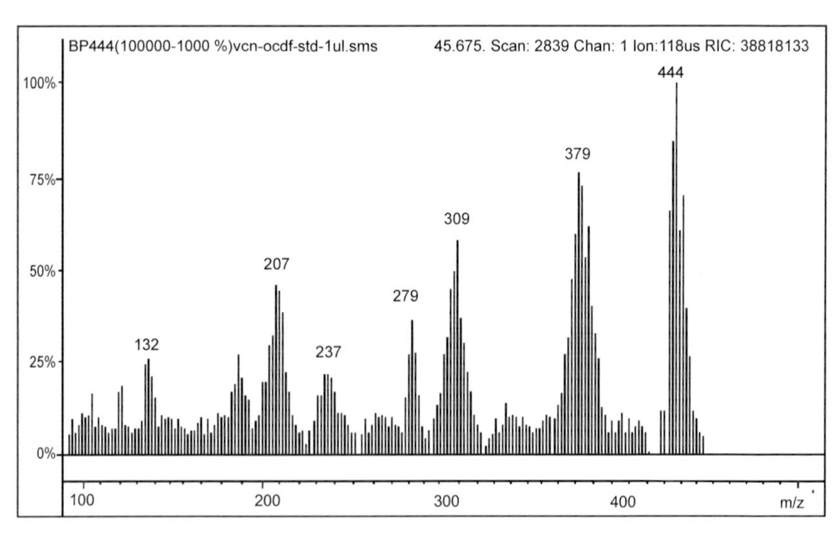

Fig. 5.11: Mass spectra of OCDF standard on DB-5 capillary column.

sample are shown in Figs. 5.10 and 5.11. A number of experiments were carried out at varied concentration ranges to find out the most efficient solvent with greater extraction efficiency and to establish the percent recovery, standard deviation, and relative standard deviation for various solvents used. The two extraction methods were compared with respect to time, economic aspect, and ease of operation.

Sample Preparation

Known amount of dry soil sample was taken in a beaker and spiked with known concentration of 2,3,7,8-TCDD, 2,3,7,8-TCDF, OCDD, and OCDF solution prepared in nonane. The soil slurry was made with acetone and the samples were thoroughly mixed with a stainless steel spatula and allowed to equilibrate at room temperature. Spiked soil samples (10 g) were weighed in 15 cm whatman filter paper and were extracted with three different solvents. Extraction was carried out using rotary shaker and soxhlet extraction apparatus.

60 simulated soil samples of 2,3,7,8-TCDD of concentration 25 ng–200 ng/10 gm of soil were prepared. The samples were divided into three sets. Samples were extracted using soxhlet (30 Nos.) and the rotary shaker method (30 Nos.); the extracting solvent used was dichloromethane (200 mL). The same procedure was repeated using hexane:acetone and toluene as the extracting solvent. The extracted samples were filtered through Whatman filter paper and subjected to various clean up processes followed by concentration and GC analysis. The extraction efficiency of each solvent was established. Each sample was analysed in duplicate for authenticity of the results.

The extraction method was performed under identical conditions following the procedure mentioned in subsequent sections. The QA/QC data in terms of % recovery, standard deviation and relative standard deviation, variance, and confidence limit of both extraction methods was calculated. GC-MS chromatogram and MS spectra of 2,3,7,8-TCDD in soil are shown in Figs. 5.12 and 5.13.

Extraction

(a) Rotary shaker

(i) The samples, 10 g each were individually placed in a separate filter paper (Whatman No. 1 of 15 cm dia). The filter papers were carefully folded to form a half circle with the sample in the centre and folded at the end of the half circle towards the centre, the total resulting length to be 70 mm; then starting at the diameter line it was rolled into and approximately cylindrical shape and inserted

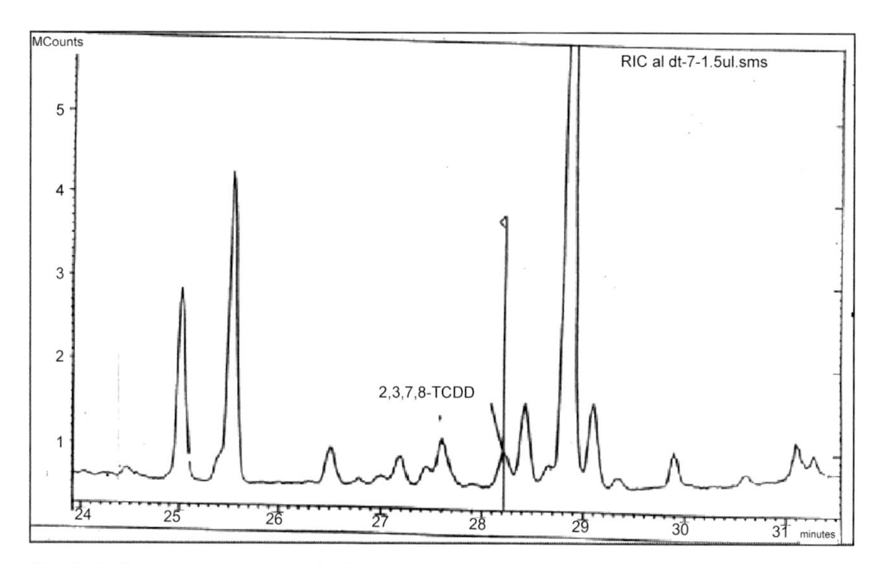

Fig. 5.12: Gas chromatogram of 2,3,7,8-TCDD in simulated soil sample on DB-5 capillary column.

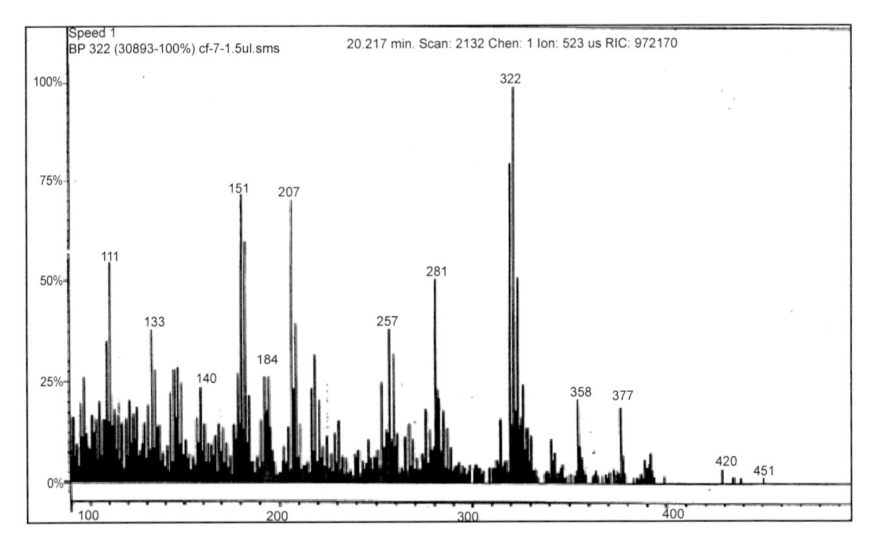

Fig. 5.13: Mass spectra of 2,3,7,8-TCDD in simulated soil sample on DB-5 capillary column.

into the (500 mL capacity) different conical flasks containing separately 100 mL each of dichloromethane, acetone-hexane, and toluene. These conical flasks were fitted to the rotatory shaker (Newtonic model) and subjected to shaking for a known period of 24 hours.

(ii) After the completion of the extraction period, the extracts were filtered and collected in 500 mL flat bottom flask. Conical flasks were rinsed with few mL of respective solvents and the rinsings-transferred to the flat bottom flasks.

(iii) Extracts were concentrated upto 2–3 mL under reduced pressure using suction pump and the reduced extract volume was transferred in each case to the evaporative concentrator or K-D evaporative tube. The evaporative tube was attached to the Snyder column and immersed about 1.5 inches into the water bath. Extracts were evaporated down to 0.1 mL, removed from the water bath, and allowed to cool down.

(iv) A soil sample reagent blank was also initiated and extracted through the identical extraction process.

(v) Concentrated extract (1 mL) of the spiked soil samples and blank were subjected to clean-up processes before GC-MS analysis.

(b) Soxhlet

(i) Approximately 10 g weighed spiked soil samples were taken in a whatman filter paper of 15 cm dia., which were folded carefully to form a half circle with the sample in the center (the total resulting length of a 70 mm) and then inserted into the extraction thimbles.

(ii) The soil sample blank was also prepared and processed through the entire extraction process.

(iii) The thimbles containing spiked soil samples and the blank were placed into the separate Soxhlet extractors. The boiling flasks were half filled with 200 mL of respective solvents, 4–5 glass beads were added, and the flasks were heated for a period of approximately 24 h on the heating mantle. There should be completion of at least 98–80 Syphon (cycles) by the Soxhlet extraction assembly, one siphon cycles normally takes 5 to 6 mins.

(iv) After completion of the extraction period (24 hours), the assembly was dissembled and the joints between the flask and condenser were rinsed with few mL of extraction solvent.

(v) Extracts were transferred from 250 mL soxhlet flask to the K-D flask, the Soxhlet flask was rinsed with 3 portions of 5 mL each of n-hexane and the rinsings were transferred to the corresponding K-D flasks or the evaporative tubes.

5.6.2 Water

Separation of 2,3,7,8-TCDD, 2,3,7,8-TCDF, OCDD, and OCDF in water is based on the principle of partition coefficient of the analytes in the organic and aqueous phase. The analysis of PCDD and PCDF congener in spiked water samples was carried out by liquid-liquid extraction followed by qualitative and quantitative estimation by GC-MS. The separating funnels of 1 L capacity were used for the extraction of the 2,3,7,8-TCDD, 2,3,7,8-TCDF, OCDD, and OCDF in water samples.

Fifty samples of simulated water of 2,3,7,8-TCDD, 2,3,7,8-TCDF, OCDD, and OCDF in duplicate were prepared in the range of 25–200 ng/L and analysed using liquid-liquid extraction. The characteristics of the tap water used for preparation of simulated samples are given in Table 5.6. Data obtained was used to establish the QA/QC data in terms of % recovery, standard deviation and relative standard deviation, variance, and confidence limit. The GC-MS chromatogram and mass spectra of 2,3,7,8-TCDD in water sample are shown in Figs. 5.14 and 5.15.

Liquid-liquid extraction

500 mL of the simulated water sample containing 2,3,7,8-TCDD, 2,3,7,8-TCDF, OCDD, and OCDF was taken in a 1 L capacity separatory funnel to which 30 mL of dichloromethane was added. The sample was extracted by shaking the separating funnel for 2 min with periodic venting. The organic layer was allowed to separate from the water phase for a minimum of 10 min and the dichloromethane layer was collected into a 250 mL concentrator flask by passing the sample extract through a column packed with a glass wool plug and 50 g of anhydrous sodium sulphate. The extraction was repeated with two additional 30 mL portions of dichloromethane, collecting each extract through anhydrous sodium sulphate column and adding it to the same concentrator flasks. After the third extraction, the sodium sulphate column was rinsed with 30 mL of dichloromethane to ensure quantitative transfer and the rinsate was added to the concentrator flask.

Concentration

Macro-concentration: Clean boiling chips were added to the flask and the extract was concentrated on heating mantle at 60–65°C under diminished pressure till the volume of the solvent reached to 5–6 mL. 50 mL of hexane was added to the extract flask and again the extract was concentrated to approximately 5–6 mL. When the apparent

Table 5.6: Physico-chemical properties of tap water.

Sr. No.	Properties	
1.	pH	6.3
2.	Conductivity (μs cm^{-1})	4
3.	TDS	2.4
4.	Total Alkalinity (mgL^{-1})	16
5.	Total Hardness (mgL^{-1})	4
6.	Calcium as Ca (mgL^{-1})	ND
7.	Magnesium as Mg (mgL^{-1})	ND
8.	Chloride as Cl (mgL^{-1})	22
9.	Sulfate as SO_4 (mgL^{-1})	ND
10.	Nitrate as NO_3 (mgL^{-1})	ND
11.	Phosphate as SO_4 (mgL^{-1})	ND
12.	Sodium as Na (mgL^{-1})	ND
13.	Potassium as K (mgL^{-1})	ND
14.	Pesticides	ND
15.	Trihalomethanes	ND
16.	Dioxins and Furans	ND

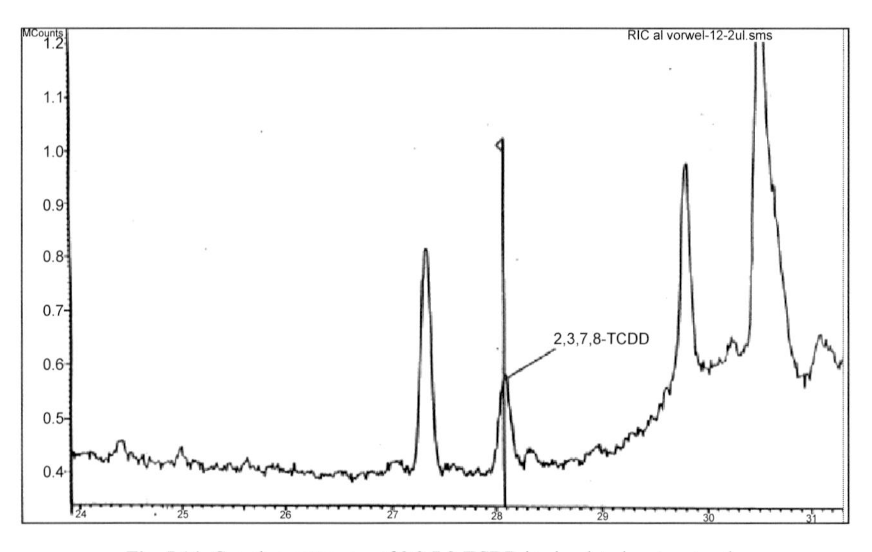

Fig. 5.14: Gas chromatogram of 2,3,7,8-TCDD in simulated water sample on DB-5 capillary column.

volume of approximately 5–8 mL is reached, the concentrated extract was transferred to the K-D evaporating tube alongwith the rinsing.

Micro-concentration: One to two clean boiling chips were added to the K-D concentrator tube and a three-ball macro Snyder column was attached to it. The column was prewet by adding approximately 1 mL of solvent through the top. The K-D apparatus was placed in a hot water bath so that the entire lower portion of the tube was bathed with steam. The temperature of the bath was adjusted (85–90°C) so as

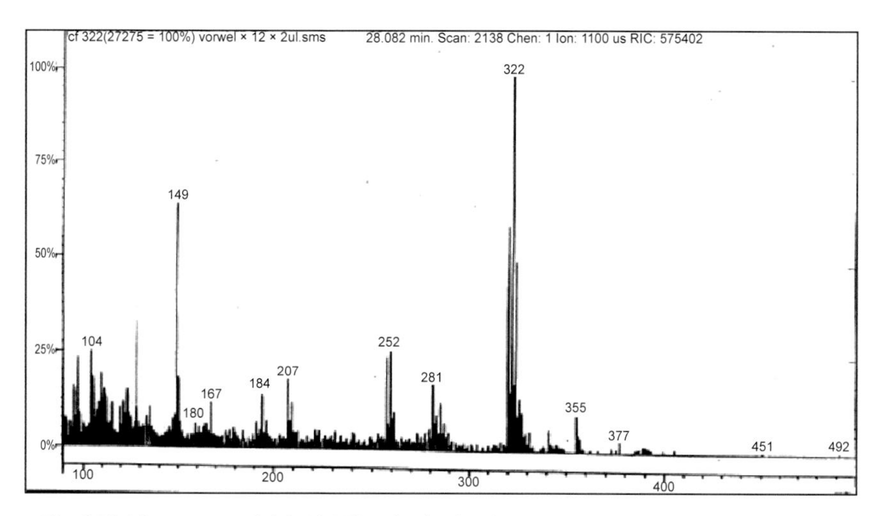

Fig. 5.15: Mass spectra of 2,3,7,8-TCDD in simulated water sample on DB-5 capillary column.

to complete the concentration in 15–20 minutes. At the proper rate of distillation, the balls of the column will actively chatter but the chambers will not flood.

When the liquid reached an apparent volume of 1 mL, the K-D apparatus was removed from the bath and the solvent was allowed to drain and cool for at least 10 minutes. The Snyder column was removed and the flask and its lower joint rinsed into the concentrator tube with 1–2 mL of solvent. The three-ball Snyder column was removed, a fresh boiling chip was added, and a two-ball micro Snyder column was attached to the concentrator tube. Again the column was prewet by adding approximately 0.5 mL of solvent through the top and the apparatus was placed in the hot water bath. The water temperature was adjusted to complete the concentration in 5–10 minutes. When the liquid reached an apparent volume of 0.5 mL, the apparatus was removed from the water bath and allowed to drain and cool for at least 10 minutes.

For clean up using silica gel, alumina, or carbon the extracts were exchanged into hexane. When the extracts were free of interferences, the extract was concentrated for injection into the GC/MS. The extract was quantitatively transferred to a conical vial for final concentration, rinsing the larger vial with hexane, and adding the rinse to the conical vial. The volume was reduced to approximately 100 μL. 10 μL of nonane was added to the vial, and the solvent evaporated to the level of the nonane. The vial was sealed and labeled with the sample number and stored in the dark at room temperature until ready for GC/MS analysis.

Clean-Up

Silica gel: A glass-wool plug was placed in a 15 mm ID chromatography column. The column was packed from bottom to top with 1 g silica gel, 4 g basic silica gel, 1 g silica gel, 8 g acidic silica gel, 2 g silica gel, and 4 g granular anhydrous sodium sulphate. The column was tapped to settle the adsorbents. The column was pre-eluted with 50–100 mL of hexane and the eluate was discarded. The column was checked for channelling. If channelling was present, the column was discarded and another column was prepared. The concentrated extract was added to the column. The receiver was rinsed twice with 1 mL portions of hexane, and applied separately to the column. All the PCDD and PCDF congeners were eluted with 100 mL hexane, and the eluate collected. The eluate was concentrated as discussed above for further cleanup or injection into GC/MS (Fig 5.16a).

Alumina: A glass-wool plug was placed in a 15 mm ID chromatography column. The column was packed by adding 6 g acid alumina. The column was tapped to settle the adsorbents. The column was pre-eluted with 50–100 mL of hexane, the stopcock was closed when the hexane was within 1 mm of the alumina, the eluate was discarded. The column was checked for channeling. If channeling was present, the column was discarded and another column was prepared. The concentrated extract was added to the column. The stopcock was opened until the extract was within 1 mm of the alumina. The receiver was rinsed twice with 1 mL portions of n-hexane, and applied separately to the column. The interfering compounds were eluted with 100 mL n-hexane and the eluate was discarded. The PCDDs/PCDFs eluted from the column with 20 mL methylene chloride:n-hexane (20:80 v/v) and the eluate was collected. The eluate was concentrated as per Section 3.5.3 for further cleanup or injection into GC/MS (Fig. 5.16b).

Carbopak/Celite: A glass-wool was plugged at one end of a 10 cm column having 0.5 cm ID, and the column packed with 0.55 g of Carbopak/Celite to form an adsorbent bed approximately 2 cm long. A glass-wool plug was inserted on top of the bed to hold the adsorbent in place. The column was pre-eluted with 5 mL of toluene followed by 2 mL of methylene chloride:methanol:toluene (15:4:1 v/v), 1 mL of methylene chloride:cyclohexane (1:1 v/v), and 5 mL of hexane. When the solvent was within 1 mm of the column packing, the sample extract along with hexane rinsates was applied to the column. The interfering compounds were eluted with two 3 mL portions of hexane, 2 mL of methylene chloride:cyclohexane (1:1 v/v), and 2 mL of methylene chloride:methanol:toluene (15:4:1 v/v). The eluate was discarded. The column was

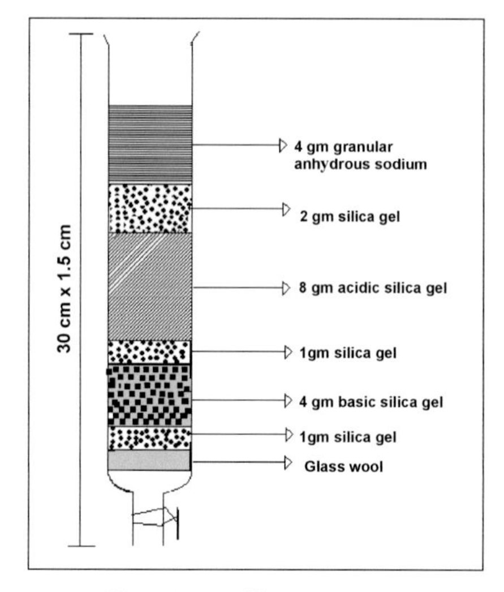

Fig. 5.16(a): Silica gel column.

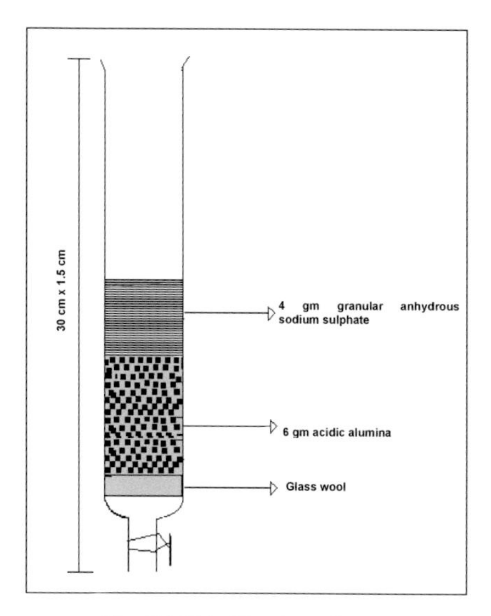

Fig. 5.16(b): Alumina column.

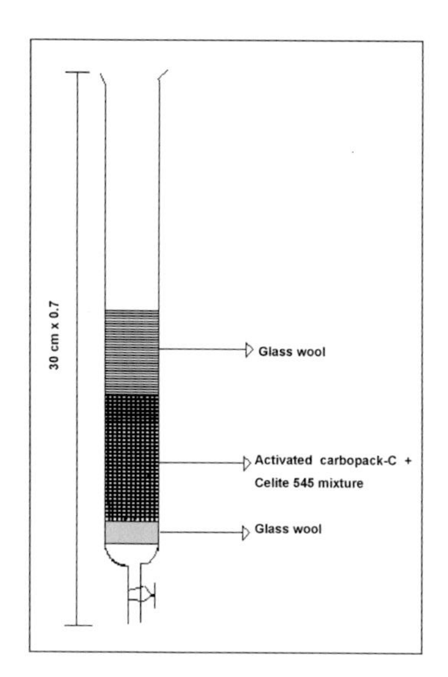

Fig. 5.16(c): Carbopack-Celite 545 column.

inverted, and the PCDDs and PCDFs eluted with 20 mL of toluene. The eluate was concentrated as detailed earlier for further injection into the GC/MS (Fig. 5.16c).

GC-MS Analysis

The operating conditions are given in Table 5.1 through 5.3. 10 μL of the appropriate internal standard solution was added to the sample extract immediately prior to injection to minimize the possibility of loss by evaporation, adsorption, or reaction. 1.0 μL or 2.0 μL of the concentrated extract containing the internal standard solution was injected, using on-column or splitless injection. The MS data was collected after the solvent peak elutes. The column was returned to the initial temperature for analysis of the next extract or standard.

Qualitative determination: Qualitative identification of a congener in the sample is made by matching the retention time with the retention time of the standard obtained under identical conditions. The signal-to-noise ratio (S/N) for the GC peak at each exact m/z must be greater than or equal to 2.5 for each PCDD and PCDF congener detected in a sample extract. The relative retention time of the peak for a congener is commonly recorded which must be within the limit. If interferences preclude identification, a new aliquot of sample must be extracted, further cleaned up, and analyzed.

Quantitative estimation: The quantity of compound present is proportional to the area of the peak and can be used to determine the concentration of the component in the sample extract.

Calculations

Soil

The concentrations of 2,3,7,8-TCDD, 2,3,7,8-TCDF, OCDD, and OCDF in soil sample were calculated using the following formula:

$$C_x = \frac{C_s \times A_x \times V_1}{A_s \times V \times B}$$

where :

C_x = quantity (ng) of the analyte present per gram
C_s = quantity (ng) of standard injected
A_s = peak area of standard
A_x = peak area of analyte
V_1 = final volume of the sample extract (μL)
V = volume of the sample extract injected (μL)
B = amount (gm) of soil taken

Water

The concentrations of 2,3,7,8-TCDD, 2,3,7,8-TCDF, OCDD, and OCDF in water sample were calculated using the formula:

$$C_x = \frac{C_s \times A_x \times V_1}{A_s \times V \times B}$$

where :

C_x = quantity (pg) of 2,3,7,8-TCDD present per μL
C_s = quantity (pg) of standard injected
A_s = peak area of standard
A_x = peak area of analyte
V_1 = final volume of the sample extract (mL)
V = volume of the sample extract injected (μL)
B = volume (mL) of water samples taken

Percent recovery

Percent Recovery = A/B × 100
A = observed concentration of the analyte
B = actual concentration of the analyte

Standard deviation

$$\text{Standard deviation } (\sigma) = \sqrt{\frac{\Sigma(x - x_1)^2}{n-1}}$$

x_1 = individual observation
x = mean observation
n = number of observation

Relative standard deviation

Relative standard deviation = σ/x x 100
σ = standard deviation
x = mean of the observations

5.7 APPLICATION OF DEVELOPED METHOD FOR FIELD SAMPLE ANALYSES

A field survey was conducted to study the extent of contamination due to PCDDs/PCDFs in various pulp and paper mills and plastic and PVC industries. Those industries in India were randomly selected which were using chlorine or chlorine based compounds in their manufacture process. Samples of final effluent and soil were collected from various pulp and paper mills and three plastic and PVC industries situated in Maharashtra, Tamilnadu, and Andhra Pradesh for validation of the developed method for PCDDs and PCDFs analysis. Four specific PCDDs and PCDFs were monitored in the contaminated samples.

Sample collection and preservation: 1 L of final effluent samples were collected in amber coloured 1 L capacity bottle with Teflon coated screw caps. No preservative was added to the samples. The samples were brought to the laboratory and refrigerated until analyses. Soil samples were collected in broad mouth glass bottles. The soil samples were ground and allowed to dry at room temperature.

Sample extraction and concentration: The soil samples and the effluent samples were extracted and concentrated as mentioned in previous sections.

Sample clean up: The concentrated extracts were subjected to silica gel, alumina, and carbopack/celite clean-up. The eluates were again concentrated and subjected to GC/MS for analysis.

GC-MS analysis: The concentrated extracts were analysed on GC-MS to identify the 2,3,7,8-substituted Dioxins and Furans and its precursors present in the samples.

5.7.1 Pulp and Paper Mills

Sampling was carried out at the pulp and paper mills using elemental chlorine and chlorine dioxide in pulp bleaching process. The industry using chlorine for bleaching process has shown contamination with 2,3,7,8-TCDD and other Dioxin congeners.

5.7.2 Plastic and PVC Industries

Final effluent and soil samples were collected from the plastic and PVC industries using either chlorine or chlorine base chemicals in their manufacturing processes and producing chlorine base products viz. PVC pipes, PVC resins, and chlorinated solvents.

5.8 RESULTS AND DISCUSSION

5.8.1 QA/QC of Rotary Shaker and Soxhlet Extraction Methods

To assure the validity and reliability of analytical procedure in estimating 2,3,7,8-TCDD, 2,3,7,8-TCDF, OCDD, and OCDF in soil, extensive internal quality assurance programme has been carried out. Number of laboratory exercises in the estimation of 2,3,7,8-TCDD in soil has been made for the selection of the most efficient and specific solvent. Two different extraction methods viz. rotary shaker and soxhlet were used for extraction of PCDDs and PCDFs from soil samples. The methods were compared for their efficiency. The simulated soil samples of 2,3,7,8-TCDD were prepared in soil having concentration in the range of 25–200 ng/10 gm. The simulated soil samples were subjected to rotary shaker extraction methods using three different extracting solvents viz. dichloromethane, hexane: acetone and toluene. The samples were extracted for nearly 24 h, after extraction the samples were filtered through Whatman filter paper and subjected to various clean up processes followed by concentration and GC analysis. The same procedure was followed for soxhlet extraction method. Thus the extraction efficiency of each method with all the three solvents was established. Experiments were conducted to establish the percent recovery, standard deviation, and extraction efficiency of the above solvents.

Nearly 50 simulated soil samples of 2,3,7,8-TCDD, 2,3,7,8-TCDF, OCDD and OCDF were prepared. The precision and accuracy of measurements were computed in terms of standard deviation and relative standard deviation. The data has been summarized in Tables 5.7 to 5.9. The analysis results were statistically analysed for finding out the most effective extraction method and the solvent having more extraction efficiency (Tables 5.10 to 5.12).

(i) Statistical evaluation for efficient procedure

The mean % recovery, standard deviation and coefficient of variation of 2,3,7,8-TCDD, for each of the three extracting solvents separately using the rotary shaker method has been calculated (Table 5.11). The mean % recovery, standard deviation, and coefficient of variation of 2,3,7,8-TCDF, OCDD, and OCDF using the rotary shaker method has been calculated (Table 5.11).

The results of response factor as % recovery of 2,3,7,8-TCDD using two different methods have been analysed using student's 't' test for comparison of the mean recovery for each of the solvents, used in extraction separately. The null hypothesis (Ho) has been formulated that there does not exist any significant difference between the mean results of % recovery of 2,3,7,8-TCDD by the methods using a particular

Table 5.7: 2,3,7,8-TCDD analysis in soil with dichloromethane using rotary shaker and Soxhlet extraction.

Sr. No.	Sample Code	Concentration (ng) Actual	Extraction Time (hrs.)	(%) Recovery	
				Rotary Shaker	Soxhlet
1	Blank	–	24	–	–
2	RS-SDC-1	50	24	25.34	17.2
3	RS-SDC-2	50	24	38.5	12.3
4	RS-SDC-3	50	24	20.05	35.5
5	RS-SDC-4	50	24	18.76	21.6
6	RS-SDC-5	50	24	23.8	20.2
7	RS-SDC-6	50	24	35.5	26.2

Table 5.8: 2,3,7,8-TCDD analysis in soil with hexane-acetone using rotary shaker and Soxhlet extraction.

Sr. No.	Sample Code	Concentration (ng) Actual	Extraction Time (hrs.)	(%) Recovery	
				Rotary Shaker	Soxhlet
1	Blank	–	24	–	–
2	SHA-1	50	24	16	25.4
3	SHA-2	50	24	17.1	10.2
4	SHA-3	50	24	10.32	17.4
5	SHA-4	50	24	19.98	11.4
6	SHA-5	50	24	10.44	15.5
7	SHA-6	50	24	10.3	12.5

Table 5.9: 2,3,7,8-TCDD analysis in soil with toluene using rotary shaker and Soxhlet extraction.

Sr. No.	Sample Code	Initial Concentration (ng)	Extraction Time (hrs.)	(%) Recovery	
				Rotary Shaker	Soxhlet
1	Blank-ST	–	24	–	
2	ST-1	50	24	33	–
3	ST-2	50	24	50.98	100
4	ST-3	50	24	60.22	68.99
5	ST-4	50	24	80.08	98.3
6	ST-5	50	24	87.8	74.0
7	ST-6	25	24	82	72.8
8	ST-7	25	24	81.72	74.70
9	ST-8	100	24	51.62	72.78
10	ST-9	100	24	70.56	77.0
11	ST-10	100	24	70.99	83.2
12	ST-11	100	24	68.78	72.8

Table 5.10: Student's 't' test for comparison of rotary shaker and Soxhlet extraction methods for 2,3,7,8-TCDD in soil.

Solvent	Number of Observations	Calculated Value of 't'	Degree of freedom (m + n – 2)	Critical Value of 't'
Dichloro-methane	m = 6 n = 6	0.128	10	2.228
n-Hexane-Acetone (1:1)	m = 6 n = 6	0.483	10	2.228
Toluene	m = 10 n = 10	0.014	18	2.101

Table 5.11: Statistical evaluation of solvents extraction efficiency for 2,3,7,8-TCDD in soil.

	Solvent	Mean Recovery (%)	Standard Deviation	Coefficient of Variation (%)
2,3,7,8-TCDD	Dichloromethane	27.00	8.15	30.20
	n-Hexane-Acetone (1:1 v/v)	14.02	4.24	30.26
	Toluene	80.03	5.45	6.81
2,3,7,8-TCDF	Toluene	74.85	6.27	8.38
OCDD		73.90	10.42	14.06
OCDF		75.38	6.85	9.09

solvent. The hypothesis has been tested against the alternative hypothesis (H_1) that the recovery results by two methods differ significantly. The 't' statistic has been calculated using the standard formula. The calculated value of 't' has then been compared with the critical value from the table for (m + n – 2) degrees of freedom where m and n are the number of observation in the rotary shaker and soxhlet extraction assembly respectively.

At 5% level of significance the calculated value of 't' is less than the critical value in each of the three solvents. It has been inferred from the data evaluation that for all the solvents the recovery percentage obtained by using rotary shaker and soxhlet extraction assembly is almost equal. Both the methods are equally good in % recovery of 2,3,7,8-TCDD and are almost comparable. Though both the methods have been found to be compatible and yield almost the same % recovery, the rotary shaker method has been preferred because of the less solvent consumption, less time consuming, and ease in carrying out the extraction.

Dichloromethane and hexane-acetone solvents have almost same coefficient of variation and average recovery (Table 5.11). In case of toluene, the average recovery is very high (80.03%) and the corresponding coefficient of variation is very low (12.35%). Thus, toluene has been found as a very efficient solvent for recovery of 2,3,7,8-TCDD and precise while using the rotary shaker method when compared with the other two solvents results.

The findings of statistical analysis show that both the methods are equally good in % recovery and are almost comparable (Table 5.12). The mean % recovery, standard deviation, and coefficient of variation of 2,3,7,8-TCDD among three extracting solvents dichloromethane, hexane-acetone, and toluene have been calculated using rotary and soxhlet extraction assembly. The most efficient average recovery of 80.03%

Table 5.12: Recovery efficiency of 2,3,7,8-TCDD using two different methods and three different solvents.

Sr. No.	Solvent	Method	Recovery Efficiency (%)											
			1	2	3	4	5	6	7	8	9	10	11	12
1.	Dichloromethane	Rotary Shaker	25.3	38.5	20.1	18.8	23.8	35.5	–	–	–	–	–	–
		Soxhlet	17.2	12.3	35.5	21.6	20.2	26.2	–	–	–	–	–	–
2.	n-Hexane + Acetone (1:1)	Rotary Shaker	16.0	17.1	10.3	20.0	10.4	10.3	–	–	–	–	–	–
		Soxhlet	25.4	10.2	17.4	11.4	15.5	12.5	–	–	–	–	–	–
3.	Toluene	Rotary Shaker	60.2	80.1	87.8	82.0	81.7	70.6	71.0	68.8	63.8	60.6	79.9	74.7
		Soxhlet	100.0	50.7	98.3	74.0	72.8	68.0	77.0	83.2	72.8	68.0	–	–

has been found in the case of toluene. However, for further experiments, the rotary shaker method and toluene was used for analysis of 2,3,7,8-TCDF, OCDD, and OCDF in soil samples.

The data analysed showed that the percent recovery of 2,3,7,8-TCDD in soil at 2.5, 5.0, 10.0, and 20.0 ng g^{-1} using rotary Shaker extraction was 80.03, 81.38, 76.31, and 82.9 respectively. The percent recovery of 2,3,7,8-TCDF in soil at 2.5, 5.0, 10.0, and 20.0 ng g^{-1} was 74.82, 82.34, 80.81, and 82.18 respectively. The percent recovery of OCDD in soil at 2.5, 5.0, 10.0, and 20.0 ng g^{-1} was 73.90, 100.86, 90.14, and 94.82 respectively. The percent recovery of OCDF in soil at 2.5, 5.0, 10.0, and 20.0 ng g^{-1} was 75.38, 90.94, 88.07 and 89.54 respectively (Tables 5.13 to 5.16). The method detection limit was calculated as 0.01 ng g^{-1}.

5.8.2 QA/QC of Liquid-Liquid Extraction Method

In the present work liquid-liquid extraction has been used for the extraction of water samples followed by extensive clean–up and GC-MS analysis. Nearly 50 simulated water samples were prepared at varying 2,3,7,8-TCDD, 2,3,7,8-TCDF, OCDD, and OCDF concentration (25, 50 and 100 ng L^{-1}) and analysed as mentioned in Section 3.5.2. The GC-MS chromatogram and mass spectra of 2,3,7,8-TCDD in water sample are shown in Figs. 5.12 and 5.13. Data obtained was used to establish laboratory bias, and the standard deviation of the percent recovery.

The precision and accuracy of measurements were computed in terms of standard deviation and relative standard deviation. The data has been summarized in Tables 5.17 to 5.20. The analysis results were statistically analysed for finding out the extraction efficiency, percent recovery and standard deviation.

The data analysed showed that the percent recovery of 2,3,7,8-TCDD in water at 25, 50, 100, and 200 ng L^{-1} was 89.86, 85.04, 93.28, and 92.09 respectively. The percent recovery of 2,3,7,8-TCDF in water at 25, 50, 100, and 200 ng L^{-1} was 84.82, 84.84, 87.80, and 88.06 respectively. The percent recovery of OCDD in water at 25, 50, 100, and 200 ng L^{-1} was 90.28, 91.71, 93.54, and 94.14 respectively. The percent

Table 5.13: QA/QC Data for 2,3,7,8-TCDD, 2,3,7,8-TCDF, OCDD, and OCDF at 2.5 ng g^{-1} in soil.

S. No.	Sample	% Recovery			
		2,3,7,8-TCDD	**2,3,7,8-TCDF**	**OCDD**	**OCDF**
1	REC-1	85.2	69.52	96.6	84.88
2	REC-2	81.8	73.00	61.2	66.90
3	REC-3	73.4	74.50	81.6	84.48
4	REC-4	84.9	80.00	63.6	74.80
5	REC-5	80.2	89.00	78.00	75.92
6	REC-6	74.00	68.00	75.00	72.30
7	REC-7	77.00	68.00	72.00	79.50
8	REC-8	72.80	75.50	69.00	80.00
9	REC-9	88.00	76.00	65.00	69.00
10	REC-10	83.00	75.00	77.0	66.00
Average Recovery		**80.03**	**74.85**	**73.90**	**75.38**
Standard Deviation		**5.45**	**6.27**	**10.42**	**6.85**
RSD		**6.81**	**8.38**	**14.06**	**9.09**

Table 5.14: QA/QC Data for 2,3,7,8-TCDD, 2,3,7,8-TCDF, OCDD, and OCDF at 5.0 ng g^{-1} in soil.

Sr. No.	Sample	% Recovery			
		2,3,7,8-TCDD	2,3,7,8-TCDF	OCDD	OCDF
1	REC-1	63.38	61.1	89.84	96.00
2	REC-2	82.3	104.2	118.0	88.50
3	REC-3	80.0	93.2	114.0	82.00
4	REC-4	85.66	87.9	106.0	94.00
5	REC-5	114.0	61.0	128.00	93.20
6	REC-6	85.07	81.5	85.00	95.00
7	REC-7	81.72	82.00	79.45	88.70
8	REC-8	70.99	79.50	86.30	79.00
9	REC-9	70.56	82.00	100.00	98.00
10	REC-10	80.08	91.00	102.00	95.00
Average Recovery		**81.38**	**82.34**	**100.86**	**90.94**
Standard Deviation		**13.54**	**13.41**	**15.90**	**6.29**
RSD		**16.64**	**16.29**	**15.76**	**6.92**

Table 5.15: QA/QC Data for 2,3,7,8-TCDD, 2,3,7,8-TCDF, OCDD and OCDF at 10.0 ng g^{-1} in soil.

Sr. No.	Sample	% Recovery			
		2,3,7,8-TCDD	2,3,7,8-TCDF	OCDD	OCDF
1	REC-1	79.87	77.00	87.00	95.00
2	REC-2	74.70	85.00	104.00	88.55
3	REC-3	72.78	82.00	95.00	91.28
4	REC-4	73.74	73.25	82.40	93.00
5	REC-5	72.99	86.30	80.00	86.00
6	REC-6	72.50	95.00	79.00	86.40
7	REC-7	87.00	84.20	98.00	81.00
8	REC-8	74.55	69.30	85.00	79.50
9	REC-9	83.00	70.00	94.00	90.00
10	REC-10	72.00	86.00	97.00	90.00
Average Recovery		**76.31**	**80.81**	**90.17**	**88.07**
Standard Deviation		**5.17**	**8.23**	**8.58**	**4.95**
RSD		**6.77**	**10.19**	**9.52**	**5.62**

recovery of OCDF in water at 25, 50, 100, and 200 ng L^{-1} was 90.14, 90.74, 90.41, and 93.78, respectively.

5.8.3 Validation of Developed Method

The field samples were analysed for validation of the developed method. Effluents and soil samples were collected from pulp and paper mills and PVC industries and analysed for specific four PCDDs and PCDFs.

In PPM-1, 0.12 ng L^{-1} of 2,3,7,8-TCDD was detected in its effluent (Table 5.21 and Fig. 5.17). 2,3,7,8-TCDD was not detected in effluents using both chlorine and chlorine dioxide and only when chlorine dioxide was used (Table 5.21). In PPM-2, in the effluent samples of bleached plant, pulp mill, recycled bleached plant, combined

Table 5.16: QA/QC Data for 2,3,7,8-TCDD, 2,3,7,8-TCDF, OCDD, and OCDF at 20.0 ng g^{-1} in soil.

Sr. No.	Sample	% Recovery			
		2,3,7,8-TCDD	**2,3,7,8-TCDF**	**OCDD**	**OCDF**
1	REC-1	87.68	85.20	78.00	89.00
2	REC-2	90.52	75.55	104.00	97.20
3	REC-3	89.22	95.25	87.20	91.28
4	REC-4	90.88	73.25	108.00	93.00
5	REC-5	73.99	77.00	82.00	90.10
6	REC-6	86.46	94.52	110.00	86.40
7	REC-7	74.22	82.00	102.00	81.00
8	REC-8	80.00	90.00	95.00	82.00
9	REC-9	81.000	70.00	93.00	87.45
10	REC-10	75.00	79.00	89.00	98.00
Average Recovery		**82.90**	**82.18**	**94.82**	**89.54**
Standard Deviation		**6.88**	**8.83**	**10.97**	**5.68**
RSD		**8.30**	**10.75**	**11.57**	**6.34**

Table 5.17: QA/QC Data for 2,3,7,8-TCDD, 2,3,7,8-TCDF, OCDD, and OCDF at 25 ng L^{-1} in water.

Sr. No.	Sample	% Recovery			
		2,3,7,8-TCDD	**2,3,7,8-TCDF**	**OCDD**	**OCDF**
1	REC-1	77.50	91.55	103.60	104.00
2	REC-2	84.68	92.50	83.40	82.00
3	REC-3	86.60	89.20	97.50	93.52
4	REC-4	89.20	85.0	86.00	89.60
5	REC-5	90.80	80.20	88.50	84.80
6	REC-6	83.90	83.20	85.3	81.7
7	REC-7	84.98	92.13	89.2	98.3
8	REC-8	91.28	70.64	79.4	92.5
9	REC-9	101.0	82.00	128	120.0
10	REC-10	108.0	86.20	109	100.0
Average Recovery		**89.86**	**85.27**	**90.28**	**90.14**
Standard Deviation		**8.95**	**6.75**	**9.89**	**9.24**
RSD		**9.96**	**7.91**	**10.96**	**10.25**

effluent from bleaching plant and pulp mill inlet to effluent treatment plant (ETP) and final treated effluent, 2,3,7,8-TCDD was below detection limit (Table 5.21) 0.01 ng g^{-1} of 2,3,7,8-TCDD was found in bleached pulp sample (Table 5.21). In another paper mill (PPM-3), the effluent samples collected before and after treatment and combined effluent of bleaching plant and pulp mill did not show 2,3,7,8-TCDD in the detectable range. In the samples of ETP sludge, garden soil, unbleached and bleached pulp, the final product writing paper and white paper, 2,3,7,8-TCDD was below detectable level. In paper mill (PPM-4) (Table 5.21). Dioxin contamination was not found in the samples of sludge from primary clarifier, agriculture soil, writing paper, and pulp mill effluent.

Table 5.18: QA/QC Data for 2,3,7,8-TCDD, 2,3,7,8-TCDF, OCDD, and OCDF at 50 ng L^{-1} in water.

Sr. No.	Sample	% Recovery			
		2,3,7,8-TCDD	**2,3,7,8-TCDF**	**OCDD**	**OCDF**
1	REC-1	87.8	88.3	106.6	106.7
2	REC-2	90.5	90.5	93.4	82.2
3	REC-3	89.2	89.2	97.3	92.2
4	REC-4	90.9	85.1	99.5	90.8
5	REC-5	73.99	81.0	78.1	74.8
6	REC-6	86.46	79.23	94.97	89.34
7	REC-7	81.72	85.10	100.20	85.65
8	REC-8	82.00	78.77	80.00	112.00
9	REC-9	80.08	91.20	77.52	79.20
10	REC-10	87.80	80.00	89.50	94.50
Average Recovery		**85.04**	**84.84**	**91.71**	**90.74**
Standard Deviation		**5.43**	**4.82**	**10.16**	**11.60**
RSD		**6.38**	**5.69**	**11.08**	**12.78**

Table 5.19: QA/QC Data for 2,3,7,8-TCDD, 2,3,7,8-TCDF, OCDD, and OCDF at 100 ng L^{-1} in water.

Sr. No.	Sample	% Recovery			
		2,3,7,8-TCDD	**2,3,7,8-TCDF**	**OCDD**	**OCDF**
1	REC-1	84.7	90.00	85.3	81.7
2	REC-2	91.3	80.00	89.2	98.3
3	REC-3	85.0	82.88	79.4	92.5
4	REC-4	101.0	85.00	128	120.0
5	REC-5	108.00	95.05	109	100.0
6	REC-6	92.00	81.86	96.20	89.55
7	REC-7	95.0	77.78	75.00	78.45
8	REC-8	93.0	100.00	116.00	75.55
9	REC-9	100.00	91.20	80.25	82.20
10	REC-10	98.30	84.20	77.00	85.80
Average Recovery		**93.28**	**87.80**	**93.54**	**90.41**
Standard Deviation		**7.44**	**7.36**	**18.31**	**13.18**
RSD		**7.98**	**8.38**	**19.58**	**14.58**

In PPM-5, in the samples of grade-II and grade-III treated effluent, sludge of grade-II and grade-III treated and soil sample from high rate transpiration system (HRTS), the Dioxin contamination was below detection limit (grade-II and grade-III were the various stages of industrial process). In paper mill (PPM-6), the samples of effluents from bleached plant, inlet to ETP, final treated, ETP sludge and ash from boiler, the Dioxin contamination was below detectable limit. In PPM-7, the samples of effluent before treatment showed the presence of 1,3,6,8-TCDD (Table 5.21).

Samples were collected from industries using vinyl chloride monomer (VCM), polyvinylchloride (PVC), ethylene dichloride (EDC), and chlorine as their raw materials. In PVC-1, the 2,3,7,8-TCDD concentration was 0.2 ng L^{-1} in the final

Table 5.20: QA/QC Data for 2,3,7,8-TCDD, 2,3,7,8-TCDF, OCDD and OCDF at 200 ng L^{-1} in water.

Sr. No.	Sample	% Recovery			
		2,3,7,8-TCDD	**2,3,7,8-TCDF**	**OCDD**	**OCDF**
1	REC-1	93.20	89.40	87.22	81.7
2	REC-2	80.00	79.00	100.00	98.3
3	REC-3	91.00	81.50	82.00	92.5
4	REC-4	92.00	88.32	105.00	120.0
5	REC-5	95.00	95.05	98.00	100.0
6	REC-6	101.00	81.86	96.20	112.00
7	REC-7	88.00	79.62	85.00	77.89
8	REC-8	94.50	98.65	105.00	84.44
9	REC-9	98.30	92.20	88.00	82.20
10	REC-10	87.88	95.00	95.00	88.78
Average Recovery		**92.09**	**88.06**	**94.14**	**93.78**
Standard Deviation		**5.92**	**7.17**	**8.21**	**13.84**
RSD		**6.43**	**8.14**	**8.72**	**14.76**

Table 5.21: 2,3,7,8-TCDD and Dioxin congeners in pulp and paper mills.

Sr. No.	Industry	Sample	2,3,7,8-TCDD Concentration (ng L^{-1} or ng g^{-1})	Dioxins & Furan Congeners
1.	PPM-1 (using Cl_2 and ClO_2)	Chlorine + Chlorine dioxide pulp mill effluent	0.12 ng L^{-1}	ND
		Final treated effluent	BDL	ND
2.	PPM-2 (using Cl_2)	Bleached plant effluent	BDL	ND
		Pulp mill effluent	BDL	ND
		Final treated effluent	BDL	Furan-2
		Bleached pulp	0.01 ng g^{-1}	ND
3.	PPM-3 (using Cl_2)	Pulp mill effluent before treatment	BDL	ND
		Final treated effluent	BDL	ND
4.	PPM-4 (using Cl_2)	Paper mill effluent before treatment	BDL	ND
		Final treated effluent	BDL	ND
5.	PPM-5 (using Cl_2 and ClO_2)	Raw effluent at inlet to clarification	BDL	ND
		Final treated effluent	BDL	ND
6.	PPM-6 (using Cl_2 and ClO_2)	Bleached plant effluent	BDL	ND
		Final treated effluent	BDL	ND
7.	PPM-7 (using ClO_2)	Effluent before treatment	BDL	1,3,6,8-TCDD
		Final treated effluent	BDL	ND

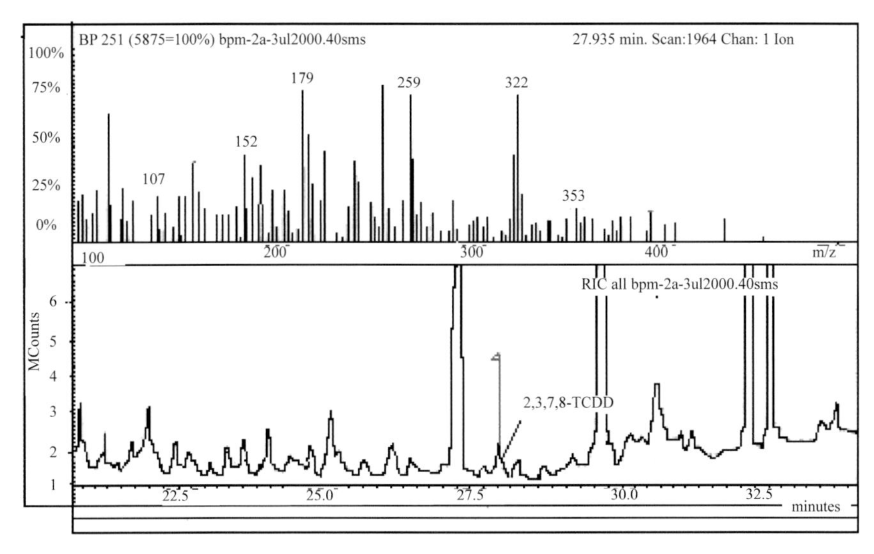

Fig. 5.17: Sample analysis of untreated effluent of paper mill (PPM-1).

Table 5.22: 2,3,7,8-TCDD in Plastic and PVC industries using chlorine based chemicals.

Sr. No.	Industry	Sample	2,3,7,8-TCDD Concentration (ng L^{-1} or ng g^{-1})	Dioxins & Furan Congeners
1.	PVC-1	Final treated effluent.	0.12 ng L^{-1}	1,2,3,4-TCDD
2.	PVC-2	Effluent before treatment (after super decanter)	BDL	1,2,3,4,7,8-HxCDD
3.	PVC-3	Effluent before treatment from VCM plant	BDL	ND
		Effluent before treatment from suspended PVC plant	0.15 ng L^{-1}	ND
		Final Treated Effluent	0.20 ng L^{-1}	ND
		Sludge	BDL	Octachloro-dibenzoFuran

BDL: Below detection limit

treated effluent. In PVC-2, the 2,3,7,8-TCDD concentration was 0.15 ng L^{-1} and 0.2 ng L^{-1} in the effluent before treatment and the final treated effluent respectively (Table 5.22). 1,2,3,4,7,8-hexachlorodibenzo-p-dioxin (HxCDD), was found in effluent before treatment (after super decanter) (Table 5.22).

CHAPTER 6

Development of Removal Techniques for Dioxins and Furans

6.1 AVAILABLE DESTRUCTION TECHNIQUES OF DIOXINS AND FURANS

A large part of POPs has been deposited in landfills along the history and many of them, especially those contained within consumer goods, are still widely deposited in industrial countries. Some relevant POPs present in landfills are Dioxins/Furans (Weber et al. 2008). Dioxins/Furans can enter the environment via escaping leachate, in fact, although there are not many studies on this subject, the presence of Dioxins/Furans has been reported in several landfill leachates samples (Casanovas et al. 1994; Pujadas et al. 2001; Choi and Lee 2006). If not treated and disposed safely, landfill leachate could be a major source of water contamination because it could percolate through soil and subsoil, causing high pollution to receiving waters (Aziz et al. 2010; Rocha et al. 2011). Therefore, the treatment of hazardous leachates is of great importance to meet the disposal standards and to reduce the negative impact on human health and environment. Dioxins enter into the environment mainly from the flue gas and formation of fly ash originating from incineration and combustion processes, and Dioxins contaminated soil due to industrial sources. The following section highlights comprehensive state-of the-art techniques for remediation, reduction, and prevention of Dioxins and Furans.

After laboratory studies showed that Dioxins broke down easily when exposed to temperature above 1200°C, EPA considered incineration the preferred technology for treating Dioxin-containing materials (USEPA 1986). A mobile research incinerator specifically designed for this purpose was built by EPA to treat recalcitrant organic chemicals in which the destruction and removal efficiency (DRE) of Dioxin in treated waste exceeded 99.9999 percent. This led EPA to adopt thermal treatment as the appropriate method for destroying Dioxin-containing waste. Extensive research has resulted in the development of several incineration technologies. The most noteworthy in relation to Dioxin treatment are rotary kilns, liquid injection, fluidized bed/circulating fluidized bed, high-temperature fluid wall destruction (advanced electric reactor), infrared thermal destruction, plasma arc pyrolysis, and supercritical water oxidation.

6.1.1 Rotary Kiln Incineration

Rotary kilns are classified into two categories: stationary and transportable. The variety of solid and liquid wastes can be treated in the kiln incinerators as shown in Fig. 6.0. The waste is fed into the rotary kiln and partially burned into inorganic ash and gases. The ash is discarded in an ash bin and the gaseous and the gaseous products containing uncombusted organic materials are sent to the secondary combustion chamber for complete destruction. The waste containing 2,3,7,8-TCDD from Seveso Italy was treated to residue levels 0.05 to 0.2 ppb. Similarly, Dioxin and Furan contaminated oils generated during the conversion of lindane waste through 2,4,5-TCP to 2,4,5-Trichlorophenoxyacetic acid which leaked out of a landfill in Hamburg, Germany were treated. Dioxins and Furans were reported to be present in the waste at levels exceeding 42,000 ppb (Harmut and Paul 1988).

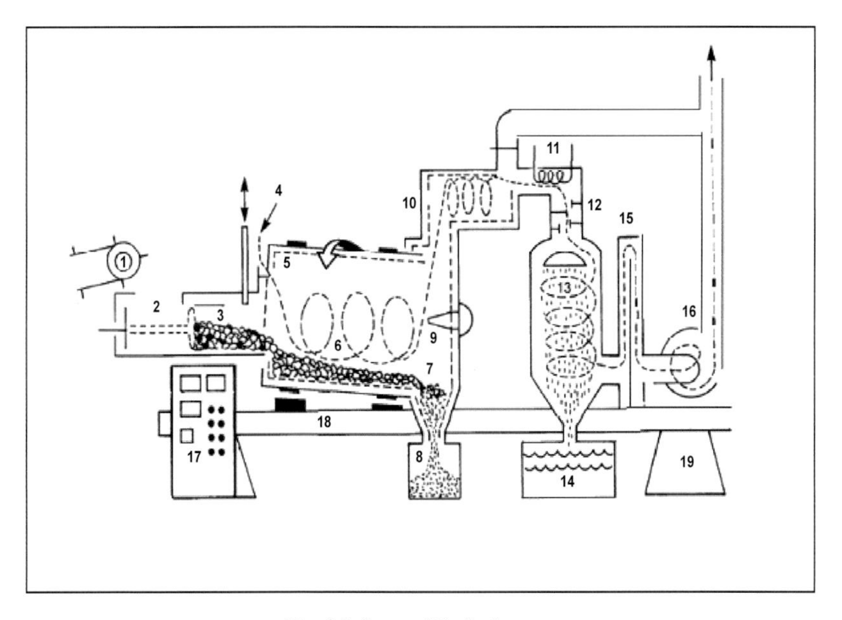

Fig. 6.0: Rotary Kiln Incinerator.

1. Material handling sytem
2. Auto-cycle feeding sytem
3. Waste to incinerator
4. Combustion air
5. Refractory-lined, rotating cylinder
6. Tumble-burning action
7. Incombustible ash
8. Ash bin
9. Auto-control burner package: programmed pilot burner
10. After burner chamber
11. Heat recuperation
12. Pre-cooler
13. Scrubber package: stainless steel, corrosion-free scrubber
14. Recycle water, fly ash sludge
15. Neutralization column
16. Exhaust fan and stack
17. Self-compensating instrumentation and controls
18. Support frame
19. Support piers

Source: Calvin R. Brunner, Incinerator System-Selection and Design (New York, NY: Van Nostrand Reinhold, 1984), p. 239.

6.1.2 Stationary Kiln Incinerators

Rollins Rotary Kiln Incinerator owned by Rollins, Inc., located in Deer Park, Texas is used to dispose hazardous waste. In this kiln, solids are fed in 55-gallon metal or fibre drums, whereas liquid waste is atomized directly into the secondary combustion chamber. The latter unit normally operates between 1300–1500°C. After being burned, combustion gases are passed to a combination venturi scrubber/absorption tower for particle removal. Fans are employed to drive scrubber gases through the stack and into the atmosphere. Maximum feed rates for the Rollins incinerator are 1440 pounds per hour for solids and 6600 pounds per hour for liquids.

6.1.3 Mobile Rotary Kiln Incinerator

Mobile incinerator are used for treating waste resulting from cleanup operations at uncontrolled hazardous waste sites; to promote the application of cost-effective and advanced technologies; and to reduce the potential risks associated with transporting waste over long distances (USEPA 1989). EPA achieved DREs of 99.9999 percent, with processed wastewater and treated soil containing Dioxins at insignificant levels.

The EPA mobile incinerator system consists of specialized incineration equipment mounted on four heavy-duty semitrailers and auxiliary pads. The first *trailer* contains: (1) a waste feed system for solids, consisting of a shredder, a conveyor, and a hopper (liquids are injected directly into the afterburner); (2) burners; and (3) the rotary kiln. Organics are burned in this portion of the system at about 1600°C. Once the waste has been incinerated, incombustible ash is discharged directly from the kiln, and the gaseous portion of the waste-now fully vaporized and completely or partially oxidized-flows into the secondary combustion chamber or *second trailer* in which it is completely oxidized at 2200°F (1200°C) and a residence time of 2 seconds. Flue gas is then cooled by water sprays to 190°F, and excess water is collected in a sump. Immediately after being cooled, the gas passes to the *third trailer* on which the pollution control and monitoring equipment is located. At this junction, gases pass through a wet electrostatic precipitator (WEP) for removal of submicron-sized particles and an alkaline mass-transfer scrubber for neutralization of acid gases formed during combustion. Cleaned gases are drawn out of the system through a 40-foot-high stack by an induced-draft fan whose other function is to keep the system under negative pressure to prevent the escape of toxic particles. Efficient and safe system performance is maintained through the use of continuous monitoring instrumentation, which includes computerized equipment and multiple automatic shutdown devices. The EPA mobile incinerator unit consumes 15 million British thermal units (Btu) per hour and handles up to 150 pounds of dry solids, 3 gallons of contaminated water, and nearly 2 gallons of contaminated fuel oil per minute. It is concurrently located at EPA's Edison Laboratory in Edison, NJ. EPA no longer plans to employ it for combustion of waste (Paul 1991).

6.1.4 Supercritical Water Oxidation

Supercritical Water Oxidation (SCWO) is a promising technology for Dioxins treatment. A system developed by MODAR, Inc., Natick, Massachusetts, is based on the oxidizing effect of water on organic and inorganic substances at 350–450°C and more than 218 atmospheric pressure or a supercritical state. Under supercritical conditions, the behaviour of water changes, and organic compounds become extremely soluble whereas inorganic salts become 'sparingly' soluble and tend to precipitate (Swallow 1990). Although SCWO can treat contaminated materials with up to 100 percent organic content, most research has focused on aqueous waste containing 20 percent organics or less. In this range, SCWO technology is said to be highly competitive and cost-effective with other available alternative treatment technologies. SCWO can be used to treat organic solids; slurries and sludge may also be treated with the addition of high-pressure pumping systems. The evaluation of SCWO on Dioxin contaminated soil, although successful, has been limited to bench-scale tests (Ralph 1991). During SCWO treatment, organic compounds are oxidised rapidly into their most basic chemical components; inorganic chemicals (salts, halogens, metals) become insoluble in the supercritical environment and descends to the bottom of the reactor where they are removed as salt or cool brine and hot aqueous and gaseous reaction products are recycled or released into the atmosphere after cooling. SCWO systems are limited by their ability to treat Dioxin-contaminated waste in liquid form.

The advantages of SCWO technology most relevant to Dioxin treatment include the following:

- The reduction of contaminants to their most basic chemical form and the harmless effluents produced eliminate the need to dispose of treated effluents;
- compounds that are difficult to dispose of are reduced to their most basic, nonhazardous forms in a process that can be adapted to a wide range of waste streams or scale of operations (Terry and Michael 1984);
- all chemical reactions occur in a totally enclosed and self-scrubbing system, thus allowing complete physical control of the waste and facilitating the monitoring of reactions throughout the process; and
- MODAR's SCWO technique can also be applied to condensates produced from the use of soil washing technologies.

6.1.5 Dechlorination Technologies

The dechlorination method is to destroy or detoxify hazardous chlorinated molecules through gradual, but progressive replacement of chlorine by other atoms (particularly hydrogen).

Alkaline Polyethylene Glycolate (APEG or KPEG) APEG-PLUS detoxifies materials contaminated with Dioxins, PCBs, pesticides, and other chlorinated hydrocarbons. The patented APEG-PLUS process consists of potassium hydroxide in a mixture of polyethylene glycol and dimethyl sulfoxide (DMSO). Once the unit has been assembled, excavated soil or sludge is conveyed to a mixer, where it is combined with reagents to form slurry. When proper mixing has been achieved and chlorinated organic compounds (PCBs, Dioxins, Furans) are extracted from the soil particles and incorporated into the mixture, the slurry is pumped into the reactor vessel and heated to 150°C. During the reaction, chlorine atoms attached to the Dioxin molecule are replaced by PEG to form a water-soluble substance (glycol ether) that can be degraded easily into nontoxic materials or washed from the soil (Peterson and New).

6.1.6 Treatment of Flue Gases

Incineration and combustion processes releases large amount of flue gases which are one of the bulk sources of Dioxin emissions in the environment. The formation of Dioxins in the flue gases of the incinerator system occurs by precursors and *de novo* synthesis at a temperature range of 300–500°C. The composition of Dioxins in the flue gases varies from 1–500 ng ITEQ/m^3. Therefore, it is important to treat the flue gas to reduce its concentration to an acceptable limit (0.1 ng I-TEQ/Nm^3) before entering into the environment. The following methods were adopted for the reduction in emission of Dioxins.

Particulate matter collection

It is possible to eliminate particle bound Dioxins with a dust collector. At temperatures below 200°C the collection of particle bound Dioxins overcomes the *de novo* synthesis. The removal of particle-bound Dioxins from the waste gas coming from an iron ore sintering plant with a cloth filter yielded a reduction of the Dioxins up to 73 percent (Ergebnisse 1996). Dioxin removal efficiencies of the electrostatic precipitator IZAYDAS Incinerator (Turkey) were examined in a trial burn. It showed the removal efficiencies of greater than 90 percent for all congeners and homologues of Dioxins (Karademir et al. 2003). A fabric filter and electrostatic precipitators (ESP) have more efficiency in the removal of particle bound Dioxins and are used as dust collectors during the incineration processes. The electrostatic precipitator, having a strong electrical field, is generally used for the collection of particulate matter or dust. A product consisting of particulate matter or dust and hydrated lime, settles to the bottom of the reactor vessel. It was observed that with the use of the combined system, Dioxins removal rates of 90–92 percent can be achieved (Kim et al. 2000). However, there are technical difficulties of removing the dust from the waste gas of incinerators at high temperatures. Some heavy metal salts because of their relatively high vapor pressure could not be removed from the waste gas in sufficient amounts.

Scrubbers or spray absorber and electrostatic precipitators

Scrubbers followed by electrostatic precipitators have been in use for many years in waste incinerator for reduction of Dioxins emissions. The lime slurry used as absorbent is atomized in the spray tower. The gas is first absorbed by the liquid phase and then by the solid phase. The lime slurry mixes with the combustion gases within the reactor. The neutralizing capacity of the lime reduces the percentage of acid gas constituents (e.g., HCl and SO_2 gas) in the reactor. It was also observed that the addition of coke made from bituminous coal in a quantity of up to 500 mg/m^3 a much higher Dioxin collection efficiency of approx. 90 percent can be achieved (Maier-Schwinning and Herden 1996).

Sorbent or flow injection process

The flow injection process is generally based on the injection of finely grained coke stemming from anthracite or bituminous coal mixed with limestone, lime, or inert material into the waste gas flow with a temperature of approx. 120°C. So the material is suspended in the flow homogeneously and subsequently settles in a layer on the surface of the cloth filter. The inert material which is added in an amount of more than 80 percent serves to take up the heat that is developed by the exothermic reactions involved in the adsorption process. It also helps to prevent ignition of the coke (Cudahy and Helsel 2000). The use of naturally and synthetically occurring zeolites is also found to be a good alternative (Abad et al. 2003). Flow injection processes are being used in Europe and USA in a number of waste incineration plants for the collection of Dioxins, HCl, HF, and SO_2.

Fluidized-bed process with adsorbent recycling

From the process engineering point of view, the fluidized-bed process lies between the flow injection process and the fixed bed as well as moving-bed adsorbent process. The advantage of the fluidized-bed process lies in the high residence times of the adsorbent and in better utilization of the sorbent because of the more favorable mass transfer conditions and longer solids retention times in the system. In this process, the flue gas passes through the grate from the bottom and forms a fluid bed of coke stemming from bituminous coal and inert material with a temperature of about 100 to 120°C. A limestone or lime can be used as inert material and the amount of coke can be higher than in the flow injection process. The adsorbent is separated from the flue gas in a dust collector and re-circulated to the fluidized bed. Usually the adsorbent can be recycled many times, so that, it is possible to collect other acid components such as HCl, HF, and SO_2. The advantages of the fluidized-bed process, lies in the high residence times of the adsorbent and in better utilization of sorbent because of the more favorable mass and heat transfer conditions and longer solids retention time in the system (Liljelind et al. 2001; Shiomitsu et al. 2002).

6.1.7 Fixed-Bed or Moving-Bed Processes

This process uses the same adsorbent as that of the fluidized bed process. However, the coke moves slowly from top to bottom while the waste gas flows in opposite direction. The activated coke takes up contaminants during its entire residence time in the reactor, which may be several 1000 operating hours. The time period during which an effective exchange of matter takes place is in fixed-bed or moving-bed processes is about 10 times longer than in the flow injection or fluidized-bed processes (Fell and Tuczek 1998). The difference between the fixed-bed and moving-bed process is that in the former, the bed of activated coke of cross-flow adsorbers is not moved during the time adsorption takes place and the spent coke is withdrawn and replaced by new coke. In moving-bed reactors the coke bed travels continuously. A very high Dioxins separation efficiency of more than 99% can be achieved with the moving-bed process (Karademir et al. 2004). Fixed bed process used for the waste gas cleaning has some problems like blocking due to moisture absorption and corrosion. Therefore in flue gas cleaning plants, the fixed-bed process has been largely replaced by the turbulent contact method applied in the moving-bed process with continuously exchanged adsorbent.

6.1.8 Catalytic Decomposition of Dioxins

A method of selective catalytic reduction for the NOx gases can be also applied for the Dioxins remediation. The catalysts used in selective reduction of the NOx in the flue gas suppressed the formation of Dioxins by 85 percent (Goemans et al. 2004). It proves that a single, effectively designed catalyst can be used in the removal of the oxides of nitrogen and Dioxins (Liljelind et al. 2001). The catalysts are mostly composed of the oxides of Ti, V, and W. Additionally, oxides of Pt and Au supported on silica-boria-alumina are found to be effective for the destruction of Dioxins at 200ºC (Everaert and Baeyens 2004). To avoid blockage of the catalyst with coarse fly ash particles and ammonium sulfate the catalyst for the destruction of Dioxins is usually applied after the cleaning stages. The advantage of selective catalytic reduction (SCR) over the other methods is the elimination of complicated disposal problems of residual matter. On the contrary, the catalyst lacks the capacity of removing as wide a spectrum of contaminants as activated coke (Andersson et al. 1998).

6.1.9 Electron Irradiation Processes

It is a new process for destruction of Dioxins compounds in the flue gas. The method has following features: (i) no possibility of secondary pollution because of the direct decomposition of Dioxins which is different from the recovery method using a filter, (ii) no need for temperature control, and (iii) a very simple process resulting in easy installation to existent incinerators. Hirota and Kojima (2005) studied the decomposition behavior of Dioxin and Furan isomers under electron-beam irradiation in incinerator gases at a temperature of 473 K. They noticed a significant decomposition for all PCDD isomers, which resulted from oxidation reactions with OH radicals yielded by electron-beam irradiation. With this process Dioxins can be reduced up to 99 percent. It involves gas-phase degradation of Dioxin molecules by OH radicals formed under the action of ionizing radiation on gas macro components (Gerasimov 2001).

6.1.10 Thermal Treatment of Fly Ash

The incineration processes of hospital, hazardous, sewage sludge, and municipal solid waste produces thick solid residues or fly ash which contains Dioxins and heavy metals. The Dioxins concentration in fly ash varies from 100–5000 ng/kg. In many countries, the environmental protection legislation classifies municipal solid waste incineration fly ash as hazardous material and further treatment is required before they are released in to the atmosphere or disposed of in landfills. Following methods were practiced for the destruction of Dioxins in fly ash; however, many of them are limited only to the laboratory stage.

Thermal treatment is a process by which heat is applied to the waste in order to sanitize it. The primary function of thermal treatment is to convert the waste to a stable and usable end product and reduce the amount that requires final disposal in landfills (Cheung et al. 2007; Lundin and Marklund 2007). It is observed that Dioxins present in fly ash can be decomposed by thermal treatment under suitable conditions. The work of Vogg and Stieglitz revealed that in an inert atmosphere, thermal treatment of Dioxins at 300ºC for 2 h resulted in 90 percent decomposition of Dioxins (Vogg and Stieglitz 1986). Further in an oxidative atmosphere, thermal treatment at 600ºC for 2 h resulted in 95 percent decomposition of Dioxins, but at lower temperatures Dioxins are formed. It is reported in a review that more than 95 percent destruction of Dioxins can be obtained using thermal treatment equipments such as electrical, oven, coke-bed melting furnace, rotary kiln with electric heater, sintering in LPG burning furnace, plasma melting furnace, etc. (Buekens and Huang 1998).

6.1.11 Non-Thermal Plasma

The application of non-thermal plasma technology on toxic substance process has been widely studied (Nifuku et al. 1997; Obata and Fujihira 1998). This process has several advantages over the conventional control devices. It performs effectively and economically at very low concentrations under ambient temperature condition and low maintenance. It doesn't require auxiliary fuel and eliminates disposal problems and sensitivity to poisoning by sulfur or halogen containing compounds. Zhou et al. (2003) applied non-thermal nanosecond plasma to destroy Dioxins contained fly ash. They found that a positive pulse discharge provides a higher destruction effect on the compounds contained than does a negative one. They reported that different isomer compounds show different toxic removal effects and the higher the toxicity of the compounds is, the higher is the destruction efficiency. Among all of the congener contained in the fly ash, the isomer 2,3,7,8-TCDD which has the highest toxicity shows the highest destruction efficiency up to 81 percent.

6.1.12 UV-Photolytic Irradiation

A photocatatlytic degradation of Dioxins using semiconductors films such as TiO_2, ZnO, CdS, and Fe_2O_3 under UV or solar light is a highly promising method, as it operates at ambient temperature and pressure with low energy photons. This process uses light to generate conduction band (CB) electrons and valence band (VB) holes (e^- and h^+) which are able to initiate redox chemical reactions on semiconductors. TiO_2 has been predominantly used as a semiconductor photocatalyst. The VB holes of TiO_2 are powerful oxidants that initiate the degradation reactions of a wide variety of organic compounds (Kim et al. 2006). It was reported that a complete degradation of 2-chlorordibenzo-p-dioxin and 2,7-dichlorodibenzo-p-dioxin was observed after 2 and 90 h, respectively, in UV illuminated aqueous suspension with no significant intermediates detection. The products obtained after the completion of process were CO_2 and HCl (Pelizzetti et al. 1988). Choi et al. (2000) in their work of photocatalytic degradation of highly chlorinated Dioxin compounds found that degradation rates of Dioxins decreased with the number of chlorine and increases with the intensity of light and the TiO_2 coating weight. The photolysis products from 2,3,7,8-TCDD do not bind to either the Ah receptor or the estrogen receptor *in vitro* (Konstantinov et al. 2000).

6.1.13 Chemical Reaction

A chemical reagent method involves use of a reagent and medium for the decomposition of polychlorinated aromatic compounds. In the past years, research was mainly focused on the removal and destruction of Dioxins and incineration was favored over the other methods. Nevertheless, the interest in the recovery of reusable materials (e.g., PCBs are present mostly in transformer oils) and the necessity to treat contaminated products with low concentration of PCBs have renewed the interest in the dechlorination methods. The dehalogenation methods mostly involve use of low-valent metal such as alkali metal in alcohol, Mg and Zn/acidic or basic solution (Krishnamurthy and Brown 1980). Mitoma et al. (2004) have studied detoxification of highly toxic polychlorinated aromatic compounds using metallic calcium in ethanol. They found that metallic calcium can be kept stable under atmospheric conditions for a long period as compared to metallic sodium since the surface is coated with

$CaCO_3$, which is formed in contact with air. Moreover, ethanol, which is one of the safe solvents for humans, acts not only as a solvent but also as an accelerator due to its ability to remove the carbonated coating. This decomposition method for Dioxins is therefore one of the most environment friendly and economic detoxification methods with respect to the energy and safety of the reagents. Concentration for each isomer of PCDDs, PCDFs and PCBs was reduced in 98–100 percent conversions by treatment in ethanol at room temperature. The TEQ for the total residues of isomers was reduced from 22000 to 210 pgTEQ at room temperature.

6.1.14 Hydrothermal Treatment

As a large amount of fly ashes are generated annually, there is a continuing interest in establishing ways in which they may be used. It is well known that fly ashes demonstrate satisfactory performance when intermixed with Portland cements. However, fly ashes contain toxic Dioxins compounds. Therefore, identification of further means to facilitate the use of fly ashes and avoid the need to dispose then as hazardous wastes is rather time relation. Fly ashes were put into water or a solution and subject to hydrothermal treatment at high pressure and temperature. An effective solution for Dioxins decomposition was found to be NaOH containing methanol; fly ashes containing 1100 ng/g total Dioxins subjected to hydrothermal treatment using this solution at 300°C for 20 min were found to have only 0.45 ng/g total Dioxins. It was suggested that the process is superior to purely thermal treatment at the same temperature and the regenerated fly ashes can be used in the cement industries (Ma and Brown 1997).

6.1.15 Supercritical Water Oxidation (SCWO)

A waste treatment process using supercritical water, which exists as a phase above the critical temperature (647.3 K) and critical pressure (22.12 MPa) has proved to be a novel way for an effective Dioxin remediation. Sako et al. (1997) applied the process for the decomposition of Dioxins in fly ashes with oxidizer such as air, pure oxygen gas, and hydrogen peroxide. They performed a reaction under the conditions of temperature 673 K, pressure 30 MPa, and time 30 min. They observed the importance of behavior of a strong oxidizer and found that the decomposition yield of Dioxins is 99.7 percent with the use of supercritical water and hydrogen peroxide. They have also successfully examined the process for dechlorination of PCBs from transformer oil (Sako et al. 1999). The same group studied a hybrid process for the destruction of Dioxins in fly ashes (Sako et al. 2004). They performed extraction of Dioxins from fly ashes using supercritical fluid (CO_2) and concentration by adsorption, and destruction by SCWO. In the extraction–adsorption process, Dioxins contained in fly ashes can be transferred and concentrated to the adsorbent (activated carbon). Then, the adsorbent containing Dioxins is completely destructed by SCWO.

6.1.16 Remediation of Soil and Sediment

Environmental problems created by forest fires, oil tanker accidents, and oil spillage from cars and trucks, leaky containers, industrial accidents, and poorly disposed of wastes are much more common cause for concern. The reservoir processes mainly contributes to the contamination of soil. Numerous tons of soil and sediment in the world were contaminated with Dioxins that need an appropriate remediation method. The common soil contaminants are petroleum based, for example, diesel fuel, gasoline polyaromatic hydrocarbons (PAH), etc. Many PAHs are known carcinogens and others are suspected problem chemicals which tend to spread through soil by diffusion and convection. Soil remediation involves two distinct classes: *in-situ*, or onsite, and *ex-situ*, or off-site. On-site cleanups are often preferred because they are cheaper. On the other hand, *ex-situ* remediation has the added bonus of taking the bulk of contaminants off-site before they can spread further. In addition, *in-situ* situations are limited because only the topside of the soil is accessible. These environmental limitations force *in-situ* remediation to fall into three categories: washing, venting, and bioremediation. Off-site facilities have the luxury of more complete control over the cleaning chemical processes. The following on- and off-site methods can be used for the remediation of soil.

Radiolytic degradation

Ionizing radiation in the form of high-energy electron beams and gamma rays is a potential non-thermal destruction technique. Theoretical and some empirical assessments suggest that these high-energy sources may be well suited to transforming Dioxin to innocuous products. Gamma radiolysis has been shown to be effective in the degradation of PCDD and PCBs in organic solvents and in the disinfection of wastewaters (Nickelsen et al. 1992; Zhao et al. 2007). Using a cobalt-60 gamma ray source, Hllarides et al. extensively studied Dioxin destruction on artificially contaminated soil (Gray and Hilarides 1995; Hilarides et al. 1994). A standard soil (EPASSM-91) was artificially contaminated with 2,3,7,8-TCDD to 100 ppb, and in the presence of 25 percent water and 2 percent surfactant (RA-40) and at a high irradiation dose (800 kGy), greater than 92 percent TCDD destruction was achieved, resulting in a final TCDD concentration of less than 7 ppb. The results of these experiments demonstrate that radiolytic destruction of TCDD bound to soil using gamma radiation can be achieved. The role of surfactant was very useful and was thought to mobilize TCDD molecule to a more favorable location in the soil, thereby modifying target size and density to make the direct effects of radiolysis more effective. The study of by-products and theoretical target theory calculations indicate that TCDD destruction proceeds through reductive dechlorination. Mucka et al. (2000) found that addition of promoters to the toxicants increases the percentage of destruction under electron beam radiation. They observed a positive influence of active carbon and Cu_2O oxide on dechlorination of PCBs in alkaline 2-propanol solution using radiolytic degradation method.

Base catalyzed dechlorination

The base-catalyzed decomposition (BCD) process is a chemical dehalogenation process (Chen et al. 1997). It involves the addition of an alkali or alkaline earth metal carbonate, bicarbonate, or hydroxide to the contaminated medium. BCD is initiated in a medium temperature thermal desorber (MTTD) at temperatures ranging from 315–426°C. Alkali is added to the contaminated medium in proportions ranging from 1 to about 20 percent by weight. A hydrogen donor compound is added to the mixture to provide hydrogen ions for reaction, if these ions are not already present in the contaminated material. The BCD process then chemically detoxifies the chlorinated organic contaminants by removing chlorine from the contaminants and replacing it with hydrogen. Pittman Jr. and Jinabo He have studied dechlorination of chlorinated hydrocarbons and pesticides. They applied Na/NH_3 to de-halogenate polychlorinated compounds from the soils and sludges. Several soils, purposely contaminated with 1,1,1-trichloroethane, 1-chlorooctane, and tetrachloroethylene, were remediated by slurring the soils in NH_3 followed by addition of sodium. The consumption of sodium per mole of chlorine removed was examined as a function of both the hazardous substrate's concentration in the soil and the amount of water present. The Na consumption per Cl removed increases as the amount of water increases and as the substrate concentration in soil decreases. PCB and Dioxin-contaminated oils were remediated with Na/ NH_3 as were PCB-contaminated soils and sludges from contaminated sites. Ca/NH_3 treatments also successfully remediated PCB-contaminated clay, sandy, and organic soils but laboratory studies demonstrated that Ca was less efficient than Na when substantial amounts of water were present (Pittman and He 2002).

Subcritical water treatment

Water which is held in liquid state above 100°C by applying a pressure is called subcritical water. It has properties similar to the organic solvents and can act as a benign medium. It has been used to extract PCBs and other organic pollutants from soil and sediment (Weber et al. 2002). Hashimoto and his co-workers (2004) examined the process of subcritical water extraction for removing Dioxins from contaminated soil. They observed 99.4 percent extraction of Dioxins at a temperature of 350°C within 30 min; however, it took a much longer time at lower temperatures. In one of the experiments, by the addition of OCDDs to the soil they found that dechlorination is a major reaction pathway. A use of zero-valent (ZVI) iron in reductive dechlorination of PCDDs and remediation of contaminated soils with subcritical water as reaction medium and extractive solvent was studied by Kluyev and co-workers (Kluyev et al. 2002). They observed that by using iron powder as a matrix, higher chlorinated congeners were practically completely reduced to less than tetra-substituted homologues. Zero-valent iron has become accepted as one of the most effective means of environmental remediation. It is inexpensive, easy to handle, and effective in treating a wide range of chlorinated compounds or heavy metals. It has been widely applied *in-situ*, *ex-situ*, or as part of a controlled treatment process in wastewater, drinking water soil amendment stabilization, and mine tailing applications.

Thermal desorption

Thermal desorption is a separation process frequently used to remediate many Superfund sites (Depercin 1995). It is an *ex-situ* remediation technology that uses heat to physically separate petroleum hydrocarbons from excavated soils. Thermal desorbers are designed to heat soils to temperatures sufficient to cause constituents to volatilize and desorb (physically separate) from the soil. Although they are not designed to decompose organic constituents, thermal desorbers can, depending upon the specific organics present and the temperature of the desorber system, cause some of the constituents to completely or partially decompose. The vaporized hydrocarbons are generally treated in a secondary treatment unit (e.g., an afterburner, catalytic oxidation chamber, condenser, or carbon adsorption unit) prior to discharge to the atmosphere. Afterburners and oxidizers destroy the organic constituents. Condensers and carbon adsorption units trap organic compounds for subsequent treatment or disposal. Kasai et al. (2000) and Harjanto et al. (2002) have proposed a thermal remediation process based on a zone combustion method for the remediation of soils contaminated by Dioxins. The process uses stable combustion of coke particles in the packed bed to soils. They removed 98.9 percent of Dioxins from the soil in a laboratory scale experiment. They also observed increase in the removal efficiency with the pre-treatment of soil such as drying, pre-granulation, and addition of limestone.

In-situ photolysis

In this method Dioxins can undergo photolysis by sunlight under proper conditions. It is cost effective and less destructive to the site. An organic solvent mixture is added to the contaminated soil and time is then allowed for Dioxin solubilization, transport, and photodegradation. For this purpose, the surface of the soil is sprayed with the low-toxicity organic solvent and allowed to photodegrade under the sunlight. Several researchers have used this approach, finding that Dioxins on the soil surface rapidly decomposed after being sprayed with various organics such as isooctane, hexane, cyclohexane, etc. (Balmer et al. 2000; Goncalves et al. 2006). Dougherty et al. (1993) found that solar-induced photolytic reactions can be a principal mechanism for the transformation of these chemicals to less toxic degradation products. Convective upward movement of the Dioxins as the volatile solvents evaporated was the major transport mechanism in these studies. The effectiveness of this process depends on a balance between two rate controlling factors: convective transport to the surface and sunlight availability for photodegradation. The *in-situ* vitrification is another developing process for onsite soil decontamination which means to make glass out of something. It involves the use of electricity to melt the waste and surrounding soil in place, then cooling it to form glass. The pollutants that cannot be destroyed by the heat are encapsulated within the glass, so they cannot leach into the surrounding soil or groundwater.

Solvent and liquefied gas extraction

Solvent extraction is a physico-chemical means of separating organic contaminants from soil and sediment, thereby concentrating and reducing the volume of contaminants that needs to be destroyed. This is an *ex-situ* process and requires the contaminated site soil to be excavated and mixed with the solvent. Eventually, it produces relatively clean soil and sediment that can be returned to the site (Silva et al. 2005). Liquefied gas solvent extraction (LG-SX) technology uses liquefied gas solvents to extract organics from soil. Gases, when liquefied under pressure, have unique physical properties that enhance their use as solvents. The low viscosities, densities, and surface tensions of these gases result in significantly higher rates of extraction compared to conventional liquid solvents. Due to their high volatility, gases are also easily recovered from the suspended solids matrix, minimizing solvent losses. Liquefied carbon dioxide and propane solvent is typically used to treat soils and sediments (Saldana et al. 2005). Contaminated solids, slurries, or wastewaters are fed into the extraction system along with solvent. Typically, more than 99 percent of the organics are extracted from the feed. After the solvent and organics are separated from the treated feed, the solvent and organic mixture passes to the solvent recovery system. Once in the solvent recovery system, the solvent is vaporized and recycled as fresh solvent. The organics are drawn off and either reused or disposed of. Treated feed is discharged from the extraction system as slurry. The slurry is filtered and dewatered. The reclaimed water is recycled to the extraction system and the filter cake is sent to disposal or reused. The U.S. Environmental Protection Agency's (EPA) evaluated a pilot scale solvent extraction process that uses liquefied propane to extract organic contaminants from soil and sediments. Approximately 1000 pounds of soil, with an average polychlorinated biphenyl (PCB) concentration of 260 mg/kg, was obtained from a remote Superfund site. Results showed that PCB removal efficiencies varied between 91.4 and 99.4 percent, with the propane-extracted soils retaining low concentrations of PCBs (19.0–1.8 mg/kg). Overall extraction efficiency was found to be dependent upon the number of extraction cycles used (Meckes et al. 1997).

Steam distillation

A distillation in which vaporization of the volatile constituents of a liquid mixture takes place at a lower temperature (than the boiling points of the either of the pure liquids) by the introduction of steam directly into the charge; steam used in this manner is known as open steam. It is an ideal way to separate volatile compounds from nonvolatile contaminants in high yield. Steam distillation is effective with microwave energy to treat contaminated soil and sediments. Microwaves are electromagnetic radiation with a wavelength ranging from 1 mm to 1 m in free space with a frequency between 300 GHz to 300 MHz, respectively. In the microwave process, heat is internally generated within the material, rather than originating from external sources. The heating is very fast as the material is heated by energy conversion rather than by energy transfer, as, in contrast, occurs in conventional techniques. Microwave radiation penetrates the sample and heats water throughout the matrix. The developing steam caused volatile and semi-volatile organic pollutants to be removed from the soil without decomposition. The temperature necessary for microwave induced steam distillation was less than 100ºC. Microwave treatments can be adjusted to individual waste streams: depending on the soil, the contaminants and their concentrations, remediation treatment can be conducted in several steps until the desired clean-up level is reached. All contaminants could be removed to non-detectable or trace levels (Windgasse and Dauerman 1992). Steam distillation was found to be effective for the removal of 2,7-dichlorodibenzo-p-dioxin (DCDD) from DCDD-applied soil. The DCDD concentration (250 μg/50 g soil) in the original soil decreased to less than 5% after steam distillation for only 20 min. The results suggest that steam distillation could be a new remedial method for soils contaminated with Dioxins (Mino and Moriyama 2001).

Mechanochemical (MC)

In this technology, the mechanical energy is transferred from the milling bodies to the solid system through shear stresses or compression, depending on the device used. A significant part of the milling energy is converted into heat and a minor part is used to induce breaks, stretches, and compression at the micro and macroscopic level or for performing a reaction. MC degradation can be easily performed using ball mills that are readily available in different sizes (treatment of materials up to several tons is possible) and constructions. The pollutants are eliminated directly inside a contaminated material, regardless of complex structure, and strong nature of the pollutant. This method has a high potential to dispose of organic wastes at any desired locations with flexible operation due to its use of a portable facility composed of a mill and a washing tank with a filter. Although this method needs a dechlorinating reagent such as CaO in the grinding operation, it does not require any heating operation. To support use of the MC dechlorination method, it would be useful to have a correlation between the dechlorination rate of organic waste and the grinding (MC) conditions to determine the optimum condition in a scaled-up MC reactor (Mio et al. 2002; Napola et al. 2006). The method offers several economic and ecological benefits: ball milling requires a low energy input only. Because of the strikingly benign reaction conditions, toxic compounds can be converted to defined and usable products. No harmful emissions to the environment have to be expected. This opened up the development of novel, innovative *ex-situ* Dioxins remediation and decontamination processes.

Biodegradation process

Bioremediation is a treatment process which uses microorganisms such as fungi and bacteria to degrade hazardous substances into nontoxic substances (Ballerstedt et al. 1997; Mori and Kondo 2002). The microorganisms break down the organic contaminants into harmless products—mainly carbon dioxide and water. Once the contaminants are degraded, the microbial population is reduced because they have used their entire food source. The extent of biodegradation is highly dependent on the toxicity and initial concentrations of the contaminants, their biodegradability, the properties of the contaminated soil and the type of microorganism selected. There are mainly two types of microorganisms: indigenous and exogenous. The former are those microorganisms that are found already living at a given site. To stimulate the growth of these indigenous microorganisms, the proper soil temperature, oxygen, and nutrient content may need to be provided. If the biological activity needed to degrade a particular contaminant is not present in the soil at the site, microorganisms from other locations, whose effectiveness has been tested, can be added to the contaminated soil. These are called exogenous microorganisms. Bioremediation can take place under aerobic and anaerobic conditions. With sufficient oxygen, microorganisms will convert many organic contaminants to carbon dioxide and water. Anaerobic conditions support biological activity in which no oxygen is present so the microorganisms break down chemical compounds in the soil to release the energy they need. A key difference between aerobic (oxidative) and anaerobic breakdown is the former predominantly used for lower chlorinated congeners and the later for high chlorinated congeners (hydrodechlorination). Sometimes, during aerobic and anaerobic processes of breaking down the original contaminants, intermediate products that are less, equally, or more toxic than the original contaminants are created. Kao and Wu (2000) have invented an *ex-situ* method in which a chemical pre-treatment (partial oxidation) in combination with bioremediation was developed to efficiently remediate TCDD-contaminated soils. In a slurry reactor, they used Fenton's Reagent as an oxidizing agent to transform TCDD to compounds more amenable for biodegradation. They observed up to 99% TCDD was transformed after the chemical pre-treatment process. The slurry reactor was then converted to a bioreactor for the biodegradation experiment. They concluded that the two-stage partial oxidation followed by biodegradation system has the potential to be developed to remediate TCDD-contaminated soils on-site. On this topic, an appealing review titled "Degradation of Dioxin like compounds by microorganisms" was presented (Wittich 1998).

6.2 DEVELOPMENT OF REMOVAL METHODS

The present work aims to have a modified and innovative approach for removal of specified Dioxins and Furans. In the present work, the removal of 2,3,7,8-TCDD, 2,3,7,8-TCDF, OCDD, and OCDF in water and soil samples has been carried out using the following methods.

- Activated carbon adsorption for water samples
- Photochemical method for soil and water samples

The simulated water and soil samples were prepared at the concentration range of 1.0 μg L^{-1} to 5.0 μg L^{-1} and 0.025 μg/10 gm to 5 μg/10 gm respectively, and subjected to various treatment processes.

6.2.1 Adsorption

Indigenous granular activated carbon (GAC), Suneeta Carbon, Mumbai and imported GAC, Sutcliffe Speakman Carbon Ltd., Leigh, Lancashire of specific grades (Tables 6.1 and 6.2) were used for the removal of 2,3,7,8-TCDD, 2,3,7,8-TCDF, OCDD, and OCDF from water. The effectiveness of adsorption was optimized with respect to contact period, GAC dose, pH, and adsorbate concentration.

Adsorption, involves nothing more than the preferential partitioning of substances from the gaseous or liquid phase onto the surface of a solid substrate. From the early days of using bone char for decolorization of sugar solutions and other foods, to the later implementation of activated carbon for removing nerve gases from the battlefield, to today's thousands of applications, the adsorption phenomenon has become a useful tool for purification and separation.

Adsorption phenomena are operative in most natural physical, biological, and chemical systems, and adsorption operations employing solids such as activated carbon and synthetic resins are used widely in industrial applications and for purification of water and wastewater.

Table 6.1: Specifications of indigenous granular activated carbon (Suneeta Carbon, Mumbai).

S. No.	Properties	
Physical		
1.	Raw material	Coconut shell
2.	Bulk density (g/ml)	0.6
3.	Particle size (mm)	1.68–2.0
4.	Pore size distribution (A°)	200–290
5.	Surface Area (m^2/g)	452
6.	Iodine number	700–800
7.	pH (water extracted)	8–10
Chemical (Proximate analysis on dry basis)		
1.	Moisture	13.3
2.	Ash	24.4
3.	Volatile	6.6
4.	Fixed carbon	55.7

Table 6.2: Specifications of imported granular activated carbon (Sutcliffe Speakman Carbon, Leigh, Lancashire).

S. No.	Properties	
Physical		
1.	Raw material	Wood based
2.	Bulk density (g/ml)	0.15
3.	Particle size (mm)	0.60–1.65
4.	Pore size distribution (A°)	50–350
5.	Surface Area (m^2/g)	830
6.	Iodine number	700–800
7.	pH (water extracted)	9
Chemical (Proximate analysis on dry basis)		
1.	Moisture	17.4
2.	Ash	3.6
3.	Volatile	7.0
4.	Fixed carbon	72

The process of adsorption involves separation of a substance from one phase accompanied by its accumulation or concentration at the surface of another. The adsorbing phase is the adsorbent, and the material concentrated or adsorbed at the surface of that phase is the adsorbate. Adsorption is thus different from absorption, a process in which material transferred from one phase to another (e.g., liquid) interpenetrates the second phase to form a "solution". The term sorption is a general expression encompassing both processes.

Physical adsorption is caused mainly by Vander Waals forces and electrostatic forces between adsorbate molecules and the atoms which compose the adsorbent surface. Thus adsorbents are characterized first by surface properties such as surface area and polarity.

A large specific surface area is preferable for providing large adsorption capacity, but the creation of a large internal surface area in a limited volume inevitably gives rise to large numbers of small sized pores between adsorption surfaces. The size of the micropores determines the accessibility of adsorbate molecules to the internal adsorption surface, so the pore size distribution of micropores is another important property for characterizing adsorptivity of adsorbents. Especially materials such as zeolite and carbon molecular sieves can be specifically engineered with precise pore size distributions and hence tuned for a particular separation.

Surface polarity corresponds to affinity with polar substances such as water or alcohols. Polar adsorbents are thus called "hydrophillic" and aluminosilicates such as zeolites, porous alumina, silica gel, or silica-alumina are examples of adsorbents of this type. On the other hand, non-polar adsorbents are generally "hydrophobic". Carbonaceous adsorbents, polymer adsorbents, and silicalite are typical nonpolar adsorbents. These adsorbents have more affinity with oil or hydrocarbons than water.

Principle

Adsorption is a surface phenomenon that is defined as the increase in concentration of a particular component at the interface between two phases. In any solid or liquid, atoms at the surface are subjected to unbalanced forces of attraction normal to the surface plane. Adsorption is distinguished into two types physical adsorption, which involves only relatively weak forces and chemisorption, which in turn, involves essentially the formation of a chemical bond between the sorbate molecule and the surface of the adsorbent.

Physical adsorption does not involve the sharing or transfer of electrons and thus always maintains the individuality of interacting species. The interaction is fully reversible, enabling desorption to occur at the same temperature although the process may be slow because of diffusion effects. This adsorption is not specific; the adsorbed molecules are free to cover the entire surface. This enables the surface area measurement of adsorbents.

Adsorption phenomenon

Adsorption from aqueous solution involves concentration of the solute on the solid surface. As the adsorption process proceeds the sorbed solute tends to desorb into the solution. Equal amounts of solute eventually are being adsorbed and desorbed simultaneously; consequently, the adsorption and desorption rate will reach the equilibrium state, called adsorptions equilibrium. At the equilibrium, no change can be observed in the concentration of the solute on the solid surface or in the bulk solution. The position of equilibrium is characteristic of the entire system, the solute, adsorbent, solvent temperature, pH, etc. Adsorbed quantities at equilibrium usually increase with the increase in the solute concentration. The presence of the amounts of solute adsorbed per unit of adsorbent as a function of the equilibrium concentration in bulk solution, at constant temperature, is termed as adsorption isotherm.

Several models can be used for the description of the adsorption data, and Langmuir's and Freundlich's adsorption isotherms are the most commonly used.

Freundlich's Adsorption Isotherms

It is the most widely used mathematical description of adsorption in aqueous system. The Freundlich's adsorption isotherms is given as

$$x/m = KC_e^{1/n} \qquad \text{eq. 1}$$

where,
x = the amount of solute adsorbed,
m = the weight of adsorbent
C_e = the solute equilibrium concentration
K and 1/n = constants characteristics of the system

For linearisation of data the Freundlich adsorption isotherms is written in logarithmic form:

$$\log x/m = \log K + 1/n \log C_e \qquad \text{eq. 2}$$

Plotting log x/m *vs* C_e, a straight line is obtained with the slope of 1/n, and log K is the intercept of log x/m at $C_e = 0$ ($C_e = 1$). The value of 1/n obtained for adsorption of most organic compounds by activated carbon is < 1. Steep slopes-that is 1/n close to 1 indicates high adsorptive capacity at high equilibrium concentrations that rapidly diminishes at lower equilibrium concentrations covered by the isotherms. Relatively flat slopes that is, $1/n << 1$ indicate that the adsorptive capacity is only slightly reduced at the lower equilibrium concentrations. As the Freundlich equation indicates, the adsorptive capacity or loading factor on the carbon, x/m, is a function of the equilibrium concentration of the solute. Therefore, higher capacities are obtained at higher equilibrium concentrations.

The Freundlich equation can be used for calculating the amount of activated carbon required to reduce any initial concentration to a predetermined final concentration. By substituting $C_o – C_e$ in eq. 1 for x, where C_o = the initial concentration:

$$\log\left(\frac{C_o - C_e}{m}\right) = \log k + \frac{1}{n}\log C_e \qquad \text{eq. 3}$$

The eq. 3 can be used for comparing different activated carbons in removal of different compounds or removal by same carbon.

Apparatus and Reagents

- GC-ECD: Perkin Elmer, Clarus 500 GC equipped with ECD and fused capillary silica column (DB-5) of length 30 m and internal diameter of 0.25 mm having film thickness of 0.25 μm was used for the analysis of PCDDs and PCDFs
- Rotary Shaker: Rotary Shaker, Thermolab and Equipment Company, Mumbai, equipped with trays of dimensions 60 × 60 cm to hold a minimum of 4 and maximum of 16 flasks with shaking rate of 100 to 260 rpm, for the duration of time as per requirement was used
- GAC: The characteristics of GACs used are given in Tables 6.1 and 6.2
- Whatman filter paper 41
- Conical Flasks of 1 L capacity
- Amber colored glass bottle of 1 L capacity
- Dichloromethane (Merck, Lichrosolv, Residue analysis grade)
- n-Hexane (Merck, Lichrosolv, Residue analysis grade)
- Standards of ^{13}C-2,3,7,8-TCDD, 2,3,7,8-TCDD, 2,3,7,8-TCDF, OCDD and OCDF

Reference standards of ^{13}C-2,3,7,8-TCDD, 2,3,7,8-TCDD, 2,3,7,8-TCDF, OCDD and OCDF of 99.99% purity were procured from Cambridge Isotope and Wellington laboratories.

Reagent grade chemicals were procured from Merck. Working standards of 2,3,7,8-TCDD, 2,3,7,8-TCDF, OCDD and OCDF were prepared in toluene or nonane from the stock solution.

Simulated samples of 2,3,7,8-TCDD, 2,3,7,8-TCDF, OCDD, and OCDF were prepared by spiking specified aliquots of working standards in 1 L of tap water.

Experimental

The application of the adsorption process for the removal of 2,3,7,8-TCDD from water was accomplished by contacting the adsorbent by simulated water in a batch type contact unit operation system.

Pretreatment of GAC

Prior to conducting the adsorption experiments it was made sure that the GAC used for the studies was properly cleaned of all the adsorbed impurities. The carbon was first washed thoroughly with double distilled deionised water till the effluent water was found free of organics and dissolved solids. The washed carbon was activated by keeping it in an oven at a temperature range of 300–400ºC for a period of 24 hours. The then pre-treated carbon was kept in closed containers and placed in a desiccator to prevent absorption of any moisture.

Determination of carbon dosage in a batch system

The first step in the evaluation of carbon dosage for a batch system was determining the contact period. This was determined by adding a known quantity of carbon to the equal volumes of simulated water samples. The samples were removed at various time intervals, and the carbon was immediately removed by filtration. The residual concentration of contaminant was subsequently determined in each sample. The data are plotted to show the change in concentration of contaminant with time. Adsorption equilibrium is achieved at the contact period where no significant change in concentration is observed with increased time.

The carbon dosage rate was determined by constructing an adsorptive isotherm. The data are plotted according to the Freundlich adsorption. According to the equation

$$\frac{x}{M} = \frac{C_o - C_f}{M} = KC_f^{1/n} \qquad \text{eq. 4}$$

The volume of x/M at C_f is read from the plot. The carbon dosage M, required at the initial concentration, C_o to the desired final concentration, C_f, is calculated by rearranging eq. 4:

$$M = \frac{C_o - C_f}{x/M}$$

6.2.1.1 *Indigenous GAC*

(a) Effect of contact period

Batch studies were conducted to evaluate the optimum dose required for the removal of 2,3,7,8-TCDD. Simulated sample of 2,3,7,8-TCDD of concentration 5.0 μg L^{-1}, 2.0 μg L^{-1} and 1.0 μg L^{-1} were prepared in tap water at neutral pH. Equal volumes of 500 mL aliquotes of samples were taken in 1 L capacity amber colored glass bottles and 500 mg of indigenous GAC was added. The samples were mixed thoroughly on a rotary shaker for varying contact time period ranging from 5 min to 60 min at a speed of 175 rpm. After mixing, the samples were allowed to settle for some time. The samples were withdrawn and immediately filtered through Whatman paper and stored in a refrigerator until analysis. The removal of 2,3,7,8-TCDD was calculated and the optimum contact period was established.

(b) Effect of GAC dose

Simulated samples of 2,3,7,8-TCDD of concentration 1 μg L^{-1}, 2 μg L^{-1} and 5 μg L^{-1} were prepared in tap water at neutral pH. 500 mL aliquotes of sample were taken in 1 L capacity amber colored glass bottles and varying doses of indigenous

GAC ranging from 25 mg L^{-1} to 1000 mg L^{-1} were added. The samples were mixed thoroughly on a rotary shaker for a time period of 60 min at a speed of 175 rpm. After mixing the samples were allowed to settle for some time. The samples were withdrawn and immediately filtered through Whatman paper and stored in refrigerator until analysis. The removal of 2,3,7,8-TCDD was calculated and the optimum dose of GAC established.

(c) Effect of pH

Simulated samples of 2,3,7,8-TCDD of concentration 5 μg L^{-1} were prepared in tap water at varying pH of 4.0, 7.0, 9.2, and 12. Equal volumes of 500 mL aliquotes of samples were taken in 1 L capacity amber colored glass bottles and 500 mg of indigenous GAC was added. The samples were mixed thoroughly on a rotary shaker for a time period of 60 min at a speed of 175 rpm. After mixing the samples were allowed to settle for some time. The samples were withdrawn and immediately filtered through Whatman paper and stored in refrigerator until analysis. The removal of 2,3,7,8-TCDD was calculated and the optimum pH was established at which maximum removal was observed.

(d) Effect of adsorbate concentration

Simulated samples of 2,3,7,8-TCDD were prepared in tap water at neutral pH and varying analyte concentration ranging from 0.5 μg L^{-1} to 5 μg L^{-1}. Equal volumes of 500 mL aliquotes of samples were taken in 1 L capacity amber colored glass bottles and 500 mg of GAC was added to each samples aliquote. The samples were mixed thoroughly on a rotary shaker for a time period of 60 min at a speed of 175 rpm. After mixing, the samples were allowed to settle for some time. The samples were withdrawn and immediately filtered through Whatman paper and stored in a refrigerator until analysis. The removal of 2,3,7,8-TCDD was calculated at varying concentrations.

6.2.1.2 Imported GAC

(a) Effect of contact period

Simulated sample of 2,3,7,8-TCDD of concentration 5 μg L^{-1} was prepared in tap water at neutral pH. Equal volumes of 500 mL aliquotes of simulated samples were taken in 1 L capacity amber colored glass bottles and 250 mg of imported GAC was added. The samples were mixed thoroughly on a rotary shaker for varying contact time period ranging from 2 min to 60 min at a speed of 175 rpm. After mixing the samples were allowed to settle for some time. The samples were withdrawn and immediately filtered through Whatman paper and stored in refrigerator until analysis. The removal was calculated and the optimum contact time was established for maximum removal of 2,3,7,8-TCDD.

(b) Effect of GAC dose

Simulated sample of 2,3,7,8-TCDD of concentration 5 μg L^{-1} was prepared in tap water at neutral pH. Equal volumes of 500 mL aliquotes of simulated sample were taken in 1 L capacity amber coloured glass bottles and varying doses of imported GAC ranging from 25 mg L^{-1} to 1000 mg L^{-1} were added. The samples were mixed thoroughly on a rotary shaker for a time period of 10 min at a speed of 175 rpm. After mixing the samples were allowed to settle for some time. The samples were withdrawn and immediately filtered through Whatman paper and stored in refrigerator until analysis. The removal of 2,3,7,8-TCDD was calculated and the optimum dose of GAC established.

(c) Effect of adsorbate concentration

Simulated samples of 2,3,7,8-TCDD were prepared in tap water at neutral pH and varying adsorbate concentration ranging from 0.5 μg L^{-1} to 5 μg L^{-1}. Equal volumes of 500 mL aliquotes of samples were taken in 1 L capacity amber colored glass bottles and 250 mg of imported GAC was added to each sample aliquote. The samples were mixed thoroughly on a rotary shaker for a time period of 10 min at a speed of 175 rpm. After mixing the samples were allowed to settle for some time. The samples were withdrawn and immediately filtered through Whatman paper and stored in refrigerator until analysis. The removal of 2,3,7,8-TCDD at varying concentration was calculated.

6.2.1.3 Results and Discussion

Indigenous-GAC

Contact period: Removal of 2,3,7,8-TCDD was evaluated at varying contact period of 0 to 60 min. The removal of 2,3,7,8-TCDD increased with the increase in contact time with GAC. The analysis results showed that adsorption of 2,3,7,8-TCDD by indigenous GAC is an instantaneous reaction. The results show that adsorption was gradual over the entire contact period.

Removal of 2,3,7,8-TCDD was obtained as 99.43% at an initial concentration of 1 μg L^{-1} after a contact period of 60 min (Fig. 6.1). Removal of 94.00% was observed at an initial concentration of 2 μg L^{-1} of 2,3,7,8-TCDD after a contact period of 60 min (Fig. 6.1). Whereas removal of 92.01% was observed at an initial concentration of 5 μg L^{-1} of 2,3,7,8-TCDD after a contact time of 60 min (Fig. 6.1). Thus an optimum contact period of 60 min was optimized for maximum removal of 2,3,7,8-TCDD. The mathematical description of adsorption of 2,3,7,8-TCDD in aqueous system is given as Freundlich adsorption isotherm [Figs. 6.2(a), 6.2(b) and 6.2(c)]. The value of slope, that is, $1/n$ obtained was < 1 which indicates good adsorption of 2,3,7,8-TCDD by GAC.

GAC dose: The effectiveness of varying doses of indigenous GAC in removal of 2,3,7,8-TCDD from water was studied. The analysis results showed the increase in removal of 2,3,7,8-TCDD with the increase in GAC dose. The removal of 2,3,7,8-TCDD increased from 9% to 91.9% with increase in the GAC dose from 25 mg L^{-1} to 1000 mg L^{-1} after 60 min of contact period at a concentration of 5 μg L^{-1}. The removal of 2,3,7,8-TCDD increased from 13% to 94.0% with increase in GAC dose from 25 mg L^{-1} to 1000 mg L^{-1} after 60 minute of contact time at a concentration of 2 μg L^{-1}. The removal of 2,3,7,8-TCDD increased from 18% to 99% with increase in GAC dose from 25 mg L^{-1} to 1000 mg L^{-1} after 60 minute of contact period at concentration of 1 μg L^{-1} [Figs. 6.3(a) and 6.3(b)]. Thus a dose of 1000 mg L^{-1} of indigenous activated GAC was optimised for maximum removal of 2,3,7,8-TCDD.

pH: The effect of pH on removal of 2,3,7,8-TCDD by indigenous GAC was studied. The analysis results showed increased removal in alkaline pH, whereas the removal was found to be less in acidic pH. Removal of 97.0%, 95.0%, 91.0%, and 76.2% was observed at pH 12, 9.2, 7.0, and 4.0 respectively at an initial concentration of 5.0 μg L^{-1} of 2,3,7,8-TCDD after the contact period of 60 min (Fig. 6.4).

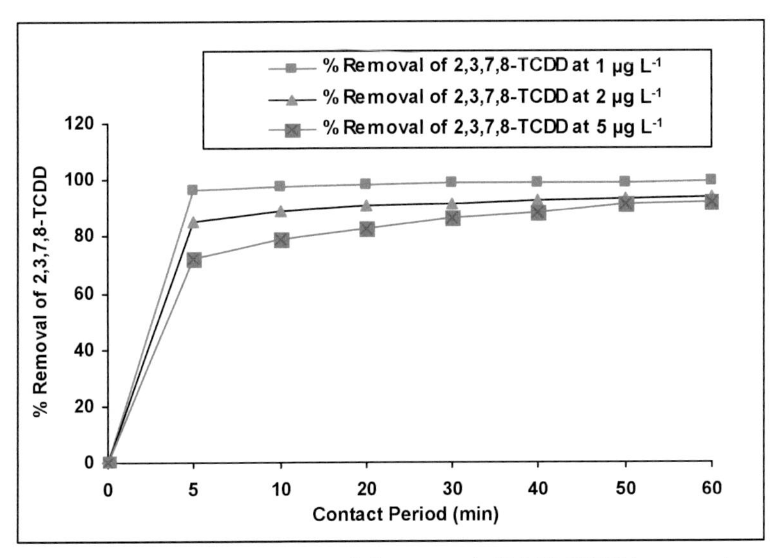

Fig. 6.1: Effect of varying contact period on removal of 2,3,7,8-TCDD in water using indigenous GAC.

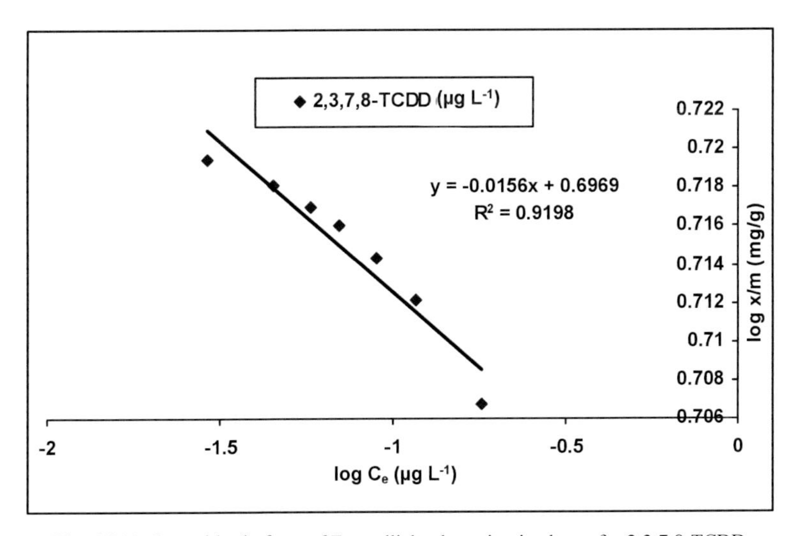

Fig. 6.2(a): Logarithmic form of Freundlich adsorption isotherm for 2,3,7,8-TCDD (5 µg L^{-1}) on GAC.

Adsorbate concentration: The removal efficiency of indigenous GAC for 2,3,7,8-TCDD at contact time of 60 min was studied by varying the 2,3,7,8-TCDD concentration. The analysis results showed decrease in removal efficiency with increase in 2,3,7,8-TCDD concentration. The removal of 99.42% was observed at 0.5 µg L^{-1} of 2,3,7,8-TCDD, which was observed to decrease to 90.20% at 5.0 µg L^{-1} of 2,3,7,8-TCDD (Fig. 6.5).

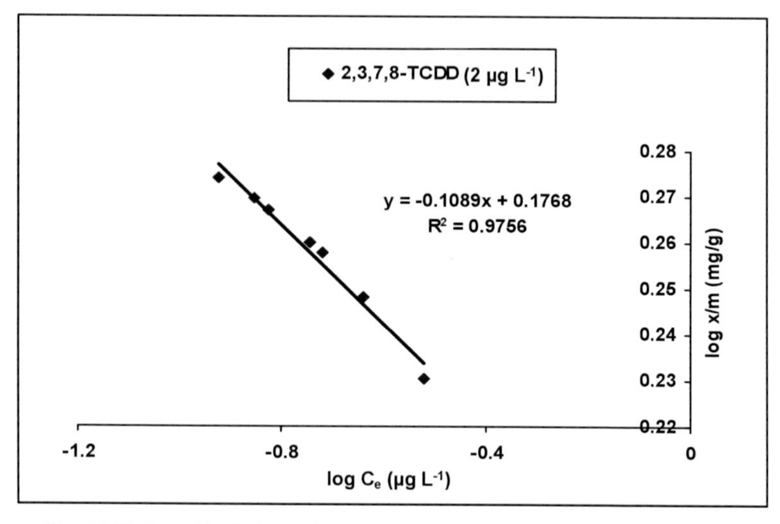

Fig. 6.2(b): Logarithmic form of Freundlich adsorption isotherm for 2,3,7,8-TCDD (2 μg L^{-1}) on GAC.

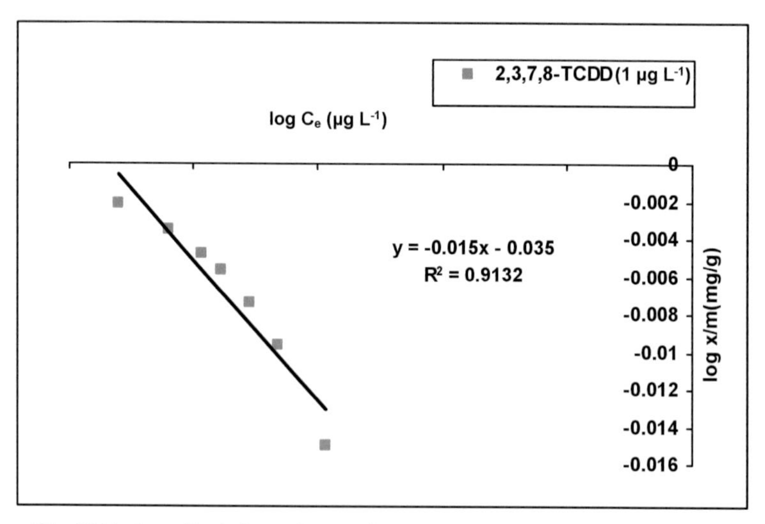

Fig. 6.2(c): Logarithmic form of Freundlich adsorption isotherm for 2,3,7,8-TCDD (1 μg L^{-1}) on GAC.

GAC-Imported

Contact period: Removal of 94% was observed for 2,3,7,8-TCDD at an initial concentration of 5 μg L^{-1} and a contact time of 2 min, which increased to 100% after the contact period of 10 min. Thus the contact period of 10 min was optimized for maximum removal of 2,3,7,8-TCDD in water (Fig. 6.6).

GAC dose: The analysis results showed increase in removal of 2,3,7,8-TCDD with increase in GAC dose. The removal increased from 70% to 100% with increase in the

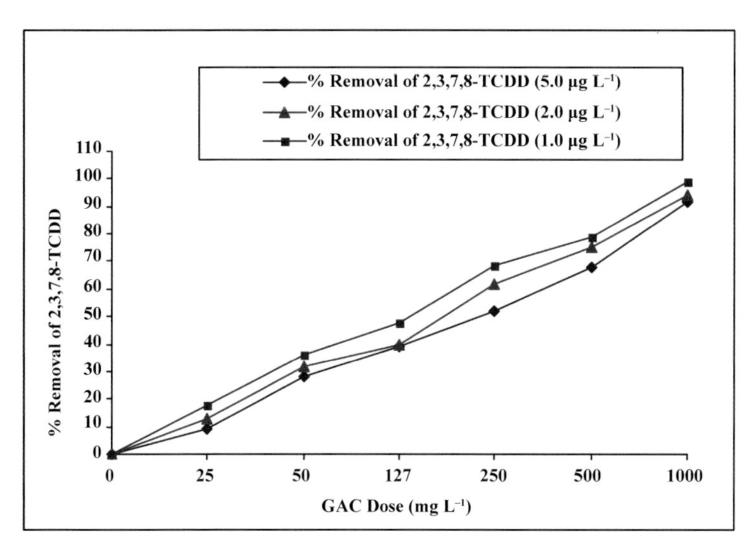

Fig. 6.3(a): Effect of varying dose of indigenous GAC on removal of 2,3,7,8-TCDD in water.

GAC dose from 25 mg L^{-1} to 1000 mg L^{-1} after 10 min of contact time (Fig. 6.7). Thus an optimum GAC dose of 250 mg L^{-1} was optimized for removal of 2,3,7,8-TCDD in water.

Adsorbate concentration: The removal efficiency of imported GAC for 2,3,7,8-TCDD at contact time of 10 min was studied by varying the 2,3,7,8-TCDD concentrations. 100% removal of 2,3,7,8-TCDD was observed in the concentration range of 0.5 μg L^{-1} to 5.0 μg L^{-1} (Fig. 6.9).

Data Evaluation

Laboratory scale experiments were conducted for evaluating the effectiveness of adsorption methods for the removal of 2,3,7,8-TCDD in simulated water samples. Adsorption studies were carried out using both indigenous (Suneeta Carbon, Mumbai) and imported GAC (Sutcliffe Speakman Carbon Ltd., Leigh, Lancasier). The adsorption phenomenon is described by the Freundlich adsorption isotherm. When log x/m was plotted against equilibrium concentration (C_e), a linear graph was obtained with slope less than 1, which follows the Freundlich isotherm. The Freundlich equation indicates the adsorptive capacity of GAC. The efficiency of indigenous GAC was evaluated with respect to its contact time and dose [Figs. 6.1 and 6.3(a)]. The imported GAC showed greater efficiency in removal as compared to indigenous GAC. This may be attributed to the greater surface area of imported GAC.

6.2.2 Photochemical Degradation

Photochemical reaction may be classified as direct or sensitized. In a direct photochemical reaction, one or more of the substances undergoing reaction absorb light. In a sensitized reaction, some substances absorbs light and initiates reactions in which it does not participate as one of the disappearing molecule. Examples of sensitized reactions are:

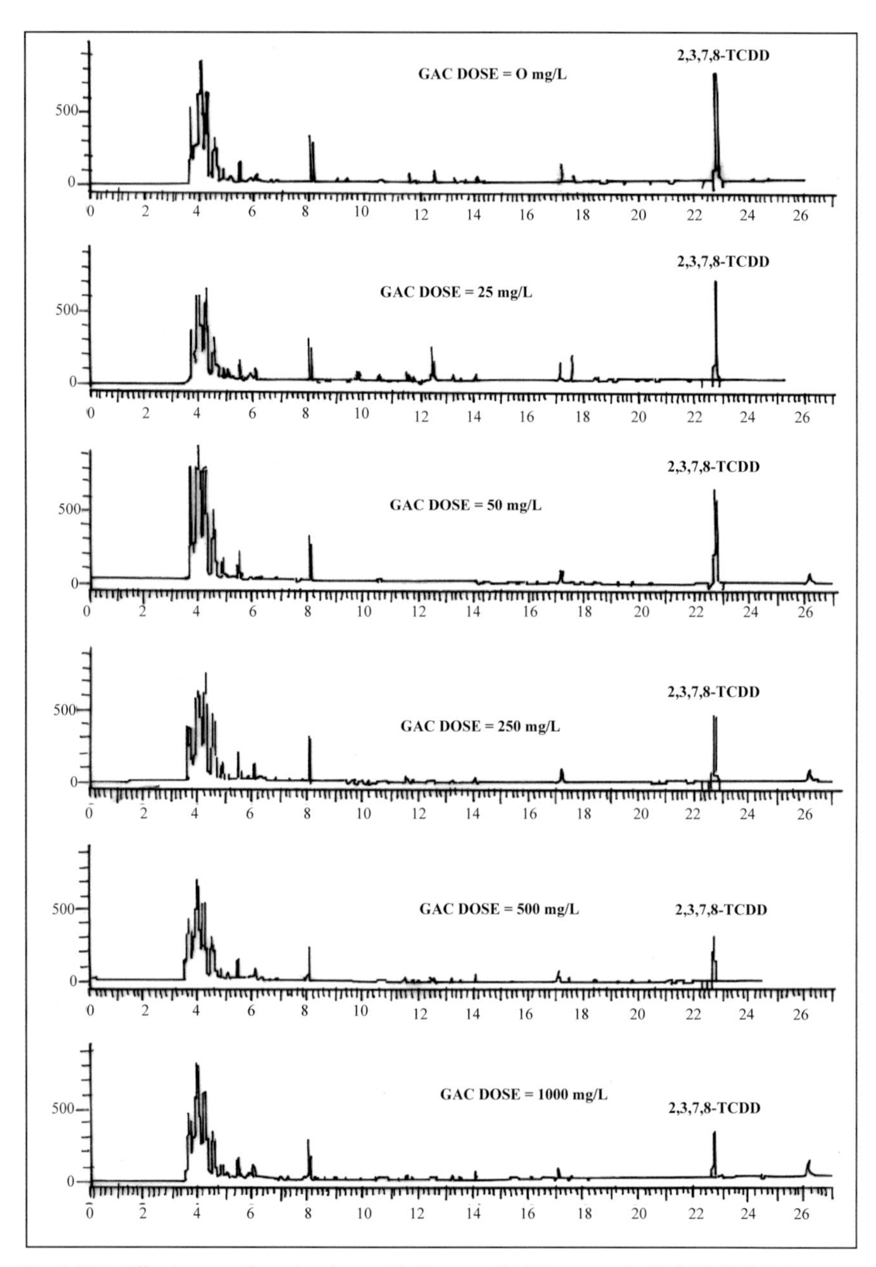

Fig. 6.3(b): Effectiveness of varying doses of indigenous GAC in removal of 2,3,7,8-TCDD in water.

(a) The hydrogenation of ethylene in the presence of mercury vapor using radiation absorbed only by the mercury vapor.

(b) The oxidation of tetrachloroethylene to trichloroacetyl chloride by oxygen in the presence of chlorine and light absorbed only by the chorine.

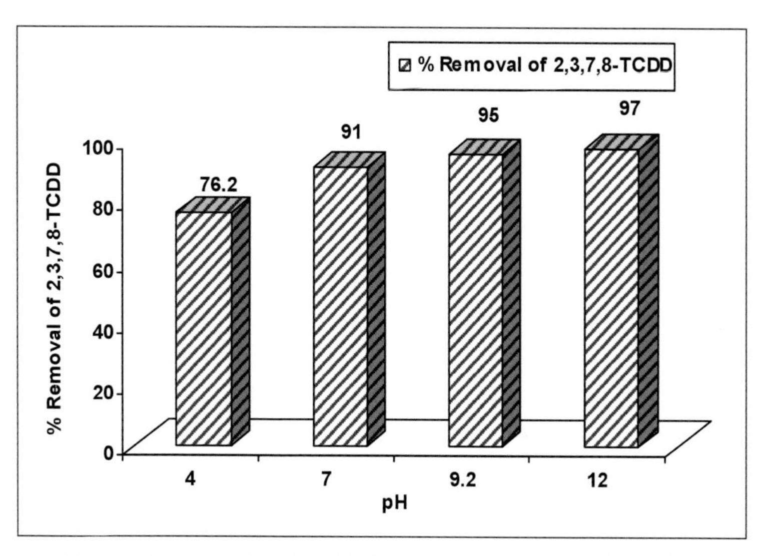

Fig. 6.4: Effect of pH on removal of 2,3,7,8-TCDD in water at 5.0 μg L^{-1} using indigenous GAC.

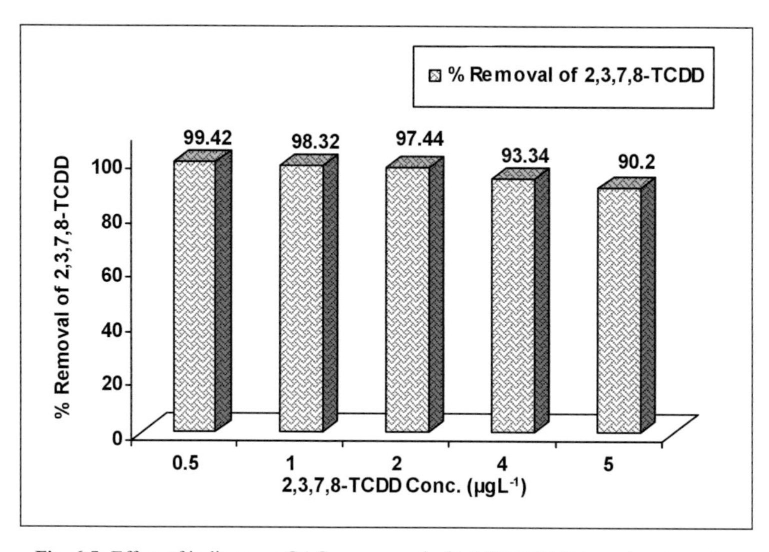

Fig. 6.5: Effect of indigenous GAC on removal of 2,3,7,8-TCDD in water at varying 2,3,7,8-TCDD concentrations.

The study of photochemical reactions may be divided into two main parts:

(a) Study of the immediate effect of the light on the absorbing molecule, that is, the primary process.

(b) Study of the reactions of the molecules, atoms, or radicals produced by the primary process.

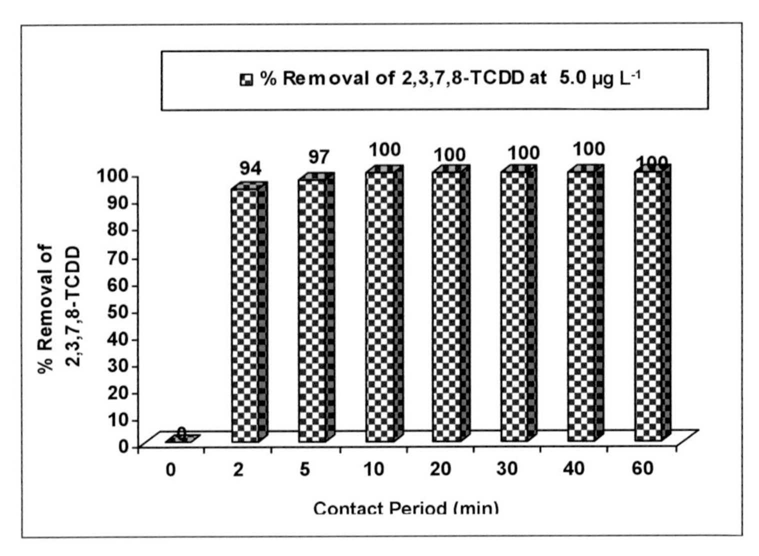

Fig. 6.6: Effect of contact period with imported GAC on removal of 2,3,7,8-TCDD in water.

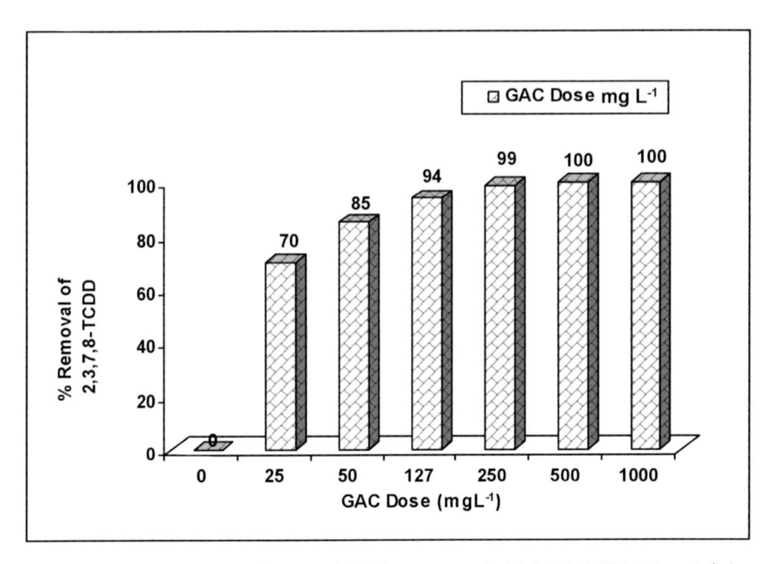

Fig. 6.7: Effect of varying dose of imported GAC on removal of 2,3,7,8-TCDD (5 µg L^{-1}) in water.

Primary process

On absorption of quantum of light energy, a molecule is placed in a situation such that it is no longer in thermodynamic equilibrium with its surroundings, and hence it must lose energy in one of the following three forms:

(a) As fluorescence or phosphorescence, that is, all or part of the energy is re-emitted as radiant energy after a time interval which is usually a small fraction of a second but which may, in some cases, be much longer

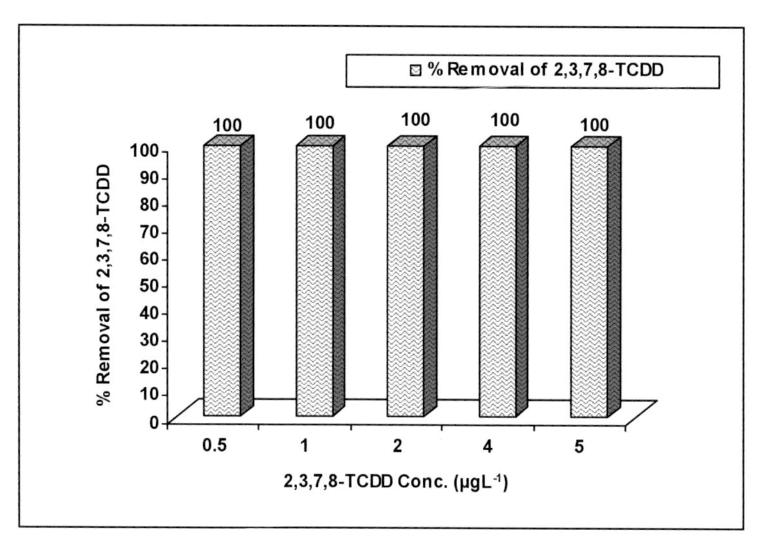

Fig. 6.8: Effect of imported GAC on removal of 2,3,7,8-TCDD in water at varying 2,3,7,8-TCDD concentrations.

(b) As thermal energy resulting in an increase in temperature of the reaction system or loss of energy to the surrounding and

(c) As chemical energy, that is, by transformation of the starting material into new compounds. The energy rich molecule which is produced on absorption of light may decompose to yield atoms or free radicals. On the other hand, it may rearrange, eventually or immediately, into a final product.

Secondary reactions

The energy rich molecules formed on absorption may lose energy by collisions with other molecules, thereby initiating chemicals reactions. In such cases the excited molecules are said to perform the role of sensitizers. In some cases the excited molecules may react directly with other molecules in the system, leading to the formation of stable products, for example, formation of anthracene peroxide from anthracene and molecular oxygen. Atoms or radicals produced in the primary process can undergo a wide variety of reactions leading to the formation of stable products.

Energies Involved: When a molecule absorbs light it receives energy in the form of discrete units, or quanta. The Einstein law of photochemical equivalence states that one molecule reacts per quantum absorbed. A quantum of radiant energy is hν ergs where h is Planck's constant (6.61×10^{-27} erg sec) and ν is the frequency of radiation. The frequency is equal to the speed of light (3×10^{10} cm per sec) divided by the wavelength in centimeters. The value of a quantum is thus seen to be:

$$1 \text{ quantum} = h\nu = 1.983 \times 10^{-8}/\text{ergs},$$

A number of quanta equal to Avogadro's number is called an Einstein, that is,

$$6.02 \times 10^{23} \times 1.98 \times 10^{-8}/\text{ergs} = 1.19 \times 10^{16}/\lambda \text{ ergs}.$$

where λ is the wavelength in angstrom units.

For a photochemical reaction to occur, the energy of the quantum adsorbed must be equal to or greater than the energy required to produce whatever effect is required in the absorbing molecule that is, either dissociation or activation.

Light Sources: Light sources are required to study the photochemical reactions. Various light sources used in photochemical reactions are as follows:

i Sunlight

ii Mercury irradiation lamps: Three types of mercury lamps are available

(a) low pressure mercury lamp

(b) medium pressure mercury lamp

(c) high pressure mercury lamp

Sunlight: The UV portion of sunlight degrades most of the organic pollutants in the environment.

Mercury lamps: Low pressure lamp (6 W or 16 W rating) emit 90% of their radiation at 254 nm. Medium pressure lamp (125 W and 400 W rating) have much more intense arcs and radiate predominantly at 365–366 nm radiation with much smaller amounts in the ultraviolet at 265, 297, 303, 313, and 334 nm as well as significantly in visible region at 404–408, 436, 546, and 577–570 nm.

The low pressure lamps need an optimum ambient temperature of 40ºC. The medium pressure lamp take a significant time (10–15 min) to attain maximum UV output. In both cases, therefore, excessive cooling of the lamps can reduce the UV output. For more accurate studies on the effect of irradiation dose on reactions, it is necessary to run a lamp in a water cooled immersion well.

These mercury lamps possess a negative volt—ampere characteristic and must be operated from reactive type supplies having a large resistance or inductance in series to provide extra power to initiate the arc and reduced power for continuous operation.

Photocatalyst

Titanium Dioxide (TiO_2) is widely used as a photocatalyst as it is chemically activated when exposed to light and decomposes organic material which come in contact with it. It has many applications as an anti-bacterial agent, self-cleaning coating, anti-fogging coating, deodorizing coating, and in water treatment. In water treatment, it has been used to remove concentrations of organic contaminants. When used to remove organic contaminants, the photocatalyst settles rapidly out of solution as is easily recovered from a solution when compared to current commercially available products.

The process of photoactive photocatalysis involves the irradiation of TiO_2 with UV irradiation below 390 nm. The photon source may be the sun or UV lamps. The radiations stimulate valence band electrons in the TiO_2 suspended water. The excited electrons jump to the conduction band leaving holes in the valence band. Possible reactions with electron/hole pairs include the production of heat and a reaction with dissolved organic or inorganic constituents in the water on recombination (Prairie et al. 1992). It has been postulated by Matthews (1986), Okamoto et al. (1985), and Fox (1983) that holes reacts with water and convert hydroxyl ions to hydroxyl radicals ($OH^{\bullet}$). The latter can completely oxidize organics including halogenated organics to water, carbon dioxide, and mineral acids. Electrons can react with dissolved oxygen to form oxide anions (O_2), which can continue to produce $OH^{\cdot}$. The probability of electron/hole pair recombination is increased by radical scavengers such as bicarbonate and carbonate ions.

6.2.2.1 Mechanism of photo-chemical degradation

Dioxins and Furans have a strong UV absorption at wavelengths shorter than 270 nm; however, their absorption band extends into the wavelengths longer than 290 nm which makes it possible to use sunlight as a radiation source for PCDD/F photolysis (Choi et al. 2000; Wagenaar et al. 1995; Tysklind et al. 1993; Choudry and Webster 1989; Dulin et al. 1986). Upon absorption of light, excitation of the PCDD/F molecule leads to the cleavage of either the C-Cl bond or the C-O bond. Cleavage of the C-Cl bond, dechlorination, accounts for less than 30% of the photodegradation products in isooctane and in aqueous solutions (Rayne et al. 2002; Kim and O'Keefe 2000; Kieatiwong et al. 1990). In hexane, dechlorination yields have typically been higher: 50% from 2,3,7,8-TCDF (Dung and O'Keefe 1994), 5 to 32% from OCDD, and 35 to 68% from OCDF (Konstantinov and Bunce 1996; Wagenaar et al. 1995). The relative yield of dechlorination products decreases with an increasing conversion of the original compound (Rayne et al. 2002; Konstantinov and Bunce 1996). Dechlorination reactions of PCDD/Fs and other aryl chlorides (ArCl) are well established (Chu and Kwan 2002; Konstantinov and Bunce 1996; Hawari et al. 1991; Bunce 1982). In the direct dechlorination mechanism (eq. 5), absorption of UV light (hν) by the ArCl molecule leads to the formation of an excited singlet state that is converted to a triplet state via intersystem crossing (ISC). Homolytic dissociation of the triplet state into radicals then takes place readily.

$$^{1}\mathrm{ArCl} \xrightarrow{[h\nu]} {}^{2}\mathrm{ArCl} \xrightarrow{[\mathrm{ISC}]} {}^{3}\mathrm{ArCl} \longrightarrow \mathrm{Ar}\bullet + \mathrm{Cl}\bullet \qquad \text{(eq. 5)}$$

In the most typical form of an indirect or sensitized mechanism, an electron is transferred from an electron donor (don-H) to the excited state of aryl chloride (ArCl*) and an intermediate aryl radical anion is formed (eq. 6).

$$\mathrm{ArCl^{*}} + \text{don-H} \longrightarrow \mathrm{ArCl^{\cdot -}} \longrightarrow \mathrm{Ar}\bullet + \mathrm{Cl^{-}} \qquad \text{(eq. 6)}$$

The generated aryl radicals (Ar•) are stabilised by hydrogen abstraction that yields the dechlorination product, ArH. Cleavage of the C-O bond yields chlorinated dihydroxybiphenyls or hydroxydiphenyl ethers (the latter compounds are also known as chlorinated phenoxyphenols) (Rayne et al. 2002; Konstantinov et al. 2000; Friesen et al. 1996; Kieatiwong et al. 1990). As yet, the reaction mechanism involving the C-O bond cleavage and a complete mass balance have only been presented for the direct photodegradation of 2,3,7,8-TCDD in aqueous solutions: Rayne and co-workers (2002) demonstrated higher than 50% conversion of 2,3,7,8-TCDD into 2,2'-dihydroxy-4,4',5,5'-tetrachlorobiphenyl. The other degradation products were chlorinated phenoxyphenols and lower chlorinated Dioxins. It was proposed that the yield of chlorinated hydroxydiphenyl ethers could be higher in better hydrogen-donor solvents. In isooctane, less than 10% of 2,3,7,8-TCDD has been found to be converted into 2,2'-dihydroxy-4,4',5,5'-tetrachlorobiphenyl and less than 10% into trichlorodibenzo-*p*-dioxin (Kieatiwong et al. 1990).

The ultimate aromatic photodegradation product of PCDDs appears to be hydroxybenzoic acid (Massé and Pelletier 1987). Some researchers have also reported on the formation of PCDD/F-solvent adducts (Hung and Ingram 1990; Kieatiwong et al. 1990) and chlorophenols (Konstantinov et al. 2000).

The data available on the influence of the chlorination degree and pattern of a PCDD/F molecule on photodegradation suggests that there is a more rapid photodegradation for lower chlorinated PCDD/Fs (Choundry and Webster 1986; Friesen et al. 1996; Kim and O'Keefe 2000; Koester and Hites 1992; Nestrick et al. 1980; Niu et al. 2003; Yan et al. 1995). PCDFs degrade more rapidly than PCDDs (Konstantinov and Bunce 1996; Niu et al. 2003; Ritterbush et al. 1994; Schuler et al. 1998; Wagenaar et al. 1995). It appears that 2,3,7,8-PCDD/Fs degrade more rapidly than non-2,3,7,8-PCDD/Fs on solid support and in the absence of photochemical sensitizers (Hosoya et al. 1995; Nestrick et al. 1980). When photochemical sensitizers are present, like in natural waters, chlorine substituents are preferably cleaved from positions 1, 4, 6, and 9, and 2,3,7,8-PCDD/Fs are more persistent (Friesen et al. 1990, 1996; Hung and Ingram 1990; Konstantinov and Bunce 1996).

6.2.3 Photolytic Degradation of Dioxins and Furans in Water

Apparatus and Reagents

Reagents

- Titanium oxide
- Anhydrous sodium sulphate
- Toluene, residue analysis grade
- Acetone
- Vegetable oil (The vegetable oil used in the experiments was refined groundnut oil. The widely used food-grade oils, was purchased from a local supermarket)

Reference standards: Reference standards of ^{13}C-2,3,7,8-TCDD, 2,3,7,8-TCDD, 2,3,7,8-TCDF, OCDD, and OCDF of 99.99% purity were procured from Cambridge Isotope Laboratories and Wellington Laboratories.

Simulated samples: Simulated samples were prepared by spiking tap water with 2,3,7,8-TCDD, 2,3,7,8-TCDF, OCDD and OCDF solution prepared in nonane and mixed in acetone. Tap water blank was checked for the presence of trace levels of PCDDs and PCDFs before preparing the simulated samples. Physico-chemical properties of tap water are given in Table 5.6.

Equipments

Photochemical Reactor: SAIC IWQ2 model immersion well photochemical reactor was used for conducting photochemical and photocatalytic experiments for removal of 2,3,7,8-TCDD, 2,3,7,8-TCDF, OCDD and OCDF. The immersion well photochemical reactor allows the solutions of reactants to be irradiated by ultraviolet or visible radiation produced by the lamp located in a cooled, double walled immersion well. The efficiency of this type of reactor is very high since the lamp is effectively surrounded by the reacting solution. The lamp is contained in a double-walled quartz borosilicate–glass immersion well through which water is passed for cooling.

Immersion well: SAIC IWQ2 immersion well, double-walled made up of quartz having outlet and inlet tubes for water cooling was used which houses the irradiation lamp. A small diameter inlct tube extends to the bottom of annular space to allow coolant flow from bottom of the well upwards.

Reaction flask: Reaction flask made of borosilicate glass and fitted with a central large ground socket (to house an immersion well) and other small sockets for a reflux condenser, sampling port, etc. was used. The reaction flask has an angled socket and one offset vertical socket and had a capacity of 500 mL.

Irradiation lamp: Medium pressure lamp of 400 W was used. The mercury lamp length is 15 cm. It operates at 220–240 volts AC power supply.

Reflux condenser: A reflux condenser fits into the vertical socket of the reaction flasks and prevents loss of vapor.

Power supply: Power supply was used to operate the UV lamp and to supply extra voltage and current required to initiate the voltage across the mercury arc.

Varian CP-3800 Gas chromatograph-Saturn 2200—Mass Spectrometer

(i) Gas Chromatograph (GC), Perkin Elmer, Clarus-500, equipped with electron capture detector (ECD), and glass lined injection port
(ii) Capillary column-DB-5, crossbond, 5% phenyl 95% polysiloxane, 30 m length, 0.25 mm id, and 0.25 μm film thickness

6.2.3.1 Set-up with UV lamp

Experimental-Batch studies were conducted using simulated samples of 2,3,7,8-TCDD, 2,3,7,8-TCDF, OCDD, and OCDF prepared in tap water. Photochemical reactions with and without catalyst TiO_2 were carried out for a varying period and at ambient temperature under controlled conditions. Experiments were carried out in a SAIC model quartz photochemical assembly with a UV lamp of 400 watt and the water jacket for cooling. The detailed specification of the assembly is given in previous section.

500 mL test water sample was taken in the photochemical reaction vessel. The sample in the reaction vessel was thoroughly mixed. The UV lamp was placed into the photochemical quartz immersion well and the inlet and the outlet tubes of the water jacket were connected to the water supply for cooling the system. The UV lamp was connected to the power supply for extra voltage and current required to initiate mercury arc. The complete assembly was placed in the wooden chamber to avoid exposure to the UV irradiations.

The samples were exposed to the UV irradiations for a fixed period and then the UV lamp was put off. The assembly was allowed to cool for some time and the samples were withdrawn. If the catalyst was added to the sample the catalyst was allowed to settle and the samples were then filtered through whatman filter paper. The filtered samples were analyzed for 2,3,7,8-TCDD, 2,3,7,8-TCDF, OCDD and OCDF. The effect of UV irradiations on the removal of 2,3,7,8-TCDD, 2,3,7,8-TCDF, OCDD and OCDF using photocatalyst and without using photocatalyst in test water was studied. The process was optimized by varying the exposure period, adsorbent, pH, and adsorbate concentration on the removal of tetra and octa congeners. The photocatalyst used was TiO_2 having a band gap of 3eV.

Control samples were also analyzed for 2,3,7,8-TCDD, 2,3,7,8-TCDF, OCDD, and OCDF where the samples were not exposed to UV radiations and were kept in dark.

(a) Effect of exposure period

Simulated samples of 2,3,7,8-TCDD, 2,3,7,8-TCDF, OCDD, and OCDF of concentration 1.0 μg L^{-1} and 2.5 μg L^{-1} were prepared in tap water at neutral pH. The samples were mixed thoroughly on a rotary shaker for a time period of 10 min, at a speed of 175 rpm. After mixing the samples, 500 mL aliquotes of sample were placed in a photoreactor cell of the photochemical assembly. The samples were exposed to UV lamp of 400 watt for varying exposure period ranging from 0 min to 20 min. After the exposure period, the samples were allowed to cool and withdrawn and stored in a refrigerator until analysis. The samples were analyzed. The removal of each congener was calculated and the optimum exposure period was established for maximum removal of each of the PCDD and PCDF congeners.

(b) Effect of analyte concentration

Simulated samples of 2,3,7,8-TCDD of varying concentration ranging from 0.025 μg L^{-1} to 2.5 μg L^{-1} were prepared in tap water at neutral pH. The samples were mixed thoroughly on a rotary shaker for a time period of 10 min, at a speed of 175 rpm. After mixing the samples, 500 mL aliquotes of samples were placed in photoreactor cell of the photochemical assembly. The samples were exposed to UV lamp of 400 watt for fixed exposure period of 20 min. After the exposure period the samples were allowed to cool, and then were withdrawn and stored in a refrigerator until analysis. The samples were analyzed and the removal of 2,3,7,8-TCDD was calculated.

(c) Effect of pH

Simulated sample of 2,3,7,8-TCDD of concentration 5.0 μg L^{-1} was prepared in deionised distilled water at neutral pH. The samples were mixed thoroughly on a rotary shaker for a time period of 10 min at a speed of 175 rpm. After mixing the samples, 500 mL aliquotes of the samples were placed in the photoreactor cell of the photochemical assembly. The samples were exposed to a UV lamp of 400 watt for 20 min at varying pH ranging from 4.0 to 12.0. After the exposure period, the samples were allowed to cool, and then were withdrawn and stored in a refrigerator until analysis. The samples were analyzed. The maximum removal of 2,3,7,8-TCDD was calculated and the optimum pH required for its removal was established.

(d) Effect of catalyst

The effect of degradation of 2,3,7,8-TCDD, 2,3,7,8-TCDF, OCDD, and OCDF using UV irradiation in conjunction with 1 gm L^{-1} of TiO_2 was studied by varying the exposure period. Simulated samples of 2,3,7,8-TCDD, 2,3,7,8-TCDF, OCDD, and OCDF at an initial concentration 1 μg L^{-1} and 2.5 μg L^{-1} were prepared in tap water at a neutral pH. The samples were mixed thoroughly on a rotary shaker at a speed of 175 rpm for a time period of 10 min. After mixing the samples, 500 mL aliquotes of simulated samples were placed in the photoreactor cell of the photochemical assembly and 1 g L^{-1} of TiO_2 was added to it. The samples were exposed to a UV lamp of

400 watt for a varying exposure period ranging from 0 to 20 min. After the exposure period, the samples were allowed to cool and withdrawn and stored in refrigerator until analysis. The samples were analyzed. The removal of 2,3,7,8-TCDD, 2,3,7,8-TCDF, OCDD, and OCDF were calculated and the optimum exposure period was established for the maximum removal of each congener.

6.2.3.2 Set-up with sunlight

Removal studies were conducted using simulated samples of 2,3,7,8-TCDD, 2,3,7,8-TCDF, OCDD, and OCDF prepared in tap water. Photochemical reactions with and without catalyst, TiO_2 were carried out for a varying exposure period. Experiments were carried out under solar radiations in 1 L capacity Borosil beakers containing 500 mL of test water samples with a fixed dose of photocatalyst and without a photocatalyst. Water samples were exposed to direct sunlight for four hrs from 11:30 am to 3:30 pm during the month of December to March when the sky was clear. The samples were exposed regularly every day for 9 days for a total time period of 34 hr. After exposure, the samples were filtered through Whatman paper if the samples were containing photocatalyst. The intensity of solar radiations was found in the range of 700–1100 lux. The filtered samples were analyzed for 2,3,7,8-TCDD, 2,3,7,8-TCDF, OCDD, and OCDF. The effect of UV irradiations on the removal of each congener using a photocatalyst and without using a photocatalyst in test water was studied. The process was optimized by varying the exposure period.

Control samples were also analyzed for 2,3,7,8-TCDD, 2,3,7,8-TCDF, OCDD, and OCDF where the samples were not exposed to sunlight and were kept in dark.

(a) Effect of exposure period

Simulated samples of 2,3,7,8-TCDD, 2,3,7,8-TCDF, OCDD, and OCDF each of initial concentration 0.05 μg L^{-1} and 1.0 μg L^{-1} were prepared in tap water at neutral pH. The samples were mixed thoroughly on a rotary shaker for a time period of 10 min at a speed of 175 rpm. After mixing the samples, 500 mL aliquotes of simulated samples were exposed to solar radiations for varying exposure period ranging from 0 to 34 hr. After the exposure period, the samples were stored in a refrigerator until analysis. The samples were analyzed. The removal of each congener was calculated and the optimum exposure period was established for maximum removal of the Dioxin congener.

(b) Effect of catalyst

Simulated samples of 2,3,7,8-TCDD of concentration 1.0 μg L^{-1} and 2.5 μg L^{-1} was prepared in tap water at neutral pH. The samples were mixed thoroughly on a rotary shaker for a time period of 10 min at a speed of 175 rpm. After mixing the samples, 1 g L^{-1} of TiO_2 was added to 500 mL aliquotes of the samples and exposed to solar radiations for varying exposure period ranging from 0 to 34 hr. After the exposure period the samples were stored in a refrigerator until analysis. The samples were analyzed. The removal of 2,3,7,8-TCDD was calculated and the optimum exposure period for maximum removal was established.

(c) Effect of pH

Simulated sample of 2,3,7,8-TCDD of concentration 1 μg L^{-1} was prepared in tap water at neutral pH. The samples were mixed thoroughly on a rotary shaker for a time period of 10 min at a speed of 175 rpm. After mixing the samples, 500 mL aliquotes of the samples were placed in the photoreactor cell of the photochemical assembly. The samples were exposed to sunlight for 34 hr at varying pH ranging from 4.0 to 12.0. After the exposure period, the samples were withdrawn and stored in refrigerator until analysis. The samples were analyzed. The removal of 2,3,7,8-TCDD was calculated and the optimum pH required for its maximum removal was established.

6.2.3.3 Results and Discussion

Photolytic degradation of PCDDs and PCDFs in water samples using 400 W UV lamp

Exposure period: The degradation of 2,3,7,8-TCDD, 2,3,7,8-TCDF, OCDD and OCDF increased with the increase in exposure period to UV irradiations. The exposure of 2,3,7,8-TCDD increased from 81.2% to 94.13% with increase in contact period from 5 min to 20 min. Maximum removal of 94.13% was observed at an initial concentration of 0.05 μg L^{-1} after an exposure period of 20 min. However, 90.20% and 85% removal of 2,3,7,8-TCDD was observed at an initial concentration of 1.0 μg L^{-1} and 2.5 μg L^{-1} respectively after an exposure period of 20 min (Fig. 6.9). Thus an optimum contact period of 20 min was established for the maximum removal of 2,3,7,8-TCDD.

The removal of 2,3,7,8-TCDF increased from 90.2% to 96.50% with increase in exposure period from 5 min to 20 min. Maximum removal of 96.50% was observed at an initial concentration of 1.0 μg L^{-1} after an exposure period of 20 min. However, 92.4% of 2,3,7,8-TCDF removal was observed at an initial concentration of 2.5 μg L^{-1} after an exposure period of 20 min (Fig. 6.10). Thus an optimum exposure period of 20 min was established for the maximum removal of 2,3,7,8-TCDF.

The removal of OCDD increased from 70.25% to 76.65% with increase in exposure period from 5 min to 20 min. Maximum OCDD removal of 76.65% was observed at an initial concentration of 1.0 μg L^{-1} after an exposure period of 20 min. However, 69.52% removal of OCDD was observed at an initial concentration of 2.5 μg L^{-1} after an exposure period of 20 min (Fig. 6.11). Thus an optimum exposure period of 20 min was established for the maximum removal of OCDD.

The removal of OCDF increased from 74.45% to 79.2% with increase in exposure period from 5 min to 20 min. Maximum removal of 79.2% was observed at an initial concentration of 1.0 μg L^{-1} after an exposure period of 20 min. However, 71.0% removal of OCDF was observed at an initial concentration of 2.5 μg L^{-1} of after an exposure period of 20 min (Fig. 6.12). Thus an optimum exposure period of 20 min was established for the maximum removal of OCDF.

Varying analyte concentrations: The effect on degradation of 2,3,7,8-TCDD using UV irradiation was studied by varying the initial concentration of 2,3,7,8-TCDD. The analysis results showed decrease in removal efficiency with increase in 2,3,7,8-TCDD concentration. The removal of 2,3,7,8-TCDD was found as 96.0% at an initial concentration of 0.025 μg L^{-1}, which was observed to decrease to 83.34% at 2.5 μg L^{-1} of 2,3,7,8-TCDD. The data analyzed showed increase in removal at lower 2,3,7,8-TCDD concentrations (Fig. 6.13).

pH: The effect of pH removal of 2,3,7,8-TCDD was studied at an initial concentration of 5.0 μg L^{-1}. The data analyzed showed that the percent removal of 2,3,7,8-TCDD increased with the increase in pH. Maximum removal of 2,3,7,8-TCDD of 94.97%

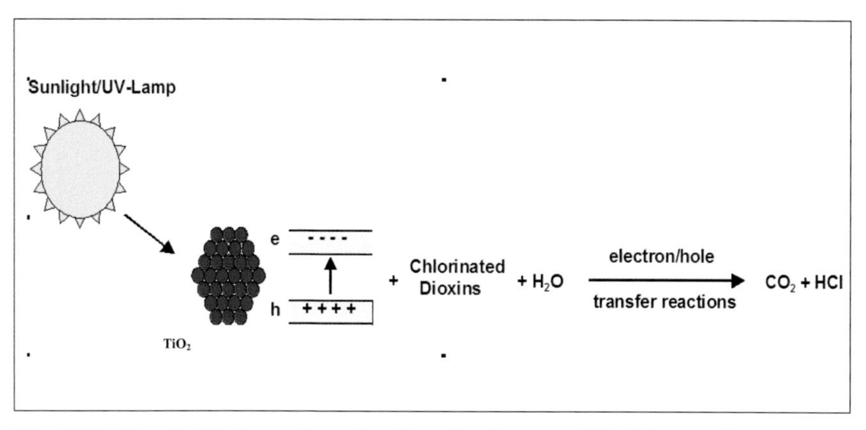

Fig. 6.9(a): Use of stable, inorganic, semiconductor band gaps to oxidize PCDDs and PCDFs using sunlight/UV-lamp.

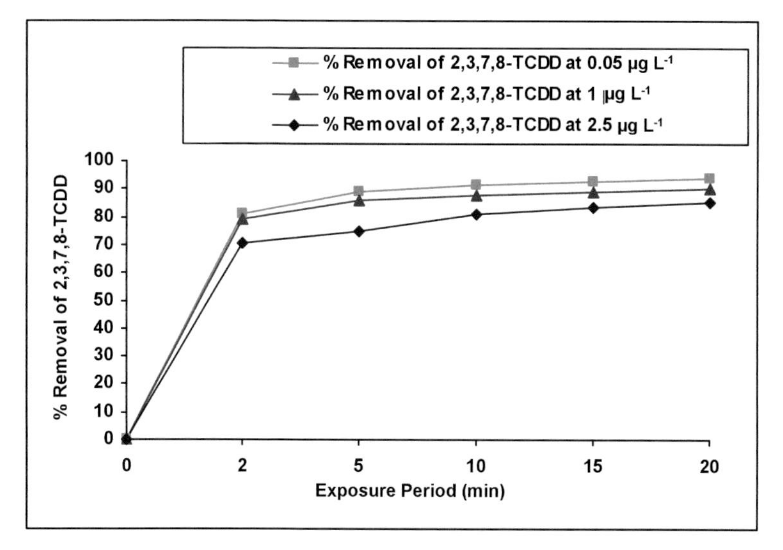

Fig. 6.9(b): Effect of photolytic UV-radiations (400 W) on removal of 2,3,7,8-TCDD in water.

was obtained at pH 12. The analysis results showed increased removal in alkaline pH, whereas the removal was found less in acidic pH. Removal of 94.97% and 66.2% was observed at pH 12 and 4.0 respectively at an initial concentration of 5.0 μg L^{-1} after the exposure period of 20 min (Fig. 6.14).

Conjunction of UV radiation and catalyst (TiO_2): The photocatalytic degradation of 2,3,7,8-TCDD, 2,3,7,8-TCDF, OCDD, and OCDF increased with the increase in exposure period to UV irradiations. The removal of 2,3,7,8-TCDD increased from 97.19% to 100% with increase in exposure period from 2 min to 20 min. Maximum removal of 100% was observed at an initial concentration of 1.0 μg L^{-1} after an exposure period of 20 min. However, at an initial concentration of 2.5 μg L^{-1} of 2,3,7,8-TCDD,

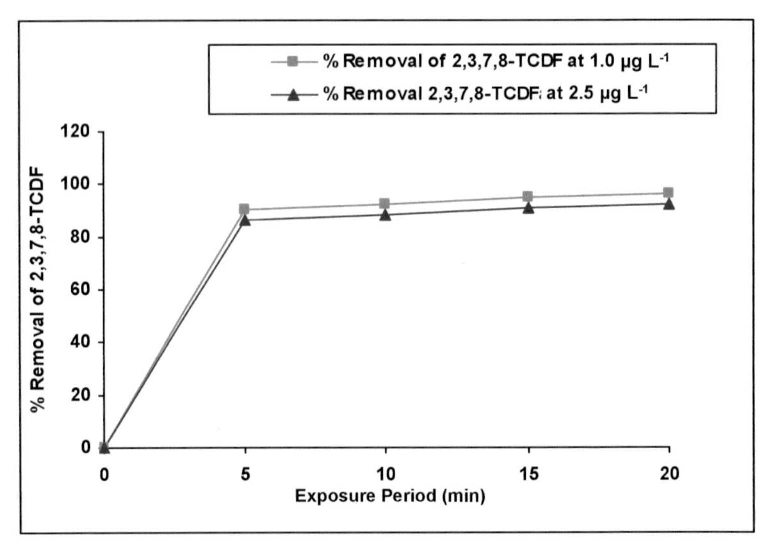

Fig. 6.10: Effect of UV-radiations (400 W) on removal of 2,3,7,8-TCDF in water.

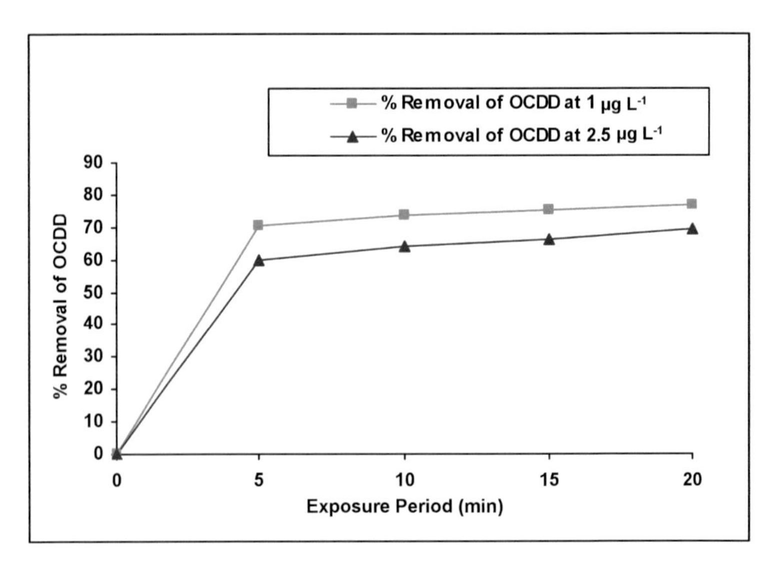

Fig. 6.11: Effect of UV-radiations (400 W) on removal of OCDD in water.

the removal increased from 90.46% to 96.90% with increase in contact time from 2 min to 20 min removal after an exposure period of 20 min (Fig. 6.15). Thus an optimum contact time of 20 min was established for maximum removal of 2,3,7,8-TCDD.

The removal of 2,3,7,8-TCDF increased from 98.42% to 100% with increase in exposure period from 2 min to 20 min. Maximum removal of 100% was observed at an initial concentration of 1.0 μg L^{-1} after an exposure period of 20 min. However, at an initial concentration of 2.5 μg L^{-1} of 2,3,7,8-TCDF, removal increased from

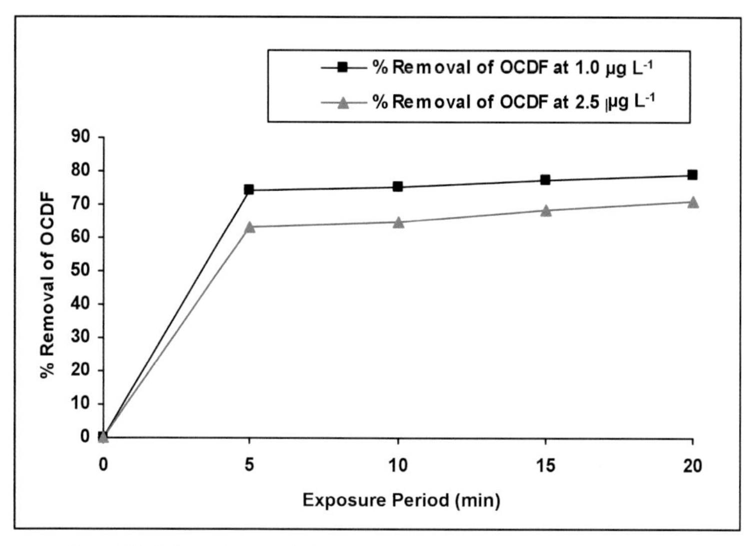

Fig. 6.12: Effect of UV-radiations (400 W) on removal of OCDF in water.

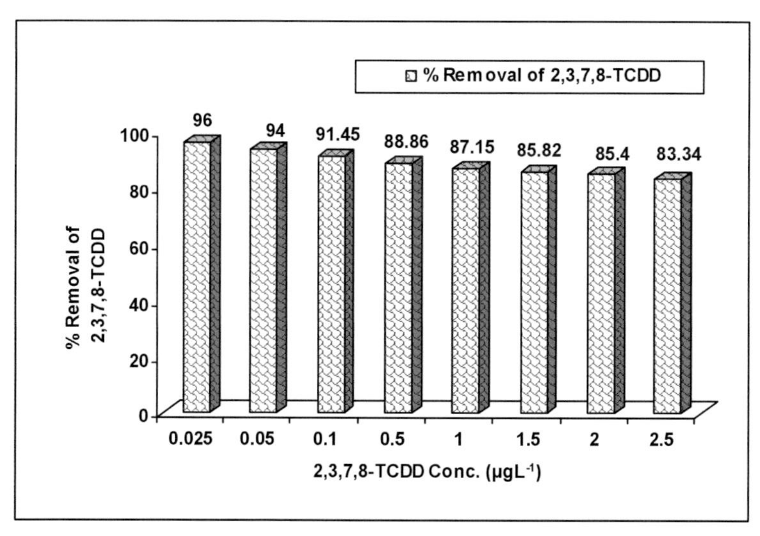

Fig. 6.13: Effect of photolysis on removal of 2,3,7,8-TCDD at varying concentrations in water.

98.00% to 100% after an exposure period of 20 min (Fig. 6.16). Thus an optimum contact period of 20 min was established for maximum removal of 2,3,7,8-TCDF.

The removal of OCDD increased from 88% to 96.75% with increase in contact time from 2 min to 20 min. Maximum removal of 96.75% was observed at an initial concentration of 1 $\mu g\ L^{-1}$ after an exposure period of 20 min. However, at an initial concentration of 2.5 $\mu g\ L^{-1}$ of OCDD, removal increased from 85.95 to 94.20% after an exposure period of 20 min (Fig. 6.17). Thus an optimum exposure period of 20 min was established for maximum removal of OCDD.

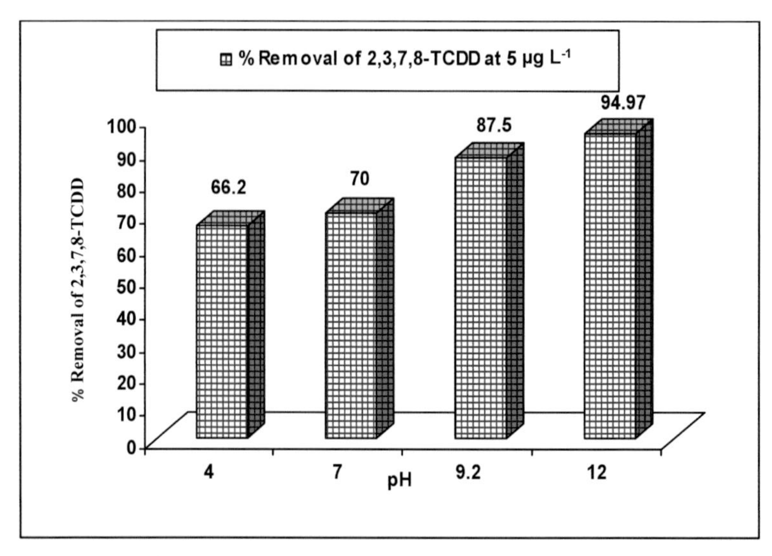

Fig. 6.14: Effect of pH on photolytic removal of 2,3,7,8-TCDD using UV radiations in water.

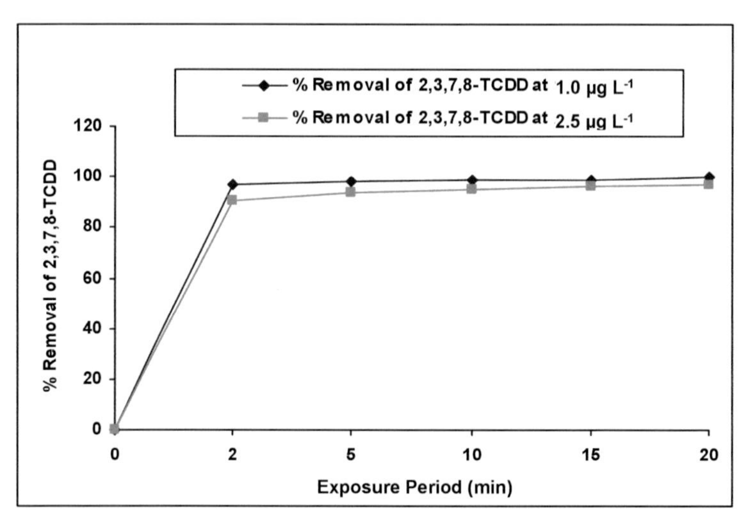

Fig. 6.15: Effect of photocatalytic UV-radiations on removal of 2,3,7,8-TCDD in water.

The removal of OCDF increased from 90.25% to 97.45% with increase in exposure period from 2 min to 20 min. Maximum removal of 97.45% was observed at an initial concentration of 1.0 µg L^{-1} after an exposure period of 20 min. However, at an initial concentration of 2.5 µg L^{-1} of OCDF, removal increased from 87.30 to 95.11% was observed after an exposure period of 20 min (Fig. 6.18). Thus an optimum exposure period of 20 min was established for maximum removal of OCDF.

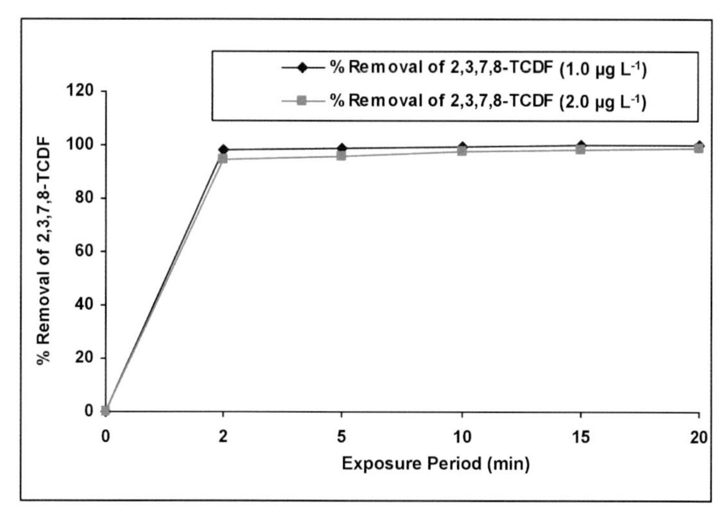

Fig. 6.16: Effect of photocatalytic UV-radiations on removal of 2,3,7,8-TCDF in water.

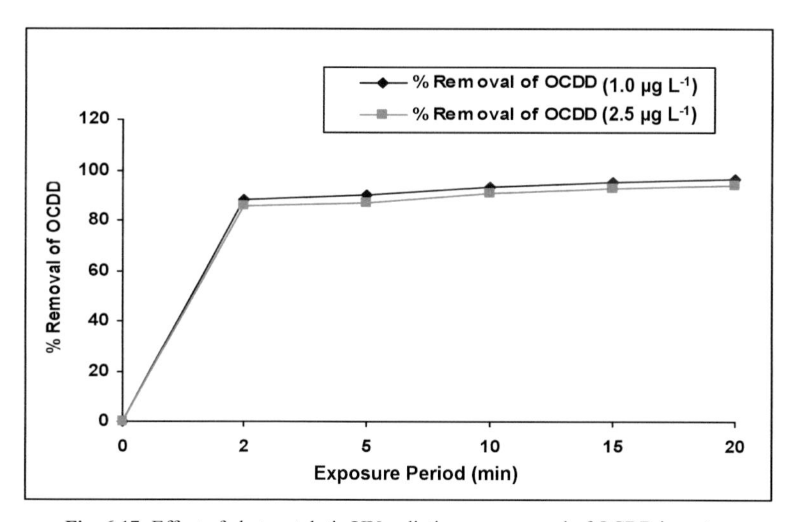

Fig. 6.17: Effect of photocatalytic UV radiations on removal of OCDD in water.

Photolytic degradation of PCDDs and PCDFs in water using sunlight

Exposure period: The effect of degradation of 2,3,7,8-TCDD using solar radiations was studied by varying the exposure period. The analysis results showed increase in removal efficiency with increase in exposure period. The removal of 30.56% was observed after 8 hr, which was observed to increase to 58.41% after 34 hr of exposure period. The data analyzed showed the maximum removal of 58.41% at an initial concentration of 1.0 μg L^{-1} of 2,3,7,8-TCDD after an exposure period of 34 hr (Fig. 6.19). However, at an initial concentration of 2.5 μg L^{-1} of 2,3,7,8-TCDD, removal increased from 25.10% to 44.2% after an exposure period of 34 hr (Fig. 6.19).

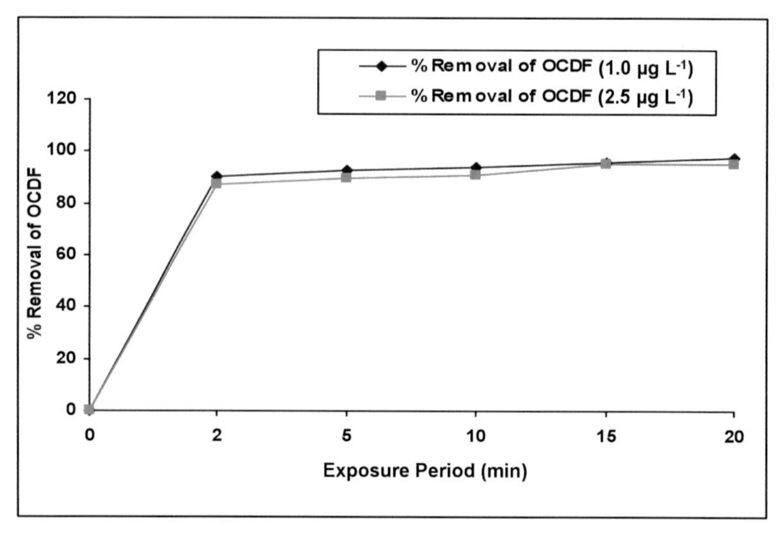

Fig. 6.18: Effect of photocatalytic UV radiations on removal of OCDF in water.

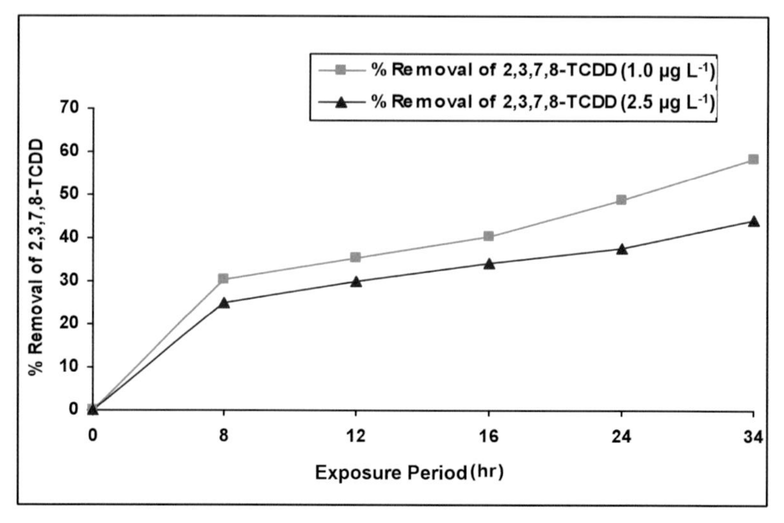

Fig. 6.19: Effect of solar radiations on removal of 2,3,7,8-TCDD in water.

pH: The effect of pH and solar irradiation on degradation 2,3,7,8-TCDD was studied at an initial concentration of 2.0 μg L^{-1}. The data analyzed showed that the removal of 2,3,7,8-TCDD increased with the increase in pH. Maximum removal of 2,3,7,8-TCDD of 60.32% was obtained at pH 12. The analysis results showed higher removal in alkaline pH, whereas the removal was found less in acidic pH. Removal of 2,3,7,8-TCDD was found as 32.05%, 44.4%, 53.2%, and 60.32% at pH 4.0, 7, 9.2, and 12, respectively, at an initial concentration of 1.0 μg L^{-1} after an exposure period of 34 hr (Fig. 6.20).

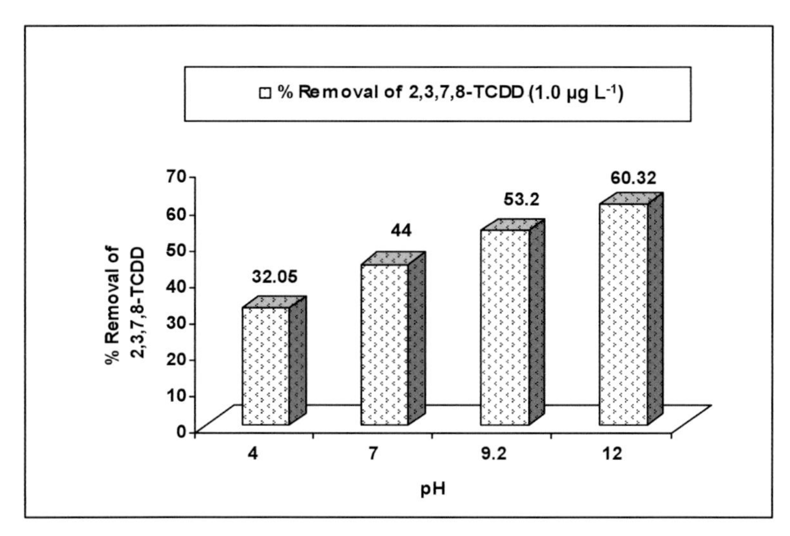

Fig. 6.20: Effect of pH on removal of 2,3,7,8-TCDD in water using photolytic solar radiations.

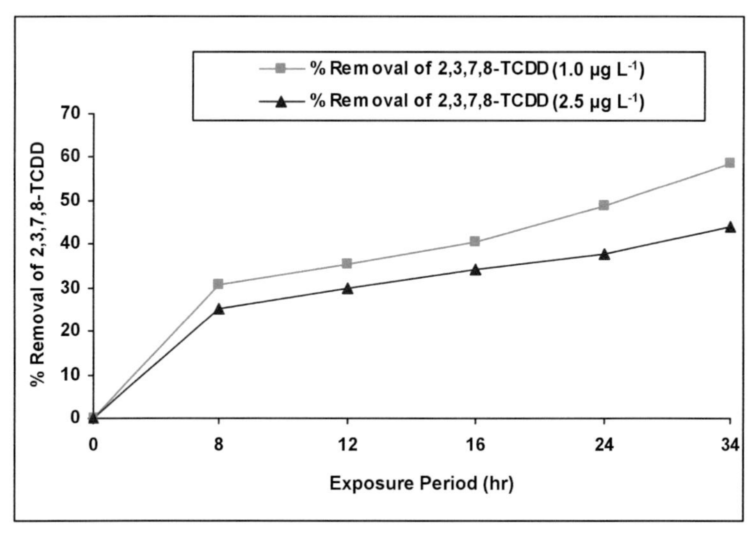

Fig. 6.21: Effect of solar radiations in conjunction with TiO_2 on removal of 2,3,7,8-TCDD in water.

Photocatalyst: The effect of photodegradation of 2,3,7,8-TCDD using solar radiations and TiO_2 was studied by varying the exposure period. The analysis results showed increase in removal efficiency with increase in exposure period. The removal of 33.56% was observed after 8 hr, which was observed to increase to 65.86% after 34 hr. The data analyzed showed the maximum removal of 65.86% at an initial concentration of 1.0 µg L^{-1} of 2,3,7,8-TCDD after an exposure period of 34 hr (Fig. 6.21). However, at an initial concentration of 2.5 µg L^{-1} of 2,3,7,8-TCDD, removal increased from 35.0% to 60.86% after an exposure period of 34 hr (Fig. 6.21).

Data Evaluation

Photochemical methods were used to study the degradation of 2,3,7,8-TCDD, 2,3,7,8-TCDF, OCDD, and OCDF. The simulated samples were exposed to high intensity UV irradiation. The UV irradiation of 400 watt lamp was able to remove 90% of the tetra isomers, and 70% of octa congeners of PCDDs and PCDFs, that is, 94.13% of 2,3,7,8-TCDD, 96.50% of 2,3,7,8-TCDF, 76.65% of OCDD, and 79.20% of OCDF. Conjunction of TiO_2 with UV irradiations showed increase in removal efficiency. However, TiO_2 alone did not remove the PCDDs and PCDFs in simulated samples. The photochemical studies were conducted at varying pH; it was observed that higher degradation was observed at alkaline pH and degradation decreased in acidic pH. Solar radiations were found to degrade PCDDs and PCDFs but to a lower extent compared to radiations of UV lamp.

6.2.4 Photolytic Degradation of Dioxins and Furans in Soil

Experimental

Batch studies were conducted using simulated samples of 2,3,7,8-TCDD, 2,3,7,8-TCDF, OCDD, and OCDF prepared in soil. Photochemical reactions were carried out for a fixed period and at ambient temperature under controlled conditions. Experiments were carried under a SAIC model quartz photochemical assembly with a UV lamp of 400 watt and the water jacket for cooling. The detailed specification of the assembly is given in earlier sections.

6.2.4.1 Set-up with UV lamp

10 g of test soil sample was taken in the photochemical reaction dish. The UV lamp was placed into the photochemical quartz immersion well and the inlet and the outlet tubes of the water jacket were connected to the water supply for cooling the system. The UV lamp was connected to the power supply for extra voltage and current required to initiate the mercury arc. The complete assembly was placed in the wooden chamber to avoid exposure to the UV irradiations. The samples were placed under the UV lamp and exposed to the UV irradiations for fixed intervals and then the UV lamp was put off. The assembly was allowed to cool for some time and the samples were withdrawn. The samples were analyzed for 2,3,7,8-TCDD, 2,3,7,8-TCDF, OCDD, and OCDF. The effect of UV irradiations on the removal of 2,3,7,8-TCDD, 2,3,7,8-TCDF, OCDD, and OCDF using oil and without using oil in test soil was studied. The process was optimized by varying the exposure period, and analyte concentration on the removal of 2,3,7,8-TCDD, 2,3,7,8-TCDF, OCDD, and OCDF. Control samples were also analyzed for 2,3,7,8-TCDD where the samples were not exposed to UV radiations and were kept in dark.

Soil samples (10 g) were dried to complete dryness and the dry weight was determined from the weight loss. The samples were spiked with known concentration of 2,3,7,8-TCDD, 2,3,7,8-TCDF, OCDD, and OCDF and then homogenized with a stainless steel spatula and extracted with toluene on a rotary shaker for 20 to 24 hr if oil was added to the soil samples then the extracts were treated with concentrated sulphuric acid to decompose the vegetable oil.

After the aforementioned pre-treatment steps, the soil samples were purified over various solid-phase columns. In brief, the samples were eluted with organic solvents or solvent mixtures through a multi-layer silica gel column, an aluminum oxide column, and an activated carbopack-Celite column. These columns were used to remove decomposed lipids and other disturbing organic matter as well as compounds that would interfere with the identification and quantification of 2,3,7,8-TCDD, 2,3,7,8-TCDF, OCDD and OCDF. For example, with an activated carbon-Celite column PCBs were separated in a different solvent fraction. Samples that were concentrated to an approximate volume of 50 to 100 μL were injected into the column of a gas chromatograph using the split technique. The ECD or MS detector was used to detect and measure the peak area of each congener.

(a) Effect of exposure period

The soil used for the removal experiments was ground, sieved and dried. Simulated soil samples of 2,3,7,8-TCDD, 2,3,7,8-TCDF, OCDD, and OCDF of initial concentration 2.5 and 5.0 ng g^{-1} were prepared. The samples were mixed and allowed to equilibrate at room temperature. After mixing the samples, 10 g aliquotes of the samples were spread evenly on a glass plate 15 × 12 cm to form soil layer of 2 mm. The samples were placed under the UV photochemical assembly. The samples were exposed to a UV lamp of 400 watt for a varying exposure period ranging from 0 to 34 h. After the exposure period the samples were allowed to cool and withdrawn and stored in an airtight container until analysis. The samples were analyzed and the removal of each Dioxin congener was calculated and the optimum exposure period for maximum removal of 2,3,7,8-TCDD, 2,3,7,8-TCDF, OCDD, and OCDF was established.

The experiment was repeated using a 125 Watt UV lamp at an initial concentration of 2.5 ng g^{-1}. The removal of each Dioxin congener was calculated and the optimum exposure period was established.

(b) Effect of analyte concentration

The soil used for the removal experiments was dried, ground, and sieved. Simulated soil samples, of 2,3,7,8-TCDD, 2,3,7,8-TCDF, OCDD, and OCDF of concentration ranging from 0.025 to 0.5 μg g^{-1} were prepared. The samples were mixed thoroughly and allowed to equilibrate at room temperature. After mixing the samples, 10 g aliquotes of simulated samples were spread evenly on a glass plate of dimensions 15 × 12 cm to form soil layer of 2 mm. The samples were placed under the photochemical assembly. The samples were exposed to UV lamp of 400 watt for fixed exposure period of 34 hr. After the exposure period, the samples were allowed to cool and withdrawn and stored in airtight container until analysis. The samples were analyzed and the removal of each Dioxin congener was calculated.

(c) Effect of groundnut oil and varying exposure period

The soil used for the removal experiments was dried, ground, and sieved. Simulated soil samples, of 2,3,7,8-TCDD, 2,3,7,8-TCDF, OCDD, and OCDF of initial concentration 2.5 ng g^{-1} and 5.0 ng g^{-1} were prepared. The samples were mixed thoroughly and allowed to equilibrate at room temperature. 20 drops of groundnut oil was added to the simulated samples. After mixing the samples, 10 g aliquotes of simulated samples were spread evenly on glass plate 15 × 12 cm to form soil layer of 2 mm. The samples were placed under the photochemical assembly. The samples were exposed to a UV lamp of 400 watt for a varying exposure period of 0 to 34 hr with periodic mixing. After the exposure period the samples were allowed to cool and withdrawn and stored in an airtight container until analysis. The samples were analysed and the removal of each Dioxin congener was calculated.

6.2.4.2 Set-up with sunlight

Batch studies were conducted using simulated soil samples of 2,3,7,8-TCDD, 2,3,7,8-TCDF, OCDD, and OCDF. Photochemical reactions were carried out for a fixed period and at ambient temperature under controlled conditions. Experiments were carried under sunlight.

A 10 g test soil sample was taken in the clean glass petri dish. The sample in the glass dish was thoroughly mixed. The samples were exposed to the solar irradiations for a fixed period. The samples were analysed for the four congeners. The effect of solar irradiations on the removal of 2,3,7,8-TCDD, 2,3,7,8-TCDF, OCDD, and OCDF in the test sample was studied. The process was optimized by varying the exposure period and analyte concentration on the removal of 2,3,7,8-TCDD, 2,3,7,8-TCDF, OCDD, and OCDF. The control samples were also analyzed for 2,3,7,8-TCDD, 2,3,7,8-TCDF, OCDD, and OCDF where the samples were not exposed to solar radiations and were kept in the dark.

(a) Effect of varying exposure period

The soil used for the removal experiments was dried, ground, and sieved. Simulated soil samples of 2,3,7,8-TCDD, 2,3,7,8-TCDF, OCDD, and OCDF of initial concentration 2.5 ng g^{-1} and 5.0 ng g^{-1} were prepared. The samples were mixed thoroughly and allowed to equilibrate at room temperature. After mixing the samples, 10 g aliquotes of samples were spread evenly on glass plate 15 × 12 cm to form soil layer of 2 mm. The samples were exposed to solar irradiations for varying time of 0 to 34 h. After the exposure period the samples were withdrawn and stored in airtight container until analysis. The samples were analyzed and removal of each Dioxin congener was calculated.

(b) Effect of varying analyte concentration

The soil used for the removal experiments was dried, ground, and sieved. Simulated soil samples of 2,3,7,8-TCDD, 2,3,7,8-TCDF, OCDD, and OCDF of varying concentrations ranging from 0.025 μg g^{-1} to 0.5 μg g^{-1} were prepared. The samples were mixed thoroughly by making slurry with acetone and allowed it to equilibrate at room temperature. After mixing the samples, 10 g aliquotes of samples were spread evenly in glass plate of dimensions 15 × 12 cm to form a soil layer of 2 mm. The samples were exposed to solar irradiations for a fixed exposure period of 34 hr. After the exposure period, the samples were withdrawn and stored in an airtight container until analysis. The samples were analyzed and the removal of each Dioxin congener was calculated.

(c) Effect of groundnut oil and varying exposure period

The soil used for the removal experiments was dried, ground, and sieved. Simulated soil samples of 2,3,7,8-TCDD, 2,3,7,8-TCDF, OCDD, and OCDF of concentration 25 ng g^{-1} were prepared. 20 drops of groundnut oil was added to the simulated samples. The samples were mixed thoroughly by making slurry with acetone and allowed to equilibrate at room temperature. After mixing the samples, 10 g aliquotes of simulated samples were spread evenly in glass plate of dimensions 15 × 12 cm to form a soil layer of 2 mm. The samples were exposed to solar irradiations for a varying exposure period of 0–34 h. After the exposure period, the samples were withdrawn and stored in an airtight container until analysis. The samples were analyzed and the removal of each Dioxin congener was calculated.

(d) Effect of varying analyte concentration using sunlight and oil

The soil used for the removal experiments was dried, ground, and sieved. Simulated soil samples of 2,3,7,8-TCDD, 2,3,7,8-TCDF, OCDD, and OCDF of varying concentration ranging from 0.025 $\mu g\ g^{-1}$ to 0.5 $\mu g\ g^{-1}$ were prepared. The samples were mixed thoroughly by making slurry with acetone and allowed to equilibrate at room temperature. The soil was sprayed with groundnut oil. After mixing the samples, 10 g aliquotes of samples were spread evenly in glass plate of dimensions 15 × 12 cm to form a soil layer of 2 mm. The samples were exposed to solar irradiations for fixed exposure period 34 h with periodic mixing. After the exposure period, the samples were withdrawn and stored in an airtight container until analysis. The samples were analyzed and removal of each Dioxin congener was calculated.

6.2.4.3 Results and Discussion

Effect of exposure period using UV lamp (400 W)

The degradation of 2,3,7,8-TCDD, 2,3,7,8-TCDF, OCDD, and OCDF increased with the increase in exposure period to UV irradiations. The removal of 2,3,7,8-TCDD increased from 5.2% to 43.61% with increase in exposure period from 2 hr to 24 hr. Maximum removal of 43.61% was observed at an initial concentration of 2.5 ng g^{-1} after an exposure period of 24 hr. However, at an initial concentration of 5.0 ng g^{-1} of 2,3,7,8-TCDD, 30% removal was observed after an exposure period of 24 hr (Fig. 6.22).

The removal of 2,3,7,8-TCDF increased from 6.0% to 49.76% with increase in exposure period from 2 hr to 24 hr. The maximum removal of 49.76% was observed at an initial concentration of 2.5 ng g^{-1} after an exposure period of 24 hr. However, at an initial concentration of 5.0 ng g^{-1} of 2,3,7,8-TCDF, 32.3% removal was observed after an exposure period of 24 hr (Fig. 6.23).

The removal of OCDD increased from 0 to 32.8% with increase in exposure period from 2 hr to 24 hr. Maximum removal of 32.8% was observed at an initial concentration of 2.5 ng g^{-1} after an exposure period of 24 hr. However, at an initial concentration of 5.0 ng g^{-1} of OCDD, 24.0% removal was observed after an exposure period of 24 hr (Fig. 6.24).

The removal of OCDF increased from 0 to 38.0% with increase in exposure period from 2 hr to 24 hr. Maximum removal of 38.0% was observed at an initial concentration of 2.5 ng g^{-1} after an exposure period of 24 hr. However, at an initial concentration of 5.0 ng g^{-1} of OCDF, 30.20% removal was observed after an exposure period of 24 hr (Fig. 6.25).

Effect of exposure period using UV lamp (125 W)

The degradation of 2,3,7,8-TCDD, 2,3,7,8-TCDF, OCDD, and OCDF increased with the increase in exposure period to UV irradiations. The removal of 2,3,7,8-TCDD increased from 0 to 15.3% with increase in exposure period from 2 hr to 24 hr.

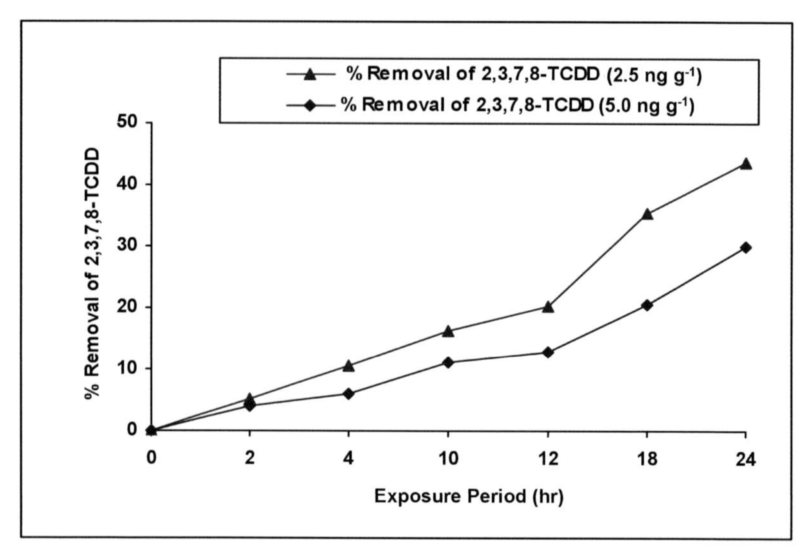

Fig. 6.22: Effect of UV radiations (400 W) on removal of 2,3,7,8-TCDD in soil.

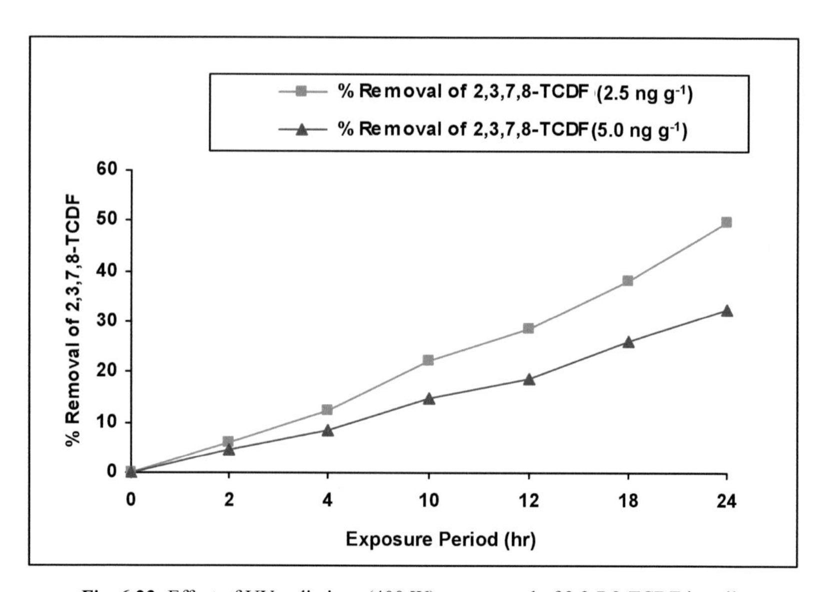

Fig. 6.23: Effect of UV radiations (400 W) on removal of 2,3,7,8-TCDF in soil.

Maximum removal of 15.3% was observed at an initial concentration of 2.5 ng g^{-1} after an exposure period of 24 hr (Fig. 6.26).

The removal of 2,3,7,8-TCDF increased from 0 to 17.25% with increase in exposure period from 2 hr to 24 hr. Maximum removal of 17.25% was observed at an initial concentration of 2.5 ng g^{-1} after an exposure period of 24 hr (Fig. 6.27).

The removal of OCDD increased from 0 to 11.22% with an increase in exposure period from 2 hr to 24 hr. Maximum removal of 11.22% was observed at an initial

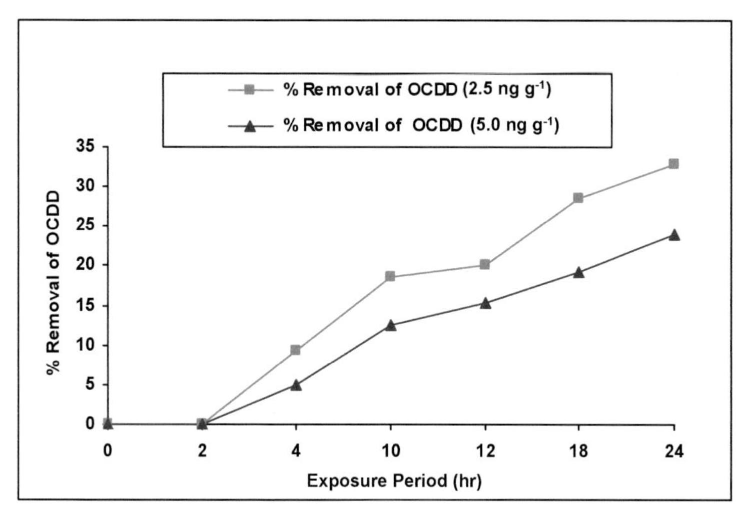

Fig. 6.24: Effect of UV radiations (400 W) on removal of OCDD in soil.

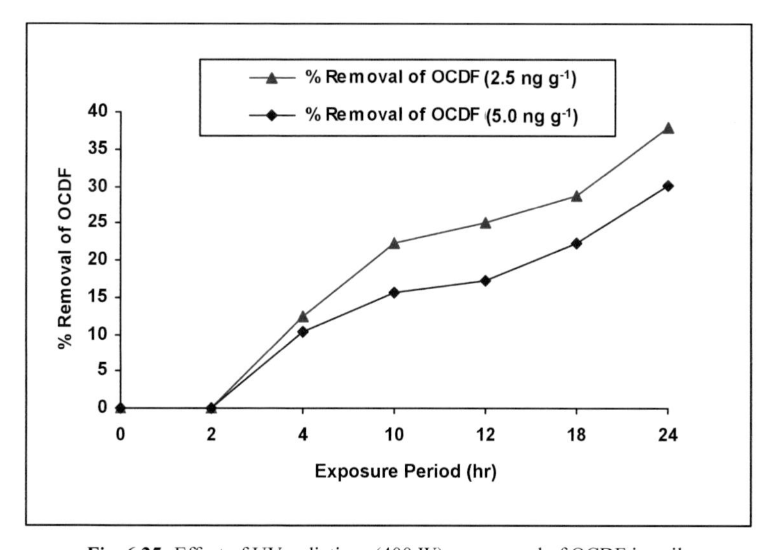

Fig. 6.25: Effect of UV radiations (400 W) on removal of OCDF in soil.

concentration of 2.5 ng g^{-1} after an exposure period of 24 hr (Fig. 6.28). The removal of OCDF increased from 0–13.25% with increase in exposure period from 2 hr to 24 hr. Maximum removal of 13.25% was observed at an initial concentration of 2.5 ng g^{-1} after an exposure period of 24 hr (Fig. 6.29).

Effect of analyte concentration using UV lamp (400 W)

The degradation of 2,3,7,8-TCDD in simulated soil sample was studied at varying concentration of 0.1 μg g^{-1}, 0.2 μg g^{-1}, 0.4 μg g^{-1}, 0.5 mg g^{-1} and 2.5 ng g^{-1} of 2,3,7,8-TCDD. Removal of 42.80% for 2,3,7,8-TCDD was observed at an initial concentration of 2.5 ng g^{-1} after an exposure period of 24 hr. Whereas removal of 38.0% for 2,3,7,8-

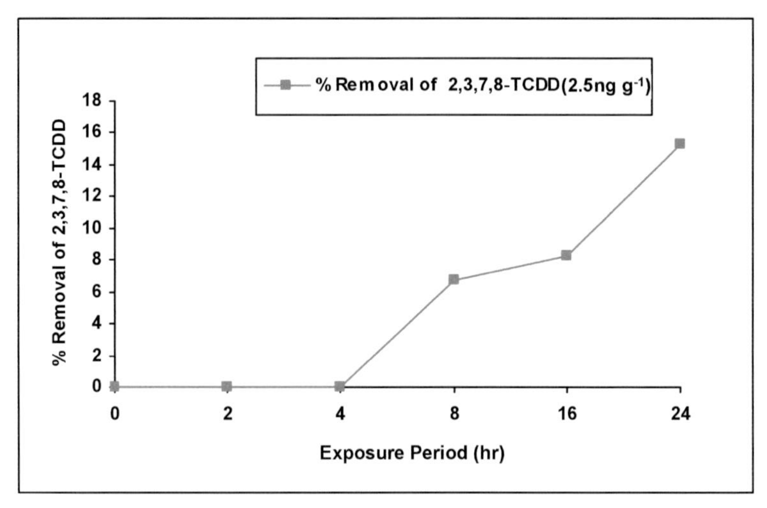

Fig. 6.26: Effect of UV radiations (125 W) on removal of 2,3,7,8-TCDD at 2.5 ng g^{-1} in soil.

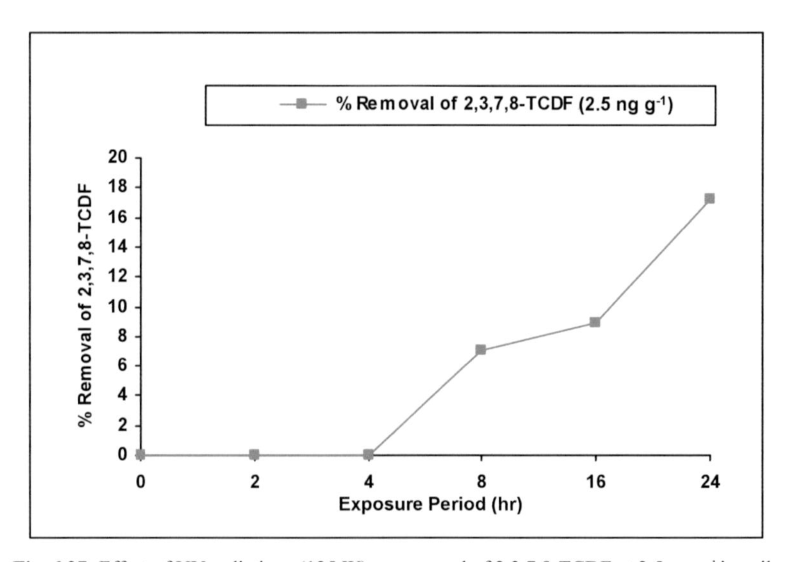

Fig. 6.27: Effect of UV radiations (125 W) on removal of 2,3,7,8-TCDF at 2.5 ng g^{-1} in soil.

TCDD was observed at an initial concentration of 0.1 μg g^{-1} after an exposure period of 24 hr. Removal of 30.0% for 2,3,7,8-TCDD was observed at an initial concentration of 0.2 μg g^{-1} after an exposure period of 24 hr. Removal of 22.0% for 2,3,7,8-TCDD was observed at an initial concentration of 0.4 μg g^{-1} after an exposure period of 24 hr. Removal of 16.0% for 2,3,7,8-TCDD was observed at an initial concentration of 0.5 μg g^{-1} after an exposure period of 24 hr (Fig. 6.30).

Effect of groundnut oil and varying exposure period using UV lamp (400 W)

The degradation of 2,3,7,8-TCDD, 2,3,7,8-TCDF, OCDD, and OCDF in a simulated soil sample was studied by varying the exposure period and addition of groundnut oil.

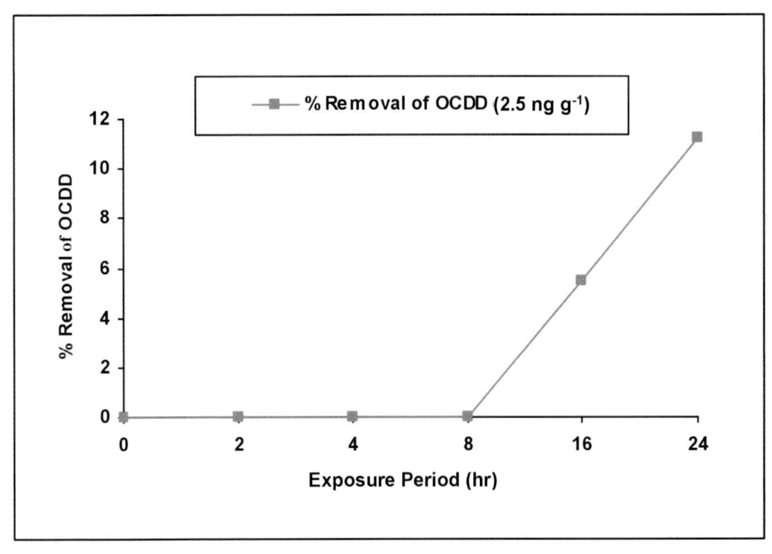

Fig. 6.28: Effect of UV radiations (125 W) on removal of OCDD at 2.5 ng g^{-1} in soil.

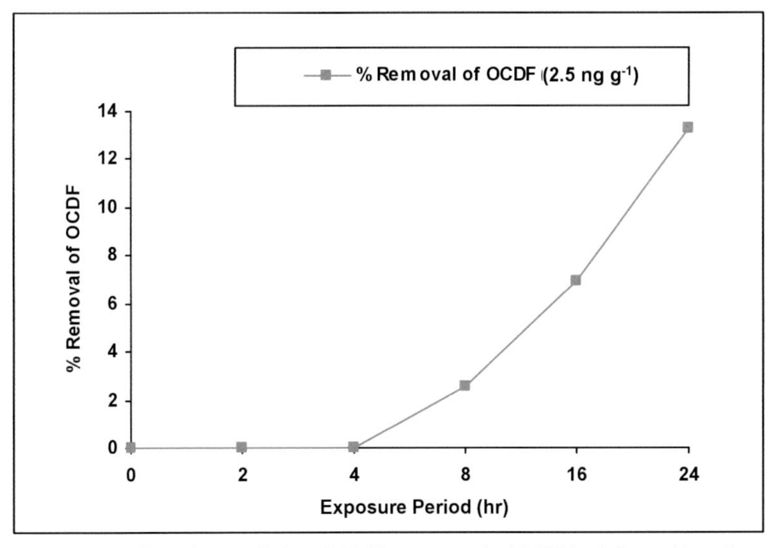

Fig. 6.29: Effect of UV radiations (125 W) on removal of OCDF at 2.5 ng g^{-1} in soil.

The degradation of 2,3,7,8-TCDD increased with the addition of the oil and increase in exposure period to UV irradiations. The removal of 2,3,7,8-TCDD increased from 13.44% to 62.3% with increase in exposure period from 2 hr to 24 hr. Maximum removal of 62.3% was observed at an initial concentration of 2.5 ng g^{-1} after an exposure period of 24 hr. However, at an initial concentration of 5.0 ng g^{-1} of 2,3,7,8-TCDD, 50.61% removal was observed after an exposure period of 24 hr (Fig. 6.31).

The removal of 2,3,7,8-TCDF increased from 15.44% to 69.24% with increase in exposure period from 2 hr to 24 hr. Maximum removal of 69.24% was observed at

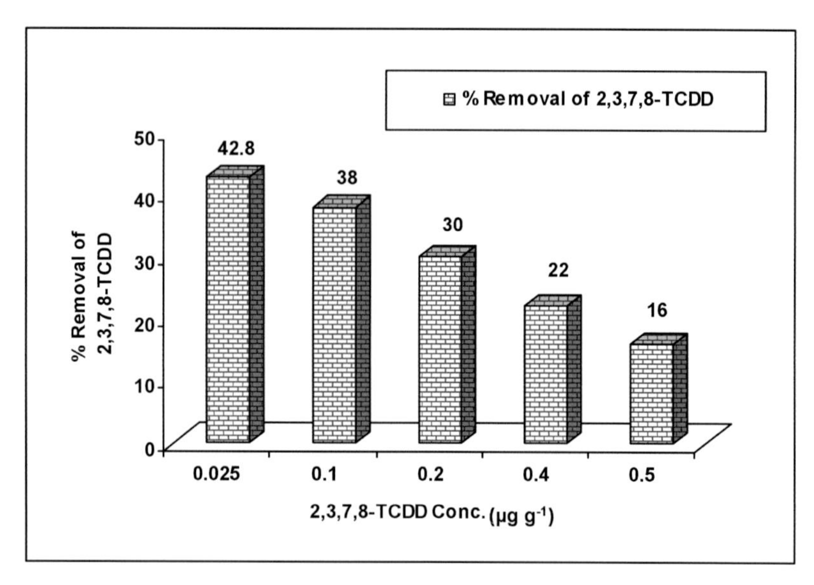

Fig. 6.30: Effect of UV radiations on removal of 2,3,7,8-TCDD in soil at varying initial concentrations.

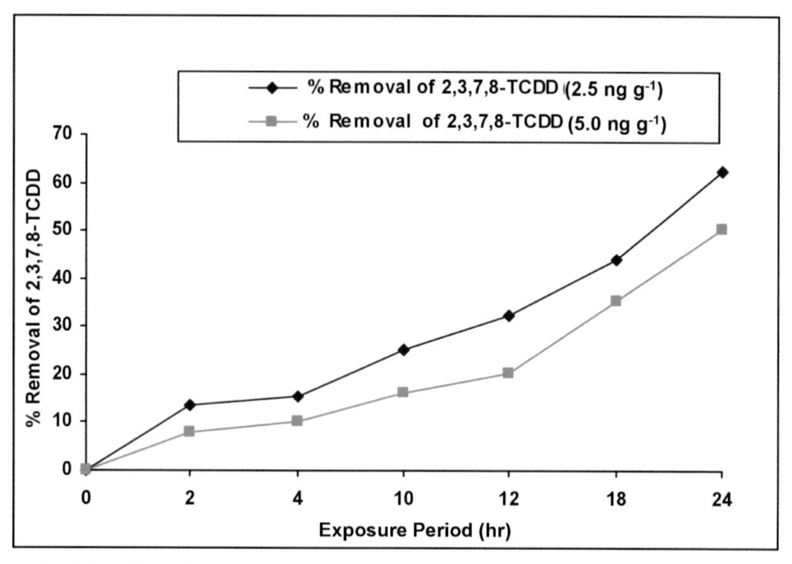

Fig. 6.31: Effect of UV radiations (400 W) and oil on removal of 2,3,7,8-TCDD in soil.

an initial concentration of 2.5 ng g^{-1} after an exposure period of 24 hr. However, at an initial concentration of 5.0 ng g^{-1} of 2,3,7,8-TCDF, 44.1% removal was observed after an exposure period of 24 hr (Fig. 6.32).

The removal of OCDD increased from 9.57% to 56.03% with increase in exposure period from 2 hr to 24 hr. Maximum removal of 56.03% was observed at an initial

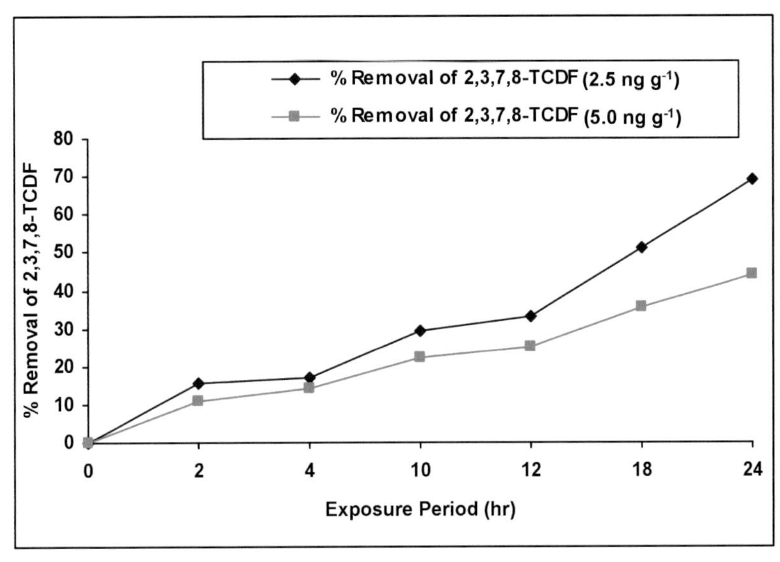

Fig. 6.32: Effect of UV radiations (400 W) and oil on removal of 2,3,7,8-TCDF in soil.

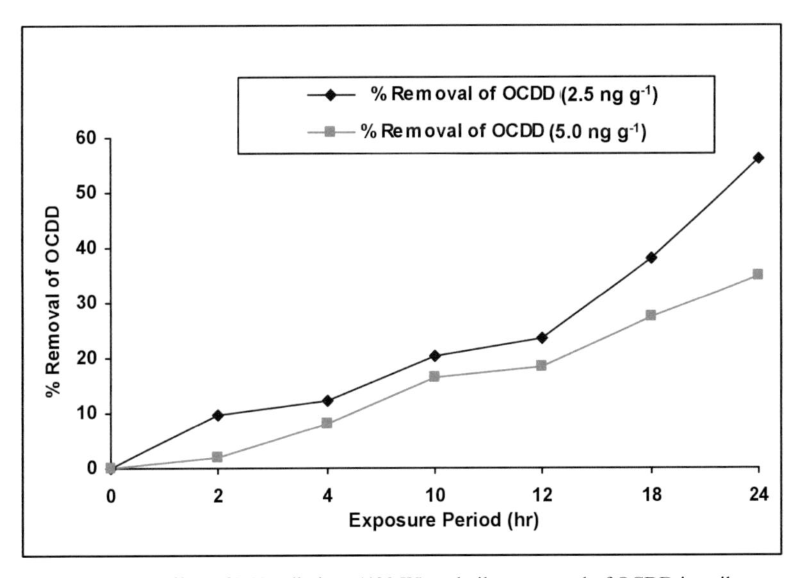

Fig. 6.33: Effect of UV radiations (400 W) and oil on removal of OCDD in soil.

concentration of 2.5 ng g^{-1} after an exposure period of 24 hr. However, at an initial concentration of 5.0 ng g^{-1} of OCDD, 35.0% removal was observed after an exposure period of 24 hr (Fig. 6.33).

The removal of OCDF increased from 11.12% to 59.64% with increase in exposure period from 2 hr to 24 hr. Maximum removal of 59.64% was observed at an initial concentration of 2.5 ng g^{-1} after an exposure period of 24 hr (Fig. 6.34).

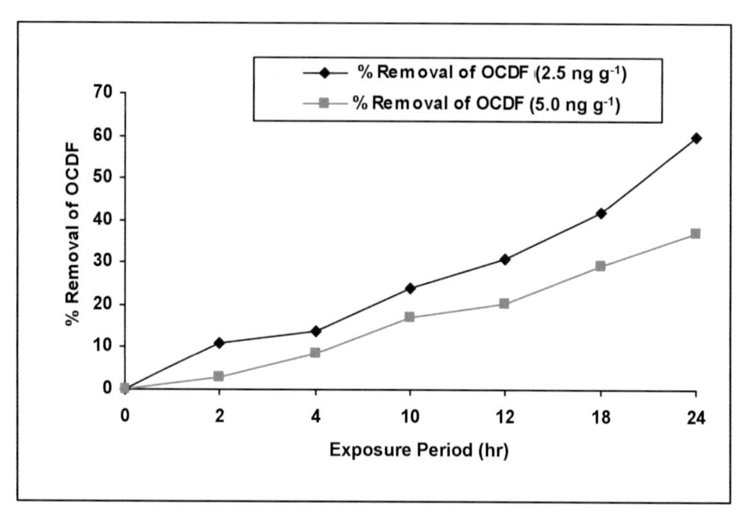

Fig. 6.34: Effect of UV radiations (400 W) and oil on removal of OCDF in soil.

Effect of groundnut oil and varying exposure period using UV lamp (125 W)

The degradation of 2,3,7,8-TCDD, 2,3,7,8-TCDF, OCDD, and OCDF increased after the addition of the oil and increase in exposure period to UV irradiations. The removal of 2,3,7,8-TCDD increased from 0 to 15.33% with increase in exposure period from 2 hr to 24 hr. Maximum removal of 15.33% was observed at an initial concentration of 2.5 ng g^{-1} after an exposure period of 24 hr (Fig. 6.35).

The removal of 2,3,7,8-TCDF increased from 0 to 17.25% with increase in exposure period from 2 hr to 24 hr. Maximum removal of 17.25% was observed at an initial concentration of 2.5 ng g^{-1} after an exposure period of 24 hr. However, at an initial concentration of 5.0 ng g^{-1} of 2,3,7,8-TCDF, 20.0% removal was observed after an exposure period of 24 hr (Fig. 6.36).

The removal of OCDD increased from 0 to 11.22% with increase in exposure period from 2 hr to 24 hr. Maximum removal of 11.22% was observed at an initial concentration of 2.5 ng g^{-1} after an exposure period of 24 hr (Fig. 6.37).

The removal of OCDF increased from 0 to 13.25% with increase in exposure period from 2 hr to 24 hr. Maximum removal of 13.25% was observed at an initial concentration of 2.5 ng g^{-1} after an exposure period of 24 hr (Fig. 6.38).

Effect of analyte concentration using groundnut oil and a UV lamp (400 W)

The degradation of a 2,3,7,8-TCDD in simulated soil sample was studied at varying concentration of 0.5 μg g^{-1}, 0.4 μg g^{-1}, 0.1 μg g^{-1}, 0.2 μg g^{-1} and 2.5 ng g^{-1} of 2,3,7,8-TCDD. Removal of 60.0% for 2,3,7,8-TCDD was observed at an initial concentration of 2.5 ng g^{-1} after a exposure period of 24 hr. Whereas 51.20%, 46.32%, 41.0%, and 36.10% removal of 2,3,7,8-TCDD was observed at an initial concentration of 0.1 μg g^{-1}, 0.2 μg g^{-1}, 0.4 μg g^{-1}, and 0.5 μg g^{-1} respectively after an exposure period of 24 hr (Fig. 6.39).

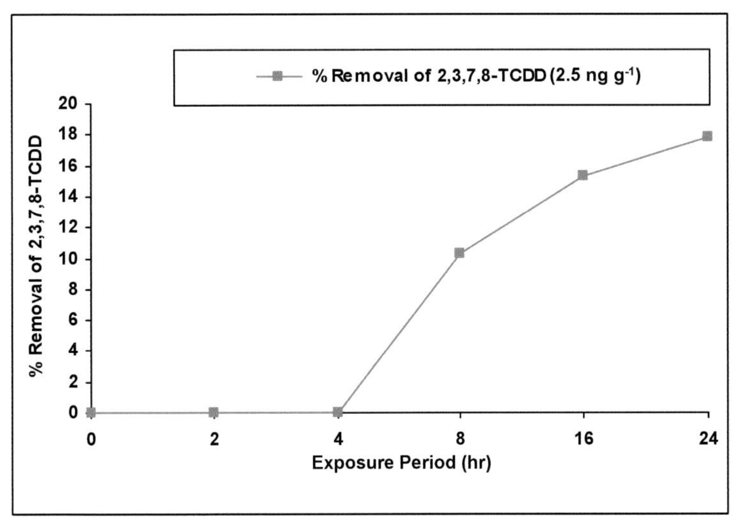

Fig. 6.35: Effect of UV radiations (125 W) and oil on removal of 2,3,7,8-TCDD at 2.5 ng g^{-1} in soil.

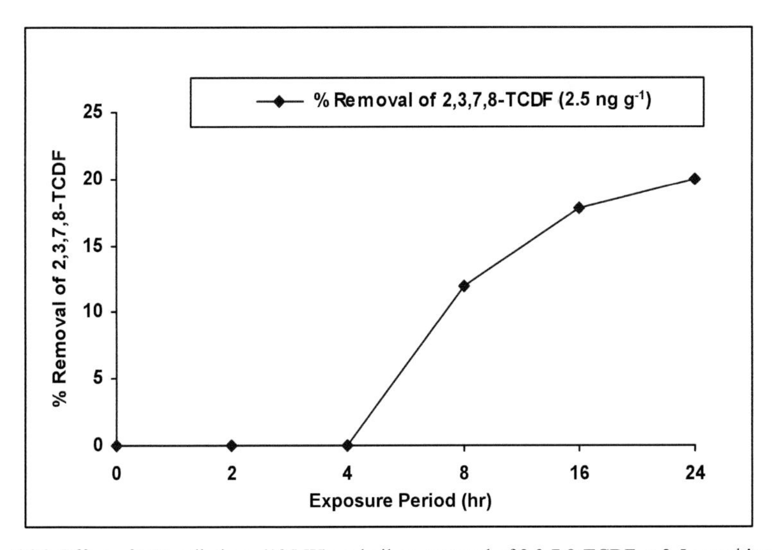

Fig. 6.36: Effect of UV radiations (125 W) and oil on removal of 2,3,7,8-TCDF at 2.5 ng g^{-1} in soil.

Effect of varying exposure period using sunlight

The degradation of 2,3,7,8-TCDD, 2,3,7,8-TCDF, OCDD, and OCDF increased with the increase in exposure period to solar irradiations. The removal of 2,3,7,8-TCDD increased from 5.08% to 22.36% with increase in exposure period from 8 hr to 34 hr. Maximum removal of 22.36% was observed at an initial concentration of 2.5 ng g^{-1} after an exposure period of 34 hr. However, 15.58% removal of 2,3,7,8-TCDD was observed at an initial concentration of 5.0 ng g^{-1} after an exposure period of 34 hr (Fig. 6.40).

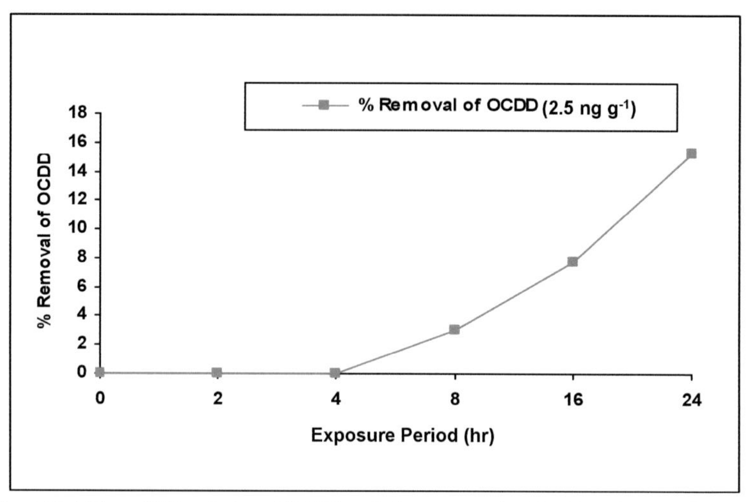

Fig. 6.37: Effect of UV radiations (125 W) and oil on removal of OCDD at 2.5 ng g^{-1} in soil.

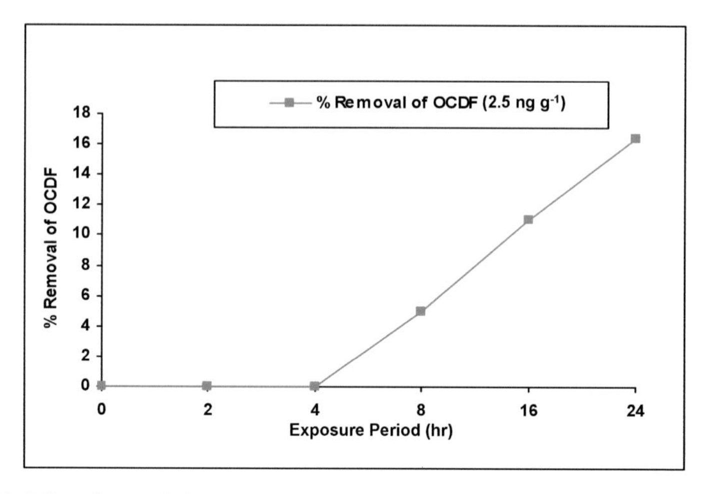

Fig. 6.38: Effect of UV radiations (125 W) and oil on removal of 2,3,7,8-TCDD at 2.5 ng g^{-1} in soil.

The removal of 2,3,7,8-TCDF increased from 6.4% to 32.2% with an increase in exposure period from 8 hr to 34 hr. Maximum removal of 32.2% was observed at an initial concentration of 2.5 ng g^{-1} after an exposure period of 34 hr. However, 17% removal of 2,3,7,8-TCDF was observed at an initial concentration of 5.0 ng g^{-1} after an exposure period of 34 hr (Fig. 6.41).

The removal of OCDD increased from 0 to 17.52% with increase in exposure period from 8 hr to 34 hr. Maximum removal of 17.52% was observed at an initial concentration of 2.5 ng g^{-1} after an exposure period of 34 hr. However, 11.65% removal of OCDD was observed at an initial concentration of 5.0 ng g^{-1} after an exposure period of 34 hr (Fig. 6.42).

The removal of OCDF increased from 0 to 18.55% with increase in exposure period from 8 hr to 34 hr. Maximum removal of 18.55% was observed at an initial

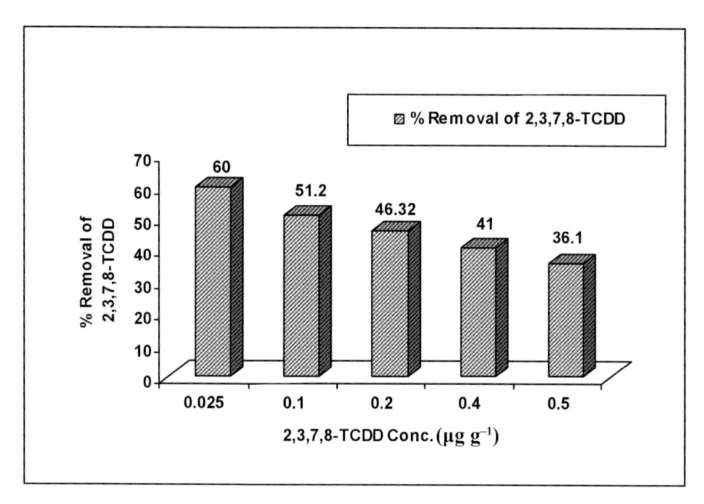

Fig. 6.39: Effect of UV radiations and oil on removal of 2,3,7,8-TCDD in soil at varying concentrations.

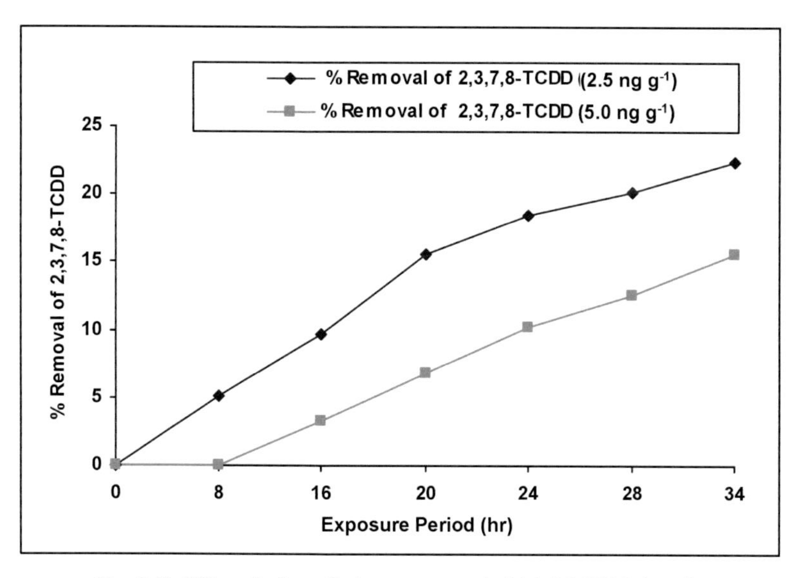

Fig. 6.40: Effect of solar radiations on removal of 2,3,7,8-TCDD in soil.

concentration of 2.5 ng g^{-1} after an exposure period of 34 hr. However, 12.05% removal of OCDF was observed at an initial concentration of 5.0 ng g^{-1} after an exposure period of 34 hr (Fig. 6.43).

Effect of varying analyte concentration using sunlight

The degradation of 2,3,7,8-TCDD in a simulated soil sample was studied at varying 2,3,7,8-TCDD concentration of 0.1 μg g^{-1}, 0.2 μg g^{-1}, 0.4 μg g^{-1}, 0.5 μg g^{-1}, and 2.5 ng g^{-1}. The removal of 22.36% for 2,3,7,8-TCDD was observed at an initial concentration of 2.5 ng g^{-1} after an exposure period of 34 hr. Whereas removal of

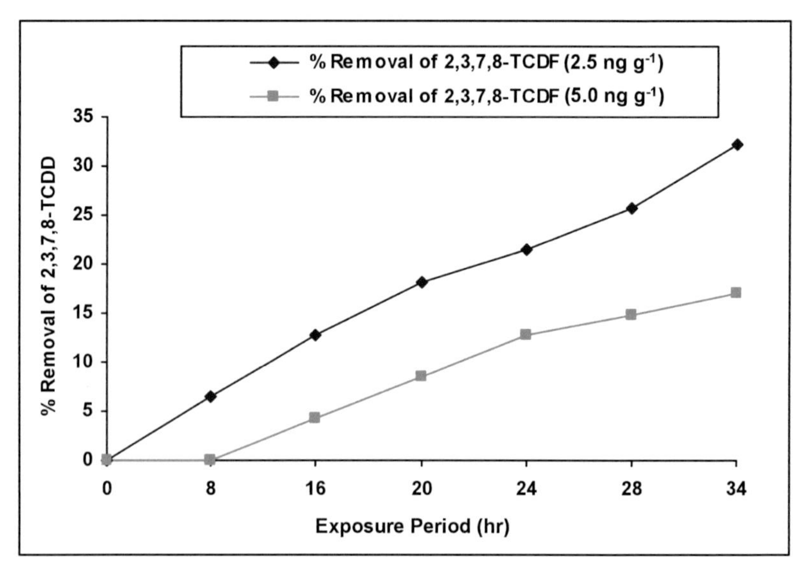

Fig. 6.41: Effect of solar radiations on removal of 2,3,7,8-TCDF in soil.

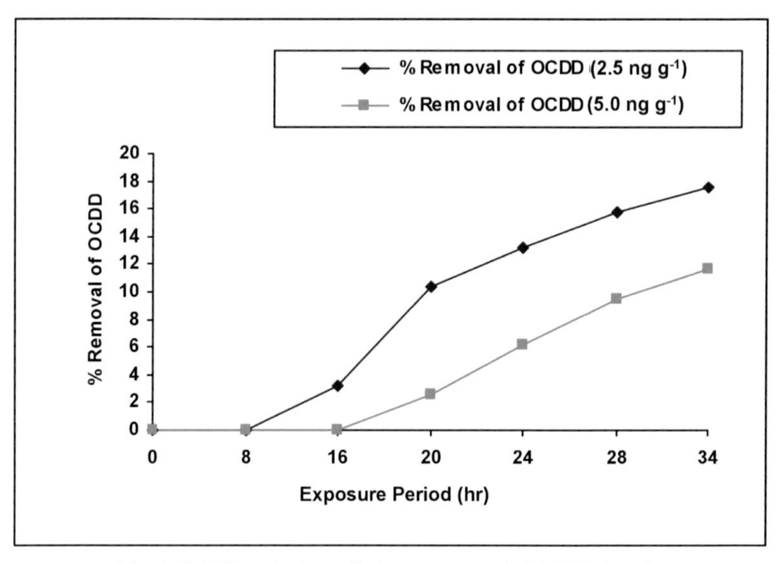

Fig. 6.42: Effect of solar radiations on removal of OCDD in soil.

16.80% for 2,3,7,8-TCDD was observed at an initial concentration of 0.1 μg g^{-1} after an exposure period of 34 hr. Removal of 8.0% of 2,3,7,8-TCDD was observed at an initial concentration of 0.5 μg g^{-1} after an exposure period of 34 hr (Fig. 6.44).

Effect of groundnut oil using sunlight

The degradation of 2,3,7,8-TCDD, 2,3,7,8-TCDF, OCDD, and OCDF increased with the increase in exposure period to solar radiations. The removal of 2,3,7,8-TCDD increased from 6.53% to 38.3% with increase in exposure period from 8 hr to 34 hr.

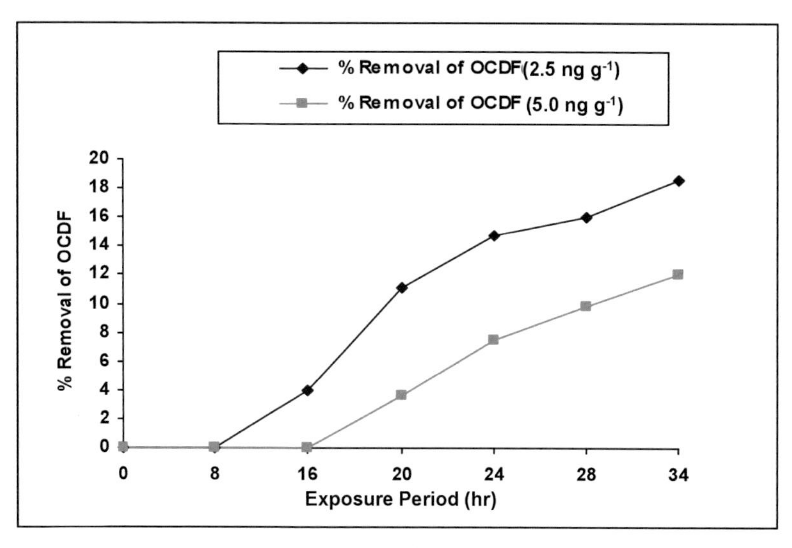

Fig. 6.43: Effect of solar radiations on removal of OCDF in soil.

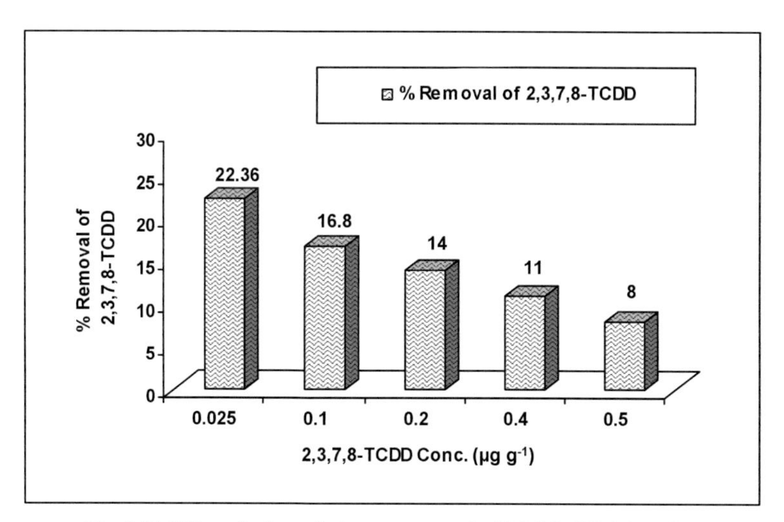

Fig. 6.44: Effect of solar radiations on removal of 2,3,7,8-TCDD in soil at varying initial concentration.

Maximum removal of 38.3% was observed at an initial concentration of 2.5 ng g^{-1} after an exposure period of 34 hr. However, 22.75% removal of 2,3,7,8-TCDD was observed at an initial concentration of 5.0 ng g^{-1} after an exposure period of 34 hr (Fig. 6.45).

The removal of 2,3,7,8-TCDF increased from 11.44% to 49.72% with increase in exposure period from 8 hr to 34 hr. Maximum removal of 49.72% was observed at an initial concentration of 2.5 ng g^{-1} after an exposure period of 34 hr. However, 26.0% removal of 2,3,7,8-TCDF was observed at an initial concentration of 5.0 ng g^{-1} after an exposure period of 34 hr (Fig. 6.46).

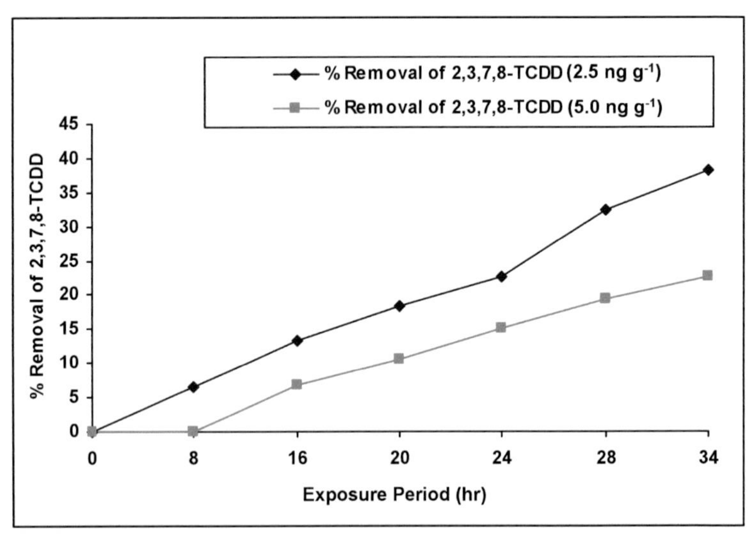

Fig. 6.45: Effect of solar radiations and oil on removal of 2,3,7,8-TCDD in soil.

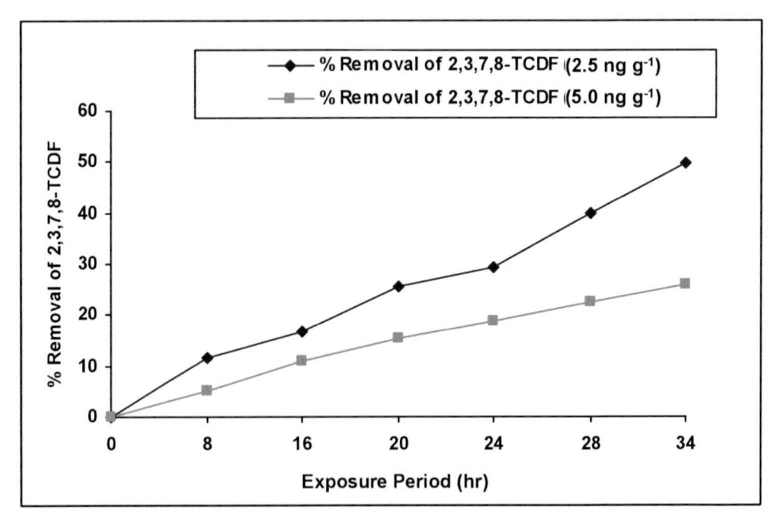

Fig. 6.46: Effect of solar radiations and oil on removal of 2,3,7,8-TCDF in soil.

The removal of OCDD increased from 3.26% to 27.0% with increase in exposure period from 8 hr to 34 hr. Maximum 27.0% removal of was observed at an initial concentration of 2.5 ng g^{-1} after an exposure period of 34 hr. However, 17.65% removal of OCDD was observed at an initial concentration of 5.0 ng g^{-1} after an exposure period of 34 hr (Fig. 6.47).

The removal of OCDF increased from 4.11% to 30.25% with an increase in exposure period from 8 hr to 34 hr. Maximum removal of 30.25% was observed at an initial concentration of 2.5 ng g^{-1} after an exposure period of 34 hr. However,

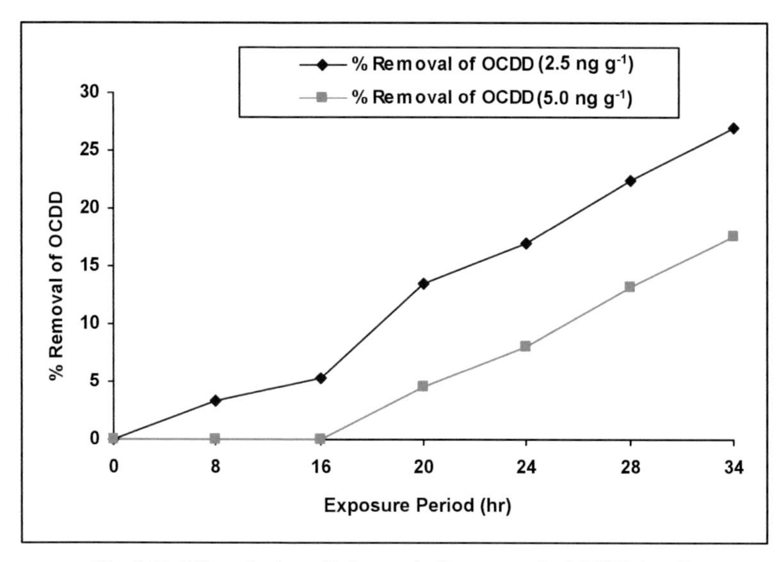

Fig. 6.47: Effect of solar radiations and oil on removal of OCDD in soil.

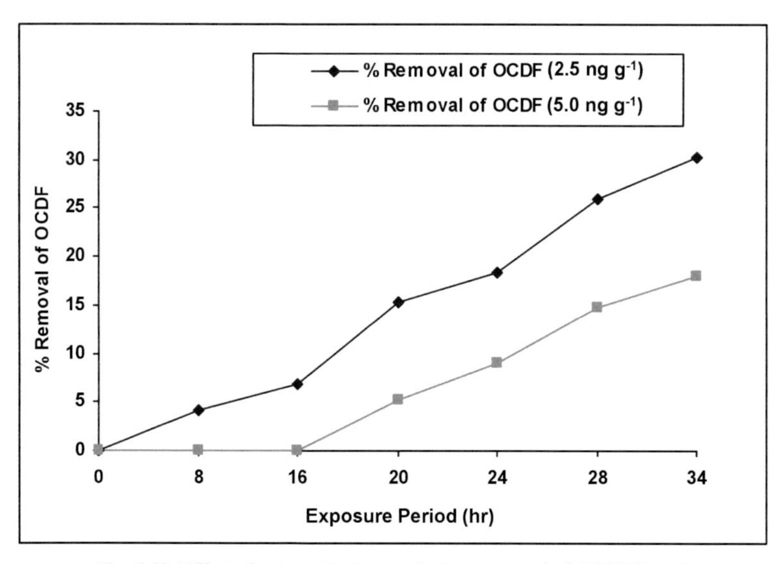

Fig. 6.48: Effect of solar radiations and oil on removal of OCDF in soil.

18.0% removal of OCDF was observed at an initial concentration of 5.0 ng g^{-1} after an exposure period of 34 hr (Fig. 6.48).

Effect of varying analyte concentration using groundnut oil and sunlight

The degradation of 2,3,7,8-TCDD in simulated soil sample was studied at varying 2,3,7,8-TCDD concentration of 0.5 μg g^{-1}, 0.4 μg g^{-1}, 0.2 μg g^{-1}, 0.1 μg g^{-1}, and 2.5 ng g^{-1}. Removal of 38.36% for 2,3,7,8-TCDD was observed at an initial

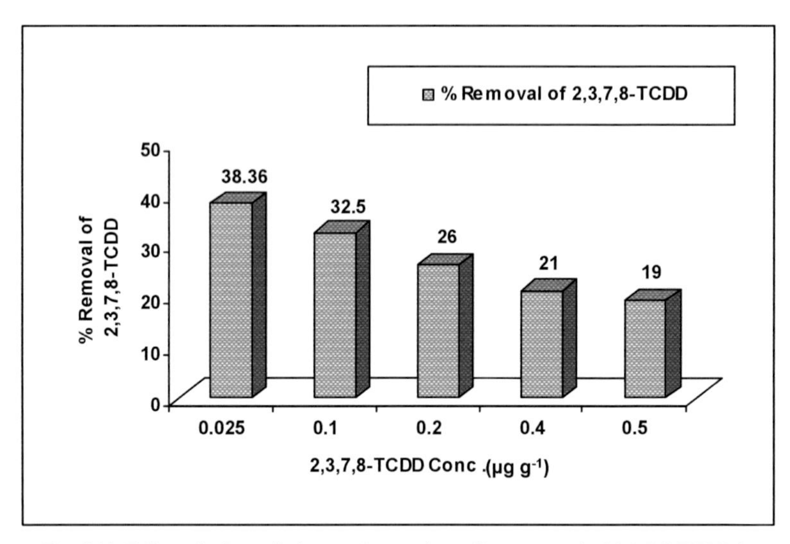

Fig. 6.49: Effect of solar radiations and groundnut oil on removal of 2,3,7,8-TCDD in soil at varying concentrations.

concentration of 2.5 ng g^{-1} after an exposure period of 34 hr. Whereas, 32.50%, 26.0%, 21.0%, and 19.0% removal of 2,3,7,8-TCDD was observed at an initial concentration of 0.1 μg g^{-1}, 0.2 μg g^{-1}, 0.4 μg g^{-1} and 0.5 μg g^{-1} respectively after an exposure period of 34 hr (Fig. 6.49).

Data Evaluation

The overall objective of this research was to evaluate the effectiveness of oil-mediated mobilization and photodegradation of chlorinated PCDDs and PCDFs as a means of soil decontamination. In order to meet this objective, a series of laboratory experiments were carried out.

A rapid photodegradation of PCDDs and PCDFs was demonstrated in vegetable oils. When contaminated soil was treated, the strong binding of PCDD/Fs to soil articles was the main barrier to photodegradation. This obstacle could be overcome by ensuring a proper solubilisation of PCDD/Fs, by using a longer exposure period and periodical mixing of the soil-vegetable oil mixture. Experiments were conducted to study the photodegradation rate of PCDDs and PCDFs in the presence of groundnut oil and to evaluate the efficacy of an optimized mobilization-photodegradation process on a laboratory scale in contaminated samples. Irradiation experiments revealed that different Dioxin congeners degrade at different rates. The data revealed an inverse relationship between degree of chlorination and rate of disappearance. The data also indicated that more toxic laterally substituted congeners degrade at a slower rate. The photodegradability of 2,3,7,8-TCDD, 2,3,7,8-TCDF, OCDD, and OCDF in vegetable oil was investigated. The degradation rates of the lower chlorinated congeners seemed to degrade more slowly than the highly chlorinated PCDDs and PCDFs.

To compare the degradability of the toxic PCDDs with toxic PCDFs, results from these experiments and congeners (mostly OCDD and OCDF) were compiled. The degradation percentage of OCDD was typically more or less the same as the degradation of OCDF. Using long irradiation times, high degradation percents can be attained for PCDDs as well as PCDFs. The results showed that PCDDs and PCDFs were effectively photodegraded at wavelengths that are available from sunlight. Irradiation with UV lamps and sunlight both resulted in significant degradation of PCDD/Fs in contaminated soils. A control sample to which only groundnut oil was added and not exposed to light proved that vegetable oil alone did not degrade PCDDs and PCDFs but a light source was also needed. The concentrations of all congeners in a PCDDs and PCDFs mixture remained unchanged when stored for four weeks in the dark at room temperature. Furthermore, concentrations analyzed in non-irradiated reference samples were equivalent to the known amounts of PCDD/Fs that had been added to the samples. In the absence of vegetable oil, sunlight irradiation degraded PCDDs and PCDFs but at a slow rate. Vegetable oils penetrated into the soil matrix and desorbed PCDDs and PCDFs molecules. The desorbing effect was intensified by UV light, because it decomposes soil organic matter (Patel-Sorrentino et al. 2004), increases the concentration gradient between soil and oil by degrading PCDDs and PCDFs, and accelerates the convective transport of PCDDs and PCDFs molecules driven by evaporation (Dougherty et al. 1993; McPeters and Overcash 1993; Overcash et al. 1991). A long-term treatment under sunlight although very slow offered certain advantages over the other treatment options, including efficient desorption of PCDDs and PCDFs from soil matrix and a free UV source.

References

Abad, E., J. Caixach and J. Rivera. 2003. Improvements in dioxin abatement strategies at a municipal waste management plant in Barcelona. Chemosphere. 50: 1175–82.

Abbasy, M.S., H.Z. Ibrahim and H.M. Abdel-Kader. 2003. Persistent organochlorine pollutants in the aquatic ecosystem of Lake Manzala, Egypt. Bull. Environ. Contam. Toxicol. 7: 1158–1164.

Abd-Allah, A.M. 1999. Organochlorine contaminants in microlayer and subsurface water of Alexandria Coast, Egypt. J. AOAC. Int. 82: 391–398.

Adams, E.M., D.D. Irish, H.C. Spencer et al. 1941. The response of rabbit skin to compounds reported to have caused acneform dermatitis. Ind. Med. 2: 1–4.

Addison, R.F. and W.T. Stobo. 2001. Trends in organochlorine residue concentrations and burdens in grey seals (Halichoerus grypus) from Sable Is., NS, Canada, between 1974–1994. Environ. Pollut. 112(3): 505–513.

Adriaens, P., Q. Fu and D. Grbic-Galic. 1995. Bioavailability and transformation of highly chlorinated dibenzo-p-dioxins and dibenzofurans in anaerobic soils and sediments. Environ. Sci. Technol. 29: 2252–2260.

Aigner, E., A. Leone and R. Falconer. 1998. Concentrations and enantiomeric ratios of organochlorine pesticides in soils from the U.S. corbel. Environ. Sci. Technol. 32: 1162–1168.

Alaluusua, S., P.L. Lukinmaa, J. Torppa, J. Tuomisto and T. Vartiainen. 1999. Developing teeth as biomarker of dioxin exposure. Lancet. 353: 206.

Albrecht, I.D., A.L. Barkovskii and P. Adriaens. 1999. Production and dechlorination of 2,3,7,8-tetrachlorodibenzop-dioxin in historically contaminated estuarine sediments. Environ. Sci. Technol. 33: 737–744.

Albro, P.W. and B.J. Corbett. 1977. Extraction and clean-up of animal tissues for subsequent determination of mixtures of chlorinated dibenzo-p-dioxins and dibenzofurans. Chemosphere. 7: 381–385.

Alcock, R.E. and K.C. Jones. 1996. Dioxins in the environment: A review of trend data. Environ. Sci. Technol. 30: 3133–3143.

Alcock, R.E., A.J. Sweetman, C.-Y. Juan et al. 1999. The intake and clearance of PCBs in humans—a generic model of lifetime exposure. Organohalogen Compounds. 44: 61–65.

Allan, R.J. and A.J. Ball. 1990. An overview of toxic contaminants in water and sediments of the Great Lakes: Part I. Water Pollution Research Journal of Canada. 25(4): 387–505.

Allen-Gil, S.M., C.P. Gubala, R. Wilson et al. 1997. Organochlorine pesticides and polychlorinated biphenyls (PCBs) in sediments and biota from four US Arctic lakes. Arch. Environ. Contam. Toxicol. 33(4): 378–387.

Al-Majed, N., H. Mohammadi and A.N. Al-Ghadban. 2000. Regional report of the state of the marine environment, ROPME. GC-10/001/1. Revised by Al-Awadi, A., Regional Organisation for the Protection of the Marine Environment (ROPME), Kuwait.

Almeida-Gonzalez, M., O.P. Luzardo, M. Zumbado et al. 2012. Levels of organochlorine contaminants in organic and conventional cheeses and their impact on the health of consumers: An independent study in the Canary Islands (Spain). Food Chem. Toxicol. 50(12): 4325–4332.

Al-Omar, M.A., F.H. Abdul-Jalil, N.H. Al-ogaily et al. 1986. A follow-up study of maternal milk contamination with organochlorine insecticide residues. Environmental Pollution. 42: 79–91.

Anderson, D.J., T.B. Bloem, R.K. Blankenbaker et al. 1999. Concentrations of polychlorinated biphenyls in the water column of the Laurentian Great Lakes: Spring 1993. J. Great Lakes Res. 25(1): 160–170.

Anderson, H.A., C. Falk, L. Hanrahan et al. 1998. Profiles of Great Lakes critical pollutants: A sentinel analysis of human blood and urine. Environmental Health Perspectives. 106(5): 279–289.

Anderson, R.L. and T.A. Bancroft. 1952. Statistical Theory in Research. Mc. Graw Hill Book Co., New York.

Andersson, P., C. Rappe, O. Maaskant et al. 1998. Low temperature catalytic destruction of PCDD/F in flue gas from waste incineration. Organohal. Compd. 36: 109–112.

Andersson, P.L., A. Blom, A. Johannisson et al. 1999. Assessment of PCBs and hydroxylated PCBs as potential xenoestrogens: *In vitro* studies based on MCF-7 cell proliferation and induction of vitellogenin in primary culture of rainbow trout hepatocytes. Arch. Environ. Contam. Toxicol. 37: 145–150.

Anezaki, K. and S. Nagahora. 2014. Characterization of polychlorinated biphenyls, pentachlorobenzene, hexachlorobenzene, polychlorinated dibenzo-p-dioxins, and dibenzofurans in surface sediments of Muroran Port, Japan. Environ. Sci. Pollut. Res. Int. 21(15): 9169–9181.

Arthur, M.R. and J.L. Frea. 1989. 2,3,7,8-Tetrachlorodibenzo-p-dioxin: Aspects of its important properties and its potential biodegradation in soils. J. Environ. Qual. 18: 1–11.

Arthur, R., J. Cain and B. Barrentine. 1977. DDT residues in air in the Mississippi delta, 1975. Pestic. Monit. J. 10: 168.

Arthur, R.D., J.D. Cain and B.F. Barrentine. 1976. Atmospheric levels of pesticides in the Mississippi Delta. Bull. Environ. Contam. Toxicol. 15: 129–134.

Aschengrau, A. and R.R. Monson. 1990. Paternal military service in Vietnam and the risk of late adverse pregnancy outcomes. Am. J. Publ. Health. 80: 1218–1223.

Atkinson, R. 1991. Atmospheric lifetimes of dibenzo-p-dioxins and dibenzofurans. Sci. Total Environ. 104(1-2): 17–33.

Atlas, E. and C.S. Giam. 1988. Ambient concentration and precipitation scavenging of atmospheric organic pollutants. Wat. Air Soil Pollut. 38: 19–36.

Atlas, E. and C.S. Giam. 1989. Sea-air exchange of high molecular weight synthetic organic compounds: results from the SEAREX Program. pp. 340–378. *In*: Riley, J.P., R. Chester, R.A. Duce (eds.). Chemical Oceanography, Vol. 10. Academic Press, London.

ATSDR. 1995. Exposure to PCBs from hazardous waste among Mohawk women and infants at Akwesasne. Atlanta, GA.: U.S. Department of Health and Human Services, Public Health Service, Agency for Toxic Substances and Disease Registry.

ATSDR, Agency for Toxic Substances and Disease Registry, Atlanta. 1998. Toxicological profile for chlorinated dibenzo-p-dioxins. Public Health Service, U.S. Department of Health and Human Services, Atlanta, GA.

Ayotte, P., E. Dewailly, J.J. Ryan et al. 1997. PCBs and dioxin-like compounds in plasma of adult inuit living in Nunavik (Arctic Quebec). Chemosphere. 34(5-7): 1459–1468.

Aziz, S.Q., H.A. Aziz, M.S. Yusoff et al. 2010. Leachate characterization in semi-aerobic and anaerobic sanitary landfills: A comparative study. J. Environ. Manage. 91: 2608–2614.

Baccarelli, A., C. Angela, Pesatori et al. 2004. Elevated dioxin levels in chloracne cases twenty years after the Seveso, Italy accident. Organohalogen Compounds. 66: 2694–2699.

Badia-Vila, M., M. Ociepa, R. Mateo et al. 2000. Comparison of residue levels of persistent organochlorine compounds in butter from Spain and from other European countries. J. Environ. Sci. Health B. 35(2): 201–210.

Baker, J.E., S.J. Eisenreich, T.C. Johnson et al. 1985. Chlorinated hydrocarbon cycling in the benthic nepheloid layer of Lake Superior. Environ. Sci. Technol. 19: 854–861.

Balfanz, E., J. Fuchs and H. Kieper. 1993. Sampling and analysis of polychlorinated biphenyls (PCB) in indoor air due to permanently elastic sealants. Chemosphere. 26(5): 871–880.

Ballerstedt, H., A. Kraus and U. Lechner. 1997. Reductive dechlorination of 1,2,3,4-tetrachlorodibenzo-p-dioxin and its products by anaerobic mixed cultures from Saale river sediment. Environ. Sci. Technol. 31: 1749–1753.

Balmer, M.E., K.U. Goss and R.P. Schwarzenbach. 2000. Photolytic transformation of organic pollutants on soil surfaces—an experimental approach. Environ. Sci. Technol. 34: 1240–1245.

Barber, J.L., A.J. Sweetman, D. van Wijk et al. 2005. Hexachlorobenzene in the global environment: Emissions, levels, distribution, trends and processes. Sci. Total Environ. 349(1-3): 1–44.

Bard, S.M. 1999. Global transport of anthropogenic contaminants and the consequences for the arctic marine ecosystem. Mar. Pollut. Bull. 38(5): 356–379.

Barkovskii, A.L. and A.L. Adriaens. 1996. Microbial dechlorination of historically present and freshly spiked chlorinated dioxins and diversity of dioxin-dechlorinating populations. Appl. Environ. Microbiol. 62: 4556–4562.

Barthel, W.F., J.C. Hawthorne, J.H. Ford et al. 1969. Pesticides in water: Pesticide residues in sediments of the lower Mississippi River and its tributaries. Pestic. Monit. J. 3: 8–34.

Basharova, G.R. 1996. Reproduction in families of workers exposed to 2,4,5-T intoxication. Organohalogen Compounds. 315–318.

Baughman, R.W. and M. Meselson. 1973. An analytical method for detecting TCDD (dioxin): Levels of TCDD in samples from Vietnam. Environ. Health Perspect. 5: 27–35.

Becher, G., G.S. Eriksen, K. Lund-Larsen et al. 1998. Dietary exposure and human body burden of dioxins and dioxin-like PCBs in Norway. Organohalogen Compounds. 38: 79–82.

Becher, H., D. Flesch-Janys, T. Kauppinen et al. 1996. Cancer mortality in German male workers exposed to phenoxy herbicides and dioxins. Cancer Causes Control. 7: 312–321.

Beck, H., K. Eckart, M. Kellert et al. 1987. Levels of PCDD's and PCDF's in samples of human origin and food in the Federal Republic of Germany. Chemosphere. 16: 1977–1982.

Beck, H., K. Eckart, W. Mathar et al. 1989a. PCDD and PCDF body burden from food intake in the Federal Republic of Germany. Chemosphere. 18: 417–424.

Beck, H., K. Eckart, W. Mathar et al. 1989c. Levels of PCDDs and PCDFs in adipose tissue of occupationally exposed workers. Chemosphere. 18: 507–516.

Beck, H., A. Dross, W.J. Kleemann et al. 1990. PCDD and PCDF concentrations in different organs from infants. Chemosphere. 20: 903–910.

Beck, H., A. Dro and W. Mathar. 1992a. PCDDs, PCDFs and related contaminants in the German food supply. Chemosphere. 25: 1539–1550.

Beck, H., A. Drob and W. Mathar. 1992b. Dependence of PCDD and PCDF levels in human milk on various parameters in Germany II. Chemosphere. 25: 1015–1220.

Berg, V., K.I. Ugland, N.R. Hareide et al. 1997. Organochlorine contamination in deep-sea fish from the Davis Strait. Mar. Environ. Res. 44: 135–148.

Berg, V., A. Polder and J.U. Skaare. 1998. Organochlorines in deep-sea fish from the Nordfjord. Chemosphere. 38: 275–282.

Bergersen, E.P. 1987. Aldrin, dieldrin, and mercury profiles in recent lake sediments at the Rocky Mountain Arsenal, Colorado. Arch. Environ. Contam. Toxicol. 16: 61–67.

Bergman, A., A. Hagman, S. Jacobsson et al. 1984. Thermal degradation of polychlorinated alkanes. Chemosphere. 13(2): 237–250.

Bernes, C. 1998. Organiska miljögifter—Ett svenskt perspektiv på ett internationellt problem. Naturvårdsverket förlag, Värnamo.

Bertazzi, P.A., C. Zocchetti, A.C. Pesatori et al. 1989b. Ten-year mortality study of the population involved in the Seveso incident in 1976. Am. J. Epidemiol. 129: 1187–1200.

Bertazzi, P.A., A.C. Pesatori, D. Consonni et al. 1993. Cancer incidence in a population accidentally exposed to 2,3,7,8-tetrachlorodibenzo-para-dioxin. Epidemiol. 4: 398–406.

Bevenue, A., J. Hylin, Y. Kawano et al. 1972. Organochlorine pesticide residues in water, sediment, algae, and fish: Hawaii, 1970–71. Pestic. Monit. J. 6(1): 56–64.

Beyer, W.N. and C. Stafford. 1993. Survey and evaluation of contaminants in earthworms and insoils derived from dredged materials at confined disposal facilities in the Great Lakes region. Environ. Monit. Assess. 24: 151–165.

Bhatnagar, V.K., J.S. Patel, M.R. Variya et al. 1992. Levels of organochlorine insecticides in human blood from Ahmedabad (rural), India. Bull. Environ. Contam. Toxicol. 48: 302–307.

Biberhofer, J. and R.J.J. Stevens. 1987. Organochlorine contaminants in ambient waters of Lake Ontario. Scientific Series—Canada, Inland Waters/Lands Directorate. 159: 1–11.

Bidleman, T., M.D. Walla, R. Roura et al. 1993. Organochlorine pesticides in the atmosphere of the southern ocean and Antarctica, January–March, 1990. Mar. Pollut. Bull. 26(5): 258–262.
Bidleman, T.E. and A. Leone. 2004. Soil-air relationships for toxaphene in the southern United States. Environ. Toxicol. Chem. 23(10): 2337–2342.
Bidleman, T.F. and C.E. Olney. 1975. Long range transport of toxaphene insecticide in the atmosphere of the western North Atlantic. Nature. 257: 475–477.
Bidleman, T.F. 1981. Interlaboratory analysis of high molecular organochlorines in ambient air. Atmos. Environ. 15: 619–24.
Bidleman, T.F., M.D. Walla, R. Roura et al. 1993. Organochlorine pesticides in the atmosphere of the Southern Ocean and Antarctica, January–March 1990. Marine Pollut. Bull. 26: 258–262.
Birmingham, A.G., C.J. Edmunds, B.W.L. Graham et al. 1989. Determination of PCDDs and PCDFs in car exhaust. Chemosphere. 19: 669–673.
Birmingham, B. 1990. Analysis of PCDD and PCDF patterns in soil samples: Use in the estimation of the risk of exposure. Chemosphere. 20(7-9): 807–814.
Birnbaum, L.S. and L.A. Couture. 1988. Disposition of octachlorodibenzo-p-dioxin (OCDD) in male rats. Toxicol. Appl. Pharmacol. 93: 22–30.
Bisanti, L., F. Bonetti, F. Caramaschi et al. 1980. Experiences from the accident of Seveso. Acta Morphologica Acad. Sci. Hung. 28: 139–157.
Bishop, F.S. 1984. Written communication (August 29) to Velsicol Chemical Company, regarding notice of intent to cancel registration: Velsicol technical endrin. EPA registration No. 876-20. Washington, DC: US Environmental Protection Agency, Office of pesticide Program, Registration Division.
Bishop, F.S. 1985. Written communication (August 29) to Velsicol Chemical Company, regarding final cancellation notice: Velsicol technical endrin. EPA registration No. 876-20. Washington, DC: US Environmental Protection Agency, Office of pesticide Program, Registration Division.
Bishop, F.S. 1986. Written communication (August 29) to Velsicol chemical Company, Chicago, IL regarding notice of intent to cancel registration of certain pesticide products: Velsicol technical endrin 1.6. EPA registration No. 876-20. Washington, DC: US Environmental Protection Agency, Office of pesticide Program, Registration Division.
Blais, J.M., D.W. Schindler, D.C.G. Muir et al. 1998. Accumulation of persistent organochlorine compounds in mountains of western Canada. Nature. 395: 585–588.
Blais, J.M., F. Wilhelm, K.A. Kidd et al. 2003. Concentrations of organochlorine pesticides and polychlorinated biphenyls in amphipods (*Gammarus lacustris*) along an elevation gradient in mountain lakes of Western Canada. Environ. Toxicol. Chem. 22: 2605–2613.
Bocio, A. and J.L. Domingo. 2005. Daily intake of polychlorinated dibenzo-p-dioxins/polychlorinated dibenzofurans (PCDD/PCDFs) in foodstuffs consumed in Tarragona, Spain: A review of recent studies (2001–2003) on human PCDD/PCDF exposure through the diet. Environ. Res. 97: 1–9.
Boden, A.R., G.E. Ladwig and E.J. Reiner. 2002. Analysis of polycyclic aromatic compounds using microbore columns. Polycyclic Aromatic Compounds. 22(3 & 4): 301–310.
Boer, F.P., M.A. Neuman, F.P. Van Remoortere et al. 1973. X-Ray diffraction studies of chlorinated dibenzo-p-dioxins. Chlorodioxins-origin and fate. Adv. Chem. Ser. 120: 14–25.
Bond, G.G., M.G. Ott, F.E. Brenner et al. 1983. Medical and morbidity surveillance findings among employees potentially exposed to TCDD. Br. J. Ind. Med. 40: 318–324.
Bond, G.G., E.A. McLaren, F.E. Brenner et al. 1989a. Incidence of chloracne among chemical workers potentially exposed to chlorinated dioxins. J. Occup. Med. 31: 771–774.
Bonn, Bernadine. 1998. Dioxins and furans in bed sediment and fish tissue of the Willamette Basin, Oregon, 1992–95. Portland. Or.: U.S. Dept. of the Interior, U.S. Geological Survey; Denver, CO: Branch of Information Services [distributor].
Booij, K., R. van Bommel, K.C. Jones et al. 2007. Air-water distribution of hexachlorobenzene and 4,4'DDE along a north-south Atlantic transect. Mar. Pollut. Bull. 54(6): 814–819.
Boon, J.P. and J.C. Duinker. 1986. Monitoring of cyclic organochlorines in the marine environment. Environ. Monit. Assess. 7: 189–208.
Bopp, R.F., M.L. Gross, H. Tong et al. 1991. A major incident of dioxin contamination: Sediments of New Jersey estuaries. Environ. Sci. Technol. 25: 951–956.

Bopp, R.F., S.N. Chillrud, E.L. Shuster et al. 1998. Trends in chlorinated hydrocarbon levels in Hudson River Basin sediments. Environ. Health Perspect. Suppl. 106(4): 1075–1081.

Bordet, F., J. Mallet, L. Maurice et al. 1993. Organochlorine pesticides and PCB congener content of French Human milk. Bull. Environ. Contam. Toxicol. 50: 425–432.

Botello, A.V., G. Diaz, L. Rueda et al. 1994. Organochlorine compounds in oysters and sediments from coastal lagoons of the Gulf of Mexico. Bull. Environ. Contam. Toxicol. 53: 238–245.

Boul, H. 1996. Effect of soil moisture on the fate of radiolabelled DDT and DDE *in vitro*. Chemosphere. 32(5): 855–866.

Breivik, E.M. and J.E. Bjerk. 1978. Organochlorine compounds in Norwegian human fat and milk. Acta Pharmacol. Toxicol. 43: 59–63.

Breivik, K., R. Alcock, Y.-F. Li et al. 2004. Primary sources of selected POPs: regional and global scale emission inventories. Environ. Pollut. 128: 3–16.

Bright, D.A., W.T. Dushenko, S.L. Grundy et al. 1995a. Effects of local and distant contaminant sources: Polychlorinated biphenyls and other organochlorines in bottom-dwelling animals from an Arctic estuary. Sci. Total Environ. 160-161: 265–283.

Bright, D.A., W.T. Dushenko, S.L. Grundy et al. 1995b. Evidence for short-range transport of polychlorinated biphenyls in the Canadian Arctic using congener signatures of PCBs in soils. Sci. Total Environ. 160-161: 251–263.

Brock, J.W., L.J. Melnyk, S.P. Caudill et al. 1998. Serum levels of several organochlorine pesticides in farmers correspond with dietary exposure and local use history. Toxicol. Ind. Health. 14(½): 275–289.

Brown, A.P. and P.E. Ganey. 1995. Neutrophil degranulation and superoxide production induced by polychlorinated biphenyls are calcium dependent. Toxicol. Appl. Pharmacol. 131: 198–205.

Brown, K.W. and K.C. Donnelly. 1988. An estimation of the risk associated with the organic constituents of hazardous and municipal waste landfill leachates. Hazardous Waste and Hazardous Materials. 5: 1–30.

Brunciak, P.A., C.L. Lavorgna, E.D. Nelson et al. 1999. Trends and dynamics of persistent organic pollutants in the coastal atmosphere of the mid-Atlantic states. Prepr. Ext. Abst. Div. Environ. Chem. Am. Chem. Soc. 39(1): 64–67.

Bryant, C.J., R.W. Hartle and M.S. Crandall. 1989. Polychlorinated biphenyl, polychlorinateddibenzo-p-dioxin, and polychlorinated dibenzofuran contamination in PCB disposal facilities. Chemosphere. 18: 569–576.

Büchert, A., T. Cederberg, P. Dyke et al. 2001. Dioxin contamination in food. Environ. Sci. & Pollut. Res. 8: 1–5.

Buckland, S.J., S.E. Scobie, M.L. Hannah et al. 1998. Concentrations of PCDDs, PCDFs, and PCBs in New Zealand retail foods and an assessment of dietary exposure. Organohalogen Compds. 38: 71–74.

Buekens, A. and H. Huang. 1998. Comparative evaluation of techniques for controlling the formation and emission of chlorinated dioxins/furans in municipal waste incineration. J. Hazard. Mater. 62: 1–33.

Bumb, R.R., W.B. Crummett, S.S. Cutie et al. 1980. Trace chemistries of fire: A source of chlorinated dioxins. Science. 210: 385–390.

Bunge, M., L. Adrian, A. Kraus et al. 2003. Reductive dehalogenation of chlorinated dioxins by an anaerobic bacterium. Nature. 421: 357–360.

Burniston, D.A., W.M. Strachan and R.J. Wilkinson. 2005. Toxaphene deposition to Lake Ontario via precipitation, 1994–1998. Environ. Sci. Technol. 39(18): 7005–7011.

Burton, M.A. and B.G. Bennett. 1987. Exposure of man to environmental hexachlorobenzene (HCB)-an exposure commitment assessment. Sci. Total Environ. 66: 137–146.

Buser, H.R. and H.P. Bosshardt. 1976. Determination of polychlorinated dibenzo-p-dioxins and dibenzofurans in commercial pentachlorophenols by combined gas chromatography—mass spectrometry. J. Assoc. Off. Anal. Chem. 59: 562–569.

Buser, H.R. and C. Rappe. 1980. High-resolution gas chromatography of the 22 tetrachlorodibenzo-p-dioxin isomers. Anal. Chem. 52: 2257–2262.

Buser, H.R., C. Rappe and P. Bergqvist. 1985. Analysis of polychlorinated dibenzofurans, dioxins, and related compounds in environmental samples. Environ. Health Perspect. 60: 293–302.

Buser, H.R. 1988. Rapid photolytic decomposition of brominated and brominated/chlorinated dibenzodioxins and dibenzofurans. Chemosphere. 17: 889–903.
Buser, H.R. 1991. Determination of polychlorodibenzothiophenes, the sulfur analogues of polychlorodibenzofurans, using various gas chromatographic/mass spectrometric techniques. Anal. Chem. 63: 1210–1217.
Bush, B., J. Snow, S. Connor et al. 1983a. Mirex in human milk in upstate New York. Arch. Environ. Contam. Toxicol. 12(6): 739–746.
California EPA. 1995. Sampling for pesticide residues in California well water: 1995 update of the well inventory data base. California Environmental Protection Agency, Department of Pesticide Regulation, 59.
Calvert, G.M., M.H. Sweeney, J.A Morris et al. 1991. Evaluation of chronic bronchitis, chronic obstructive pulmonary disease, and ventilatory function among workers exposed to 2,3,7,8-tetrachlorodibenzo-p-dioxin. Am. Rev. Resp. Dis. 144: 1302–1306.
Calvert, G.M., R.W. Hornung, M.H. Sweeney et al. 1992. Hepatic and gastrointestinal effects in an occupational cohort exposed to 2,3,7,8-tetrachlorodibenzo-p-dioxin. J. Am. Med. Assoc. 267: 2209–2214.
Campbell, C. and L. Friedman. 1966. Chemical assay and isolation of chick edema factor in biological materials. J. Am. Assoc. Agri. Chem. 49: 824–828.
Carey, A.E., J.A. Gowen, H. Tai et al. 1978. Pesticide residue levels in soils and crops, 1971 National Soils Monitoring Program (III). Pestic. Monit. J. 12: 117–136.
Carey, A.E., P. Douglas, H. Tai et al. 1979a. Pesticide residue concentrations of five United States cities, 1971-Urban soils monitoring program. Pestic. Monit. J. 13: 17–22.
Carey, A.E., J.A. Gowen, H. Tai et al. 1979. Pesticide residue levels in soils and crops from 37 states, 1972 National Soils Monitoring Program (IV). Pestic. Monit. J. 12: 209–229.
Carey, A.E., H.S.C. Yang, G.B. Wiersma et al. 1980. Residual concentration of propanil, TCAB and pther pesticides in rice-growing soils in the Unites states, 1972. Pestic. Monit. J. 14: 23–25.
Carey, A.E. and F.W. Kutz. 1985. Trends in ambient concentrations of agochemicals in humans and the environment of the United States. Environmental Monitoring and Assessment. 5: 155–163.
Carsch, S., H. Thoma and O. Hutzinger. 1986. Leaching of polychlorinated dibenzo-p-dioxins and polychlorinated dibenzofurans from municipal waste incinerator fly ash by water and organic solvents. Chemosphere. 15: 1927–1930.
Casanovas, J., R. Muro, E. Eljarrat et al. 1994. PCDF and PCDD levels in different types of environmental samples in Spain. Fresenius. J. Anal. Chem. 348: 167–170.
Cavanagh, J.E., K.A. Burns, G.J. Brunskill et al. 1999. Organochlorine pesticide residues in soils and sediments of the Herbert and Burdekin River regions, North Queensland—implications from contamination of the Great Barrier Reef. Mar. Pollut. Bull. 39: 367–375.
CDC. 1987. Comparison of serum levels of 2,3,7,8-TCDD with indirect estimates of Agent Orange exposure in Vietnam veterans: Final Report Agent Orange Projects. Atlanta, Ga: Centers for Disease Control, Public Health Service, U.S. Department of Health and Human Services.
CDC. 1988. Serum 2,3,7,8-tetrachlorodibenzo-p-dioxin levels in US Army Vietnam-era veterans. Centers for Disease Control. JAMA. 260: 1249–1254.
Chan, C.H. and L.H Perkins. 1989. Monitoring of trace organic contaminants in atmospheric precipitation. J. Great Lakes Res. 15(3): 465–475.
Chan, C.H., G. Bruce and B. Harrison. 1994. Wet deposition of organochlorine pesticides and polychlorinated biphenyls to the Great Lakes. J. Great Lakes Res. 20(3): 546–560.
Chang, K.J., K.H. Hsieh, T.P. Lee et al. 1981. Immunologic evaluation of patients with polychlorinated biphenyl poisoning: Determination of lymphocyte subpopulations. Toxicol. Appl. Pharmacol. 61: 58–63.
Chang, K.J., K.H. Hsieh, T.P. Lee et al. 1982a. Immunologic evaluation of patients with polychlorinated biphenyl poisoning: Determination of phagocyte Fc and complement receptors. Environ. Res. 28: 329–334.
Chang, R.R., W.M. Jarman, C.C. King et al. 1990. Bioaccumulation of PCDDs and PCDFs in food animals. III. A rapid cleanup of biological material using reverse-phase adsorbent columns. Chemosphere. 20: 881–886.

Charnley, G. and J. Doull. 2005. Human exposure to dioxins from food, 1999–2002. Food Chem. Toxicol. 43(5): 671–679.

Chen, P.H., C.K. Wong, C. Rappe et al. 1985b. Polychlorinated biphenyls, dibenzofurans and quaterphenyls in toxic rice-bran oil and in the blood and tissues of patients with PCB poisoning Yu-Cheng) in Taiwan. Environ. Health Perspect. 59: 59–65.

Chen, P.H. and S.-T. Hsu. 1986. PCB poisoning from toxic rice-bran oil in Taiwan. pp. 27–38. *In*: Waid, J.S. (ed.). PCBs and the Environment, Vol. 3. Boca Raton, FL: CRC Press.

Chen, A.S.C., A.R. Gavaskar, B.C. Alleman et al. 1997. Treating contaminated sediment with a two-stage base-catalyzed decomposition (BCD) process: bench-scale evaluation. J. Hazard. Mater. 56: 287–306.

Cheung, W.H., C. Lee VKC and G. McKay. 2007. Minimizing dioxin emissions from integrated MSW thermal treatment. Environ. Sci. Technol. 41: 2001–2007.

Chevreuil, M., M. Garmouma, M.J. Teil et al. 1996. Occurrence of organochlorine (PCBs, pesticides) and herbicides (triazines, phenylureas) in the atmosphere and in the fallout from urban and rural stations of the Paris area. Sci. Total Environ. 182: 25–37.

Chevreuil, M., M. Blanchard, M.J. Teil et al. 1998. Polychlorobiphenyl behaviour in the water/sediment system of the Seine river, France. Water Res. 32(4): 1204–1212.

Chia, L.G. and F.L. Chu. 1984. Neurological studies on polychlorinated biphenyl (PCB)-poisoned patients. Am. J. Ind. Med. 5: 117–126.

Chia, L.G. and F.L. Chu. 1985. A clinical and electrophysiological study of patients with polychlorinated biphenyl poisoning. J. Neurol. Neurosurg. Psychiatry. 48: 894–901.

Chiu, C., R.S. Thomas, J. Lockwood et al. 1983. Polychlorinated hydrocarbons from power plants, wood burning, and municipal incinerators. Chemosphere. 12: 607–616.

Choi, K.Y. and D.H. Lee. 2006. PCDD/DF in leachates from Korean MSW landfills. Chemosphere. 63: 1353–1360.

Choi, S.D. and F. Wania. 2011. On the reversibility of environmental contamination with persistent organic pollutants. Environ. Sci. Technol. 45(20): 8834–8841.

Choi, W., S.J. Hong, Y.S. Chang et al. 2000. Photocatalytic degradation of polychlorinated dibenzo-p-dioxins on TiO_2 film under UV or solar light irradiation. Environ. Sci. Technol. 34: 4810–4815.

Choudhry, G.G. and G.R.B. Webster. 1989. Environmental photochemistry of PCDDs: 2. Quantum yields of the direct phototransformation of 1,2,3,7-tetra-, 1,3,6,8-tetra-, 1,2,3,4,6,7,8-hepta-, and 1,2,3,4,6,7,8,9-octachlorodibenzo-p-dioxin in aqueous acetonitrile and their sunlight half-lives. J. Agric Food Chem. 37: 254–261.

Christmann, W., K.D. Kloeppel, H. Partscht et al. 1989b. PCDD/PCDF and chlorinated phenols in wood preserving formulations for household use. Chemosphere. 18: 861–865.

Cirnies-Ross, C., B. Stanmore and G. Millar. 1996. Dioxins in diesel exhaust. Nature. 381: 379.

Cleeman, M., F. Riget, G.B. Paulsen et al. 2000a. Organochlorines in Greenland marine fish, mussels and sediments. Sci. Total Environ. 245: 87–102.

Clement, R.E., S.A. Suter, E. Reiner et al. 1989a. Concentrations of chlorinated dibenzo-p-dioxins and dibenzofurans in effluents and centrifuged particles from Ontario pulp and paper mills. Chemosphere. 19: 649–954.

Clement, R.E., S.A. Suter and H.M. Tosine. 1989c. Analysis of large volume water samples near chemical dump sites using the aqueous phase liquid extractor (APLE). Chemosphere. 18: 133–140.

Clement, R.E., H.M. Tosine and B. Ali. 1985. Levels of polychlorinated dibenzo-p-dioxin and dibenzofuran in woodburning stoves, fireplaces. Chemosphere. 14: 815.

Clevenger, T.E., D.D. Hemphill, K. Roberts et al. 1983. Chemical composition and possible mutagenicity of municipal sludges. J. Water Pollut. Control. Fed. 55(12): 1470–1475.

Cole, R.H., R.E. Frederick, R.P. Healy et al. 1984. Preliminary findings of the priority pollutant monitoring project of the nationwide urban runoff program. J. Water Pollut. Control. Fed. 56: 898–908.

Cole, J.G., D. Mackay, K.C. Jones et al. 1999. Interpreting, correlating, and predicting the multimedia concentrations of PCDD/Fs in the United Kingdom. Environ. Sci. Technol. 33: 399–405.

Commission of the European Communities. 2001. Community strategy for dioxins, furans and polychlorinated biphenyls. Communication from the Commission to the Council, the European Parliament and the Economic and Social Committee. COM (2001) 593, final. Brussels, Belgium.

Cook, R.R., G.G. Bond, R.A. Olson et al. 1986. Evaluation of the mortality experience of workers exposed to the chlorinated dioxins. Chemosphere. 15: 1769–1776.

Cook, R.R., G.G. Bond, R.A. Olson et al. 1987b. Update of the mortality experience of workers exposed to chlorinated dioxins. Chemosphere. 16: 2111–2116.

Cooper, C.M., F.E. Dendy, J.R. McHenry et al. 1987. Residual pesticide concentrations in Bear Creek, Mississippi, 1976–1979. J. Environ. Qual. 16: 69–72.

Cordle, F., R. Locke and J. Springer. 1982. Risk assessment in a federal regulatory agency: An assessment of risk associated with the human consumption of some species of fish contaminated with polychlorinated biphenyls (PCBs). Environ. Health Perspect. 45: 171–182.

Corrigan, P.J. and P. Seneviratna. 1989. Pesticide residues in Australian meat. Vet. Ret. 125(8): 180–181.

Corsolini, S., S. Focardi, C. Leonzio et al. 1999. Heavy metals and chlorinated hydrocarbon concentrations in the red fox in relation to some biological parameters. Environ. Monit. Assess. 54: 87–100.

Corsolini, S., T. Romeo, S. Ademollo et al. 2001. POPs in key species of marine Antarctic ecosystem. Microb. J. 73: 187–193.

Cortes, D.R., I. Basu, C.W. Sweet et al. 1998. Temporal trends in gas-phase concentrations of chlorinated pesticides measured at the shores of the Great Lakes. Environ. Sci. Technol. 32: 1920–1927.

Cotham, W.E. and T.F. Bidleman. 1991. Estimating the atmosphere deposition of organochlorine contaminants to the Arctic. Chemosphere. 2: 165–188.

Courtney, K.D., D.W. Gaylor, M.D. Hogan et al. 1970. Teratogenic evaluation of 2,4,5-T. Science. 168: 864–866.

Covaci, A., C. Hura and P. Schepens. 2001a. Selected organochlorinated pollutants in Romania. Sci. Total Environ. 280: 143–152.

Covaci, A., C. Hura and P. Schepens. 2001b. Determination of selected persistent organochlorine pollutants in human milk using solid-phase disk extraction and narrow bore capillary GC/MS. Chromatographia. 54: 247–252.

Cramer, G., M. Bolder, S. Henry et al. 1991. USFDA assessment of exposure to 2,3,7,8-TCDD and 2,3,7,8-TCDF from foods contacting bleached paper products. Chemosphere. 23(8-10): 1537–1550.

Crosby, D.G., A.S. Wong, J.R. Plimmer et al. 1971. Photodecomposition of chlorinated dibenzo-*p*-dioxins. Science. 195: 748–749.

Crosby, D.G., K.W. Moilanen and A.S. Wong. 1973. Environmental generation and degradation of dibenzodioxins and dibenzofurans. Environ. Health Perspect. 5: 259–266.

Cross, J.N., J.T. Hardy, J.E. Hose et al. 1987. Contaminant concentrations and toxicity of sea-surface microlayer near Los Angeles, California. Mar. Environ. Res. 23: 307–323.

Crow, K.D. 1978. Chloracne—an up to date assessment. Ann. Occup. Hyg. 21: 297–298.

Crummett, W.B. 1982. Environmental chlorinated dioxins from combustion—the trace chemistries of fire hypothesis. Pergamon. Ser. Environ. Sci. 5: 253–263.

Crummett, W.B., T.J. Nestrick and L.L. Lamparski. 1985. Analytical methodology for the determination of PCDDs in environmental samples: an overview and critique. pp. 57–83. *In*: Kamrin, M.A. and P.W. Rodgers (eds.). Dioxins in the Environment, Washington, DC, Hemisphere Publishing.

Cudahy, J.J. and R.W. Helsel. 2000. Removal of products of incomplete combustion with carbon. Waste Manage. 20: 339–345.

Currier, M.F., C.D. McClimans and G. Barna-Lloyd. 1980. Hexachlorobenzene blood levels and the health status of men employed in the manufacture of chlorinated solvents. J. Toxicol. Environ. Health. 6: 367–377.

Czaja, K., J.K. Ludwicki, K. Goralczy et al. 1997a. Effect of age and number of deliveries on mean concentration of organochlorine compounds in human breast milk in Poland. Bull. Environ. Contam. Toxicol. 59: 407–413.

Czaja, K., J.K. Ludwicki, K. Goralczy et al. 1997b. Organochlorine in pesticides, HCB, and PCBs in human milk in Poland. Bull. Environ. Contam. Toxicol. 58: 769–775.

Czaja, K., J.K. Ludwicki, K. Goralczy et al. 1999a. Effect of changes in excretion of persistent organochlorine compounds with human breast milk on related exposure of breast-fed infants. Arch. Environ. Contam. Toxicol. 36: 498–503.

Czuczwa, J., V. Katona, G. Pitts et al. 1989. Analysis of fog samples for PCDD and PCDF. Chemosphere. 18: 847–850.

Czuczwa, J.M. and R.A. Hites. 1984. Environmental fate of combustion-generated polychlorinated dioxins and furans. Environ. Sci. Technol. 18: 444–450.

Czuczwa, J.M. and R.A. Hites. 1986a. Airborne dioxins and dibenzofurans: Sources and fates. Environ. Sci. Technol. 20: 195–200.

Czuczwa, J.M. and R.A. Hites. 1986b. Sources and fates of PCDD and PCDF. Chemosphere. 15: 1417–1420.

Dannenberg, D. and A. Lerz. 1999. Occurrence and transport of organic micro-contaminants in sediments of the Odra river estuarine system. Acta Hydroch. Hydrob. 27: 303–307.

Davies, K. 1988. Concentrations and dietary intake of selected organochlorines, including PCBs, PCDDs and PCDFs in fresh food composites grown in Ontario, Canada. Chemosphere. 17(2): 263–276.

Davis, B.D. and R.C. Morgan. 1986. Hexachlorobenzene in hazardous waste sites. IARC Sci. Publ. 77: 23–30.

de Boer, J. and P.G. Wester. 1993. Determination of toxaphene in human milk from Nicaragua and in fish and marine mammals from the Northeastern Atlantic and the North Sea. Chemosphere. 27: 1879–1890.

de Brito, A.P.X., D. Ueno, S. Takahashi et al. 2002a. Organochlorine and butyltin residues in walleye pollock (*Theraga chalcogramma*) from Bering Sea, Gulf of Alaska and Japan Sea. Chemosphere. 46: 401–411.

de Brito, A.P.X., I.M.R.D.A. Bruning and I. Moreira. 2002b. Chlorinated pesticides in mussels from Guanabara Bay, Rio de Janeiro, Brazil. Mar. Pollut. Bull. 44: 71–81.

de Brito, A.P.X., S. Takahashi, D. Ueno et al. 2002c. Organochlorine and butyltin residues in deep-sea organisms collected from the western North Pacific, off Tohoku, Japan. Mar. Pollut. Bull. 45: 348–361.

de March, B.G.E., C.A. de Wit and D.C.G. Muir. 1998. Persistent organic pollutants. pp. 183–372. *In*: Wilson, S.J., J.L. Murray, H.P. Huntigton (eds.). AMAP Assessment Report: Arctic Pollution Issues. Arctic Monitoring and Assessment Programme (AMAP). Oslo, Norway.

de Mora, S. and M.R. Sheikholeslami. 2002. Final Report: Interpretation of Caspian Sea Sediment Data. ASTP: Contaminant Screening Programme.

Depercin, P.R. 1995. Application of thermal-desorption technologies to hazardous waste sites. J. Hazard. Mater. 40: 203–209.

De Peyster, A., R. Donohoe and D.J. Slymen. 1993. Aquatic biomonitoring of reclaimed water for potable use: The San Diego health effects study. Journal of Toxicology and Environment Health. 39: 121–142.

De Vault, D., W. Dunn, P.A. Bergqvist et al. 1989. Polychlorinated dibenzofurans and polychlorinated dibenzo-p-dioxins in Great Lakes fish: A baseline and interlake comparison. Environmental Toxicology and Chemistry. 8: 1013–1022.

de Wit, C.A. 2002. An Overview of brominated flame retardants in the environment. Chemosphere. 46: 583–624.

Delaplane, K.S. and J.P. LaFage. 1990. Variable chlordane residues in soil surrounding house foundations in Louisiana USA. Bull. Environ. Contam. Toxicol. 45: 675–680.

Denison, M.S., J.M. Fisher and J.P. Whitlock Jr. 1989. Protein-DNA interactions at recognition sites for the dioxin-Ah receptor complex. J. Biol. Chem. 264: 16478–16482.

Des Rosiers, P.E. 1989. Chemical detoxification of dioxin contaminated wastes using potassium polyethylene glycolate. Chemosphere. 18: 343–353.

DeVault, D.S. 1985. Contaminants in fish from Great Lakes harbors and tributary mouths. Arch. Environ. Contam. Toxicol. 14: 587–594.

Dewailly, E., J.P. Weber, S. Gingras et al. 1991. Coplanar PCBs in human milk in the province of Quebec, Canada: Are they more toxic than dioxin for breast fed infants. Bull. Environ. Contam. Toxicol. 47: 491–498.

Dewailly, E., P. Ayotte, S. Bruneau et al. 1993. Inuit exposure to organochlorines through the aquatic food chain in arctic Quebec. Environmental Health Perspectives. 101: 618–620.

Dewailly, E., G. Mulvad, H.S. Pedersen et al. 1999. Concentration of organochlorines in human brain, liver, and adipose tissue autopsy samples from Greenland. Environ. Health Perspect. 107(10): 823–828.

DiDomenico, A., G. Viviano and G. Zapponi. 1982. Environmental persistence of 2,3,7,8-TCDD at Seveso. pp. 105–113. *In*: Hutzinger, O. et al. (eds.). Chlorinated Dioxins and Related Compounds, Impact on the Environment. Elmsford, NY: Pergamon Press.

Dimich-Ward, H., C. Hertzman, K. Teschke et al. 1996. Reproductive effects of paternal exposure to chlorophenate wood preservatives in the sawmill industry. Scand. J. Work Environ. Health. 22: 267–273.

Dionex. 1999. Dionex Application Note 323: Extraction of polychlorinated dibenzo-p-dioxins and polychlorinated dibenzofurans from environmental samples using accelerated solvent extraction (ASE). Dionex, Sunnyvale, CA.

Djordjevic, M.V., D. Hofmann and J. Fan. 1994. Assessment of chlorinated pesticides and polychlorinated biphenyls in adipose breast tissue using a supercritical fluid extraction method. Carcinogenesis. 15(11): 2581–2585.

Dobbs, A.J. and C. Grant. 1979. Photolysis of highly chlorinated dibenzo-p-dioxins by sunlight. Nature. 278(5700): 163–165.

Dobbs, A.J. and N. Williams. 1983. Indoor air pollution from pesticides used in wood remedial treatments. Environ. Pollution (Series B). 6: 271–296.

Domingo, J.L., M. Schuhmacher, S. Granero et al. 1999. PCDDs and PCDFs in food samples from Catalonia, Spain. An assessment of dietary intake. Chemosphere. 38: 3517–3528.

Dougherty, E.J., M.R. Overcash and R.G. Carbonell. 1991. Diffusivity of 2,3,7,8-tetrachlorodibenzo-p-dioxin in organic solvents. Hazard Waste Hazard Mater. 8: 43–53.

Dougherty, E.J., A.L. McPeters, M.R. Overcash et al. 1993. Theoretical-analysis of a method for *in situ* decontamination of soil containing 2,3,7,8-tetrachlorodibenzo-p-dioxin. Environ. Sci. Technol. 27: 505–515.

Dougherty, E.J., A.L. McPeters, M.R. Overcash et al. 1994. Sorption processes of 2,3,7,8-tetrachlorodibenzo-p-diozin on soil in the presence of organic liquids. J. Hazard Mater. 38: 405–421.

Downs, T.J., E. Cifuentes-Garcia and I.M. Suffet. 1999. Risk screening for exposure to groundwater pollution in a wastewater irrigation district of the Mexico City region. Environ. Health Perspect. 107(7): 553–561.

Dulin, D., H. Drossman and T. Mill. 1986. Products and quantum yields for photolysis of chloroaromatics in water. Environ. Sci. Technol. 20: 72–77.

Dung, M.H. and W. O'Keef. 1994. Comparative rates of photolysis of polychlorinated dibenzofurans in organic solvents and in aqueous solutions. Environ. Sci. Technol. 28: 549–554.

Durham, R.W. and B.G. Oliver. 1983. History of Lake Ontario (Canada, USA) contamination from the Niagara river by sediment radiodating and chlorinated hydrocarbon analysis. Journal of Great Lakes Research. 9(2): 160–168.

Dyke, P.H., C. Foan, M. Wenborn et al. 1997. A review of dioxin releases to land and water in the UK. Sci. Tot. Environ. 207: 119–131.

Dyke, P.H. and G. Amendola. 2007. Dioxin releases from US chemical industry sites manufacturing or using chlorine. Chemosphere. 67(9): 125–134.

EarthFax. 2003. http://www.earthfax.com/WhiteRot/Dioxin.htm

Edgerton, S.A., J.M. Czuczwa, J.D. Rench et al. 1989. Ambient air concentrations of polychlorinated dibenzo-p-dioxins and dibenzofurans in Ohio: Sources and health risk assessment. Chemosphere. 18: 1713–1730.

Eduljee, G.H. 1987. Volatility of TCDD and PCB from soil. Chemosphere. 16: 907–920.

Egeland, G.M., M.G. Sweeney, M.A. Fingerhut et al. 1994. Total serum testosterone and gonadotropins in workers exposed to dioxins. Am. J. Epidemiol. 139: 272–281.

Eisenreich, S.J., B.B. Looney and J.D. Thornton. 1981. Airborne organic contaminants in the Great Lakes ecosystem. Environ. Sci. Technol. 15(1): 30–38.

Eisenreich, S.J., P.D. Capel, J.A. Robbins et al. 1989. Accumulation and digenesis of chlorinated hydrocarbons in lacustrine sediments. Environ. Sci. Technol. 23: 1116–1126.

Eitzer, B.D. and R.A. Hites. 1989a. Atmospheric transport and deposition of polychlorinated dibenzopdioxins and dibenzofurans. Environmental Science and Technology. 23: 1396–1401.

Eitzer, B.D. and R.A. Hites. 1989a. Dioxins and furans in the ambient atmosphere: A baseline study. Chemosphere. 18: 593–598.

Eitzer, B.D. and R.A. Hites. 1989b. Polychlorinated dibenzo-p-dioxins and dibenzofurans in the ambient atmosphere of Bloomington, Indiana. Environ. Sci. Technol. 23: 1389–1395.

Eitzer, B.D. 1993. Comparison of point and nonpoint sources of polychlorinated dibenzo-p-dioxins and polychlorinated dibenzofurans to sediments of the Housatonic River. Environ. Sci. Technol. 27(8): 1632–1637.

Elferink, C.J. and J.P. Whitlock Jr. 1990. 2,3,7,8-Tetrachlorodibenzo-p-dioxin-inducible, Ah receptor mediated bending of enhancer DNA. J. Biol. Chem. 265: 5718–5721.

EPA. 1980a. Ambient water quality criteria for aldrin/dieldrin. Washington, DC: U.S. Environmental Protection Agency, Criteria and Standards Division. PB81-11730/OWRS.

EPA. 1981a. Aquatic fate process data for organic priority pollutants. U.S. Environmental Protection Agency, Office of Water Regulations and Standards, Washington, DC. (authors: Mabey et al.). EPA-440/4-81-014.

EPA. 1981b. The potential atmospheric impact of chemicals released to the environment: Proceedings of four workshops. Washington, DC: U.S. Environmental Protection Agency. Document No. PB82-119447.

EPA. 1984a. Ambient water quality criteria document for 2,3,7,8-tetrachlorodibenzo-p-dioxin. Cincinnati, OH: U.S. Environmental Protection Agency, Environmental Criteria and Assessment Office. EPA 440/5-84/194/1.

EPA. 1986f. Hazardous waste management system: Land disposal restriction. U.S. Environmental Protection Agency. Federal Register. 51: 40572–40623.

EPA. 1986i. Superfund record of decision (EPA Region 10): Toftdahl Drums, Brush Prairie, Clark County, Washington, DC: U.S. Environmental Protection Agency. EPA/ROD/R10-86-009.

EPA. 1987i. Superfund record of decision (EPA Region 4): Gallaway Ponds Site, Gallaway, Tennessee, September 1986. Washington, DC: U.S. Environmental Protection Agency. EPA/ROD/R0486-013. NITS PB87-189080.

EPA. 1987k. 2,3,7,8-TCDD: Environmental chemistry. Athens, GA: Environmental Protection Agency, Office of Research and Development, Environmental Research Laboratory. EPA600/D-87/086.

EPA. 1988f. Land disposal: Waste specified prohibitions—dioxin-containing wastes. U.S. Environmental Protection Agency. Code of Federal Regulations. 40 CFR 268.31.

EPA. 1989. Pesticides in ground water data base: 1988 interim report. Washington, DC: U.S. Environmental Protection Agency, Office of Pesticide Programs. EPA-540/09-89-036.

EPA. 1989a. Interim methods for development of inhalation reference doses. Washington, DC: U.S. Environmental Protection Agency, Office of Health and Environmental Assessment. EPA 600/8-88-066F.

EPA. 1990a. Characterization of municipal waste combustion ash, ash extracts, and leachates. Coalition on Resource Recovery and the Environment. Washington, DC: U.S. Environmental Protection Agency, Office of Solid Waste and Emergency Response. Contract No. 68-01-7310. EPA 530-SW-90-029A.

EPA. 1990a. Interim methods for development of inhalation reference concentrations. Washington, DC: U.S. Environmental Protection Agency, Office of Health and Environmental Assessment, Office of Research and Development, Environmental Criteria and Assessment Office. EPA 600-890-066A.

EPA. 1990b. Standards for owners and operators of hazardous waste treatment, storage, and disposal facilities. Environmental Protection Agency. Code of Federal Regulations. 40 CFR 264.342.

EPA. 1990c. Dibenzo-para-dioxins/dibenzofurans in bleached wood pulp and paper product referral for action. U.S. Environmental Protection Agency. 55 Federal Register 248:53047-53049.

EPA. 1991d. Atmospheric transport and deposition of polychlorinated dibenzo-p-dioxins and dibenzofurans. Washington, DC: U.S. Environmental Protection Agency. EPA600/3-91/002 Order No. PB91-144667.

EPA. 1998a. U.S. Environmental Protection Agency. Code of Federal Regulations. 40 CFR 401.15.

EPA. 1998b. Source category listing for section 112 (d) (2) rulemaking pursuant to section 112 (c) (6) requirements. U.S. Environmental Protection Agency (EPA) Federal Register 63 FR 17838-17851 (April 10, 1998).

EPA. 2010a. Method 8276. Toxaphene and toxaphene congeners by gas chromatography/negative ion mass spectrometry (GC/NIMS). Revision 0, March 2010. U.S. Environmental Protection Agency. http://www.epa.gov/osw/hazard/testmethods/sw846/new_meth.htm. September 09, 2010.

Ergebnisse. Ergebnisse von Dioxin-Emissionsmessungen an Industrieanlagen in NRW; Dioxinmeßprogramm Nordrhein-Westfalen. In: Landesumweltamt Nordrhein-Westfalen, editor. Materialien 1996, Vol. 30. 1996. 102ps Essen.

Erickson, J.D., J. Mulinare, P.W. McClain et al. 1984. Vietnam veterans' risks for fathering babies with birth defects. JAMA. 252: 903–912.

Ernst, W. 1986. Hexachlorobenzene in the marine environment: distribution, fate and ecotoxicological aspects. pp. 211–222. *In*: Morris, C.R. and J.R.P. Cabral (eds.). Hexachlorobenzene: Proceedings of an International Symposium. IARC Sci. Publ. 77. Lyon.

Esposito, M.T., T.O. Tiernan and F.E. Dryden. 1980. Dioxins, USEPA Report EPA-600/2-80-197; U.S. Government Printing Office: Washington, DC, p. 257.

Evans, R.G., K.B. Webb, A.P. Knutsen et al. 1988. A medical follow-up of the health effects of long-term exposure to 2,3,7,8-TCDD. Arch. Environ. Health. 43: 273–278.

Everaert, K. and J. Baeyens. 2004. Catalytic combustion of volatile organic compounds. J. Hazard. Mater. 109: 113–139.

Exner, J. 1987. Perspective on hazardous waste problems related to dioxins. pp 1–10. *In*: Exner, J. (ed.). Solving Hazardous Waste Problems: Learning from Dioxins. ACS Symposium Series 338, American Chemical Society, Washington, DC, USA.

Falandysz, J., K. Kannan, S. Tanabe et al. 1994. Organochlorine pesticides and polychlorinated biphenyls in cod-liver oils: North Atlantic, Norwegian Sea, North Sea and Baltic Sea. Ambio. 23: 288–293.

Falandysz, J., D.D. Danisiewicz, L. Strandberg et al. 1997. Pentachlorobenzene, HCB and DDT in a pelagic food chain in the Baltic Sea. Organohalogen Compds. 32: 370–373.

Fan, Z., F.X. Casey, G.L. Larsen et al. 2006. Fate and transport of 1278-TCDD, 1378-TCDD, and 1478-TCDD in soil-water systems. Sci. Tot. Environ. 371(1-3): 323–333.

Faqi, A.S., P.R. Dalsenter, H.J. Merker et al. 1998. Reproductive toxicity and tissue concentrations of low doses of 2,3,7,8-tetrachlorodibenzo-*p*-dioxin in male offspring rats exposed throughout pregnancy and lactation. Toxicol. Appl. Pharmacol. 150: 383–392.

Faust, S.D. and O.M. Aly. 1964. Water pollution by organic pesticides. J. Amer. Water Works Assoc. 56: 267–274.

Faust, S.D. and I.H. Suffet. 1966. Recovery, separation, and identification of organic pesticides from natural and potable waters. Residue Rev. 15: 44–116.

FDA. 1988. Food and Drug Administration Pesticide Program. Residues in foods-1987. Food and Drug Administration Program. J. Assoc. Off. Anal. Chem. 71(6): 156A–174A.

FDA. 1989. Residues in foods-1988. Food and Drug Administration Pesticide Program. J. Assoc. Off. Anal. Chem. 72: 133A–152A.

FDA. 1990. Residues in foods 1989. Food and Drug Administration. J. Assoc. Off. Anal. Chem. 73: 127A–146A.

FDA. 1991. Residues in foods-1990 (4th Annual FDA Pesticide Residue Monitoring Program Report). J. Assoc. Off. Anal. Chem. 74(5): 121A–140A.

FDA. 1992. Residue monitoring-1991 (5th Annual FDA Pesticide Residue Monitoring Program Report). J. of AOAC International. 75(5): 135A–157A.

FDA. 1993. FDA Monitoring Program. Food and Drug Administration. J. AOAC Int. 76(5): 127A–148A.

FDA. 1993. Residue monitoring 1992 (6th Annual FDA Pesticide Residue Monitoring Program Report). J. of AOAC International. 76(5): 127A–147A.

FDA. 1994a. Residue monitoring 1993 (7th Annual FDA Pesticide Residue Monitoring Program Report). J. of AOAC International. 77(5): 163A–185A.

FDA. 1994b. 302 and 303 Methods for nonfatty foods. *In*: Pesticides Analytical Manual, 3rd edition, Vol. 1: Multiresidue methods. U.S. Department of Health and Human Services, Food and Drug Administration.

FDA. 1994c. 304: Method for fatty foods. *In*: Pesticide Analytical Manual, 3rd edition Vol. 1: Multiresidue Methods. U.S. Department of Health and Human Services, Food and Drug Administration.

FDA. 1994c. Pesticide program: Residue monitoring. Food and Drug Administration. J. AOAC Int. 77(5): 163A–185A.

FDA. 1995. FDA Monitoring Program. Food and Drug Administration. J. AOAC Int. 78(5): 119A–141A.

FDA. 1995. Residue monitoring-1994 (8th Annual FDA Pesticide Residue Monitoring Program Report). J. of AOAC International. 78(5): 119A–142A.

Fell, H.J. and M. Tuczek. 1998. Removal of dioxins and furans from flue gases by nonflammable adsorbents in a fixed bed. Chemosphere. 37: 2327–2334.

Feltz, H.R. 1980. Significance of bottom material data in evaluation water quality. pp. 271–287. *In*: Baker, R.A. (ed.). Contaminants and Sediments. Vol. 1. Fate and Transport, Case Studies, Modeling, Toxicity. Ann Arbor, MI: Ann Arbor Science.

Ferrario, J., C. Byrne, D. McDaniel et al. 1996. Determination of 2,3,7,8,-chlorine substituted dibenzo-p-dioxins and furans at the part per trillion level in United States beef fat using high resolution gas chromatography/high resolution mass spectrometry. Anal. Chem. 68: 647–652.

Ferrario, J., C. Byrne, M. Lorber et al. 1997. A statistical survey of dioxin-like compounds in United States poultry fat. Organohalogen Compounds. 32: 64–70.

Ferrario, J., C. Byrne and J. Schaum. 2004. An assessment of dioxin levels in processed ball clay from the United States. Organohalogen Compounds. 66: 1639–1644.

Ferrario, J.B., C.J. Byrne and D.H. Cleverly. 2000. 2,3,7,8-dibenzo-*p*-dioxins in mined clay products from the United States: Evidence for possible natural origin. Environ. Sci. Technol. 4524–4532.

Fiedler, H., O. Hutzinger and C. Timms. 1990. Dioxins: Sources of Environmental Load and Human Exposure. Toxicol. Environ. Chem. 29: 157–23.

Fiedler, H., K.W. Schramm and O. Hutzinger. 1990. Dioxin Emissions to the Air:Mass Balance for Germany Today and in the Year 2000. Organohalogen Compd. 4: 395–400.

Fiedler, H. 1995. EPA DIOXIN-Reassessment: Implications for Germany. Organohalogen Compd. 22: 209–228.

Fiedler, H. 1998. Thermal formation of PCDD/PCDF—A survey. Environ. Eng. Sci. 15/1: 49–58.

Fillmann, G., J.W. Readman, I. Tolosa et al. 2002. Persistent organochlorine residues in sediments from the Black Sea. Mar. Pollut. Bull. 44: 122–133.

Fingerhut, M.A., M. Haring-Sweeney, D.G. Patterson Jr et al. 1989. Levels of 2,3,7,8-tetrachlorodibenzop-dioxin in the serum of U.S. chemical workers exposed to dioxin contaminated products: Interim results. Chemosphere. 19: 835–840.

Fingerhut, M.A., W.E. Halperin, D.A. Marlow et al. 1991. Cancer mortality in workers exposed to 2,3,7,8-tetrachlorodibenzo-p-dioxin. N. Engl. J. Med. 324: 212–218.

Firestone, D. 1978. The 2,3,7,8-tetrachlorodibenzo-para-dioxin problem: A review proceedings of a conference on chlorinated phenoxyacids and their dioxins, Stockholm, 1977. Ecol. Bull. 27: 39–52.

Firestone, D., R.A. Niemann, L.F. Schneider et al. 1986. Dioxin residues in fish and other foods. pp. 355–365. *In*: Rappe, C., G. Choudhary and L.H. Keith (eds.). Chlorinated Dioxins and Dibenzofurans in Perspective. Chelsea, MI: Lewis Publishers, Inc.

Flesch-Janys, D., J. Berger, P. Gurn et al. 1995. Exposure to polychlorinated dioxins and furans (PCDD/F) and mortality in a cohort of workers from a herbicide-producing plant in Hamburg, Federal Republic of Germany. Am. J. Epidemiol. 142: 1165–1175.

Fletcher, C. and W.A. McKay. 1993. Polychlorinated dibenzo-p-dioxins PCDDs and dibenzofurans PCDFs in the aquatic environment a literature review. Chemosphere. 26(6): 1041–1069.

Focant, J.-F., G. Eppe, C. Pirard et al. 2002. Levels and congener distributions of PCDDs, PCDFs and non-ortho PCBs in Belgium foodstuffs assessment of dietary intake. Chemosphere. 48: 167–179.
Ford, W.M. and E.P. Hill. 1991. Organochlorine pesticides in soil sediments and aquatic animals in the upper steele bayou watershed of Mississippi. Arch. Environ. Contam. Toxicol. 20: 161–167.
Foster, G.D., P.M. Gates and W.T. Foreman. 1993. Determination of dissolved-phase pesticides in surface water from the Yakima River Basin, Washington, using the goulden large-sample extractor and gas chromatography/mass spectrometry. Environ. Sci. Tech. 27(9): 1911–1917.
Fox, M.E., R.M. Khan and P.A. Thiessen. 1996. Loadings of PCBs and PAHs from hamilton harbour to Lake Ontario. Water Qual. Res. J. Can. 31(3): 593–608.
Frank, R., J. Rasper, M.S. Smout et al. 1988. Organochlorine residues in adipose tissues blood and milk from Ontario, Canada residents 1976–1985. Can. J. Public Health. 79(3): 150–158.
Franz, T.P. and S.J. Eisenreich. 1993. Wet deposition of polychlorinated biphenyls to Green Bay, Lake Michigan. Chemosphere. 26(10): 1767–1788.
Franz, T.P. and S.J. Eisenreich. 1998. Snow scavenging of polychlorinated biphenyls and polycyclic aromatic hydrocarbons in Minnesota. Environ. Sci. Technol. 32: 1771–1778.
Freeman, R.A. and J.M. Schroy. 1989. Comparison of the rate of TCDD transport at Times each and at Eglin AFB. Chemosphere. 18(1-6): 1305–1312.
Friesen, K.J., D.C.G. Muir and G.R.B. Webster. 1990a. Evidence of sensitized photolysis of polychlorinated dibenzo-p-dioxins in natural waters under sunlight conditions. Environ. Sci. Technol. 24: 1739–1744. `
Friesen, K.J., J. Vilk and D.C.G. Muir. 1990b. Aqueous solubilities of selected 2,3,7,8-substituted polychorinated dibenzofurans (PCDFs). Chemosphere. 20(1-2): 7–32.
Friesen, K.J., M.M. Foga and M.D. Loewen. 1996. Aquatic photodegradation of polychlorinated dibenzofurans: Rates and photoproduct analysis. Environ. Sci. Technol. 30: 2504–2510.
Froese, K.L. and O. Hutzinger. 1993. Polychlorinated benzene and polychlorinated phenol in heterogeneous combustion reactions of ethylene and ethane. Environ. Sci. Technol. 27: 121–129.
Froese, K.L. and O. Hutzinger. 1996. Polychlorinated benzene, phenol, dibenzo-p-dioxin and dibenzofuran in heterogeneous combustion reactions of acetylene. Environ. Sci. Technol. 30: 998–1008.
Furr, A.K., A.W. Lawrence, S.C.S. Tong et al. 1976. Multielement and chlorinated hydrocarbon analysis of municipal sewage sludges of American cities. Environ. Sci. Tech. 10(7): 683–687.
Fürst, P., Chr. Kruger, H.A. Meemken et al. 1989. PCDD and PCDF levels in human milk-dependence on the period of lactation. Chemosphere. 18: 439–444.
Fürst, P., C. Fürst and W. Groebel. 1990. Levels of PCDDs and PCDFs in foodstuffs from the Federal Republic of Germany. Chemosphere. 20: 787–792.
Fürst, P., C. Fürst and K. Wilmers. 1994. Human milk as a bioindicator for body burden of PCDDs, PCDFs, organochlorine pesticides, and PCBs. Environ. Health Perspect. 102(1): 187–193.
Furubayashi, M., R. Shinohara and K. Nagai. 2001. Study of dioxin removal by activated carbon. Kagaku Kogaku Ronbunshu. 27(2): 241–242.
Fytianos, K., G. Vasilikiotis, L.Weil et al. 1985. Preliminary study of organochlorine compounds in milk products, human milk, and vegetables. Bull. Environ. Contam. Toxicol. 34: 504–508.
Gardner, A.M. and K.D. White. 1990. Polychlorinated dibenzofurans in the edible portion of selected fish. Chemosphere. 21: 215–222.
Gartrell, M., J. Craun, D. Podrebarac et al. 1985. Pesticides, selected elements, and other chemicals in adult total diet samples, October 1979–September, 1980. J. Assoc. Off. Anal. Chem. 68(5): 1184–1197.
Gartrell, M., J. Craun, D. Podrebarac et al. 1986a. Pesticides, selected elements, and other chemicals in adult total diet samples, October, 1980–March, 1982. J. Assoc. Off. Anal. Chem. 69(1): 146–161.
Gartrell, M., J. Craun, D. Podrebarac et al. 1986b. Pesticides, selected elements, and other chemicals in infant and toddler total diet samples, October 1980–March 1982. J. Assoc. Off. Anal. Chem. 69(1): 123–145.
Gaus, C., G.J. Brunskill, R. Weber et al. 2001. Historical PCDD inputs and their source implications from dated sediment cores in Queensland (Australia). Environ. Sci. Technol. 35: 4597–4603.
Gerasimov, G.Y. 2001. Degradation of dioxins in electron-beam gas cleaning of sulphur and nitrogen dioxides. Radiat. Chem. 35: 427–431.

Germain, A. and C. Langlois. 1988. Pollution of the water and suspended sediments of the St. Lawerence River (Ontario, Quebec, Canada) by organochlorine pesticides, polychlorinated biphenyls, and other priority pollutants. Water Pollution Research Journal of Canada. 23(4): 602–614.

Geyer, H., I. Scheunart and F. Korte. 1986. Bioconcentration potential of organic environmental chemicals in humans. Regul. Toxicol. Pharmacol. 6: 313–347.

Geyer, H., I. Scheunart and F. Korte. 1987. Correlation between the bioconcentration potential of organic environmental chemicals in humans and their n-octanol/water partition coefficients. Chemosphere. 16: 239–252.

Giddings, J.C. 1962. Theory of minimum time operation in gas chromatography. Anal. Chem. 34(3): 314.

Gierthy, J.F. and D. Crane. 1985. Development of *in vitro* bioassays for chlorinated dioxins and dibenzofurans. pp. 267–284. *In*: Keith, L.H., C. Rappe and G. Choudhary (eds.). Chlorinated Dioxins and Dibenzofurans in the Total Environment II, Stoneham, Maine, Butterworth Publishers.

Giesy, J.P., D.A. Verbrugge, R.A. Othout et al. 1994. Contaminants in fishes from the Great lakes-influenced sections and above dams of three Michigan rivers. I: Concentrations of organo chlorine insecticides, polychlorinated biphenyls, dioxin equivalents, and mercury. Arch. Environ. Contam. Toxicol. 27: 202–212.

Gizzi, F., R. Reginato, E. Benfenati et al. 1982. PCDDs and PCDFs in emissions from an urban incinerator: I. Average peaks and values. Chemosphere. 11: 577–583.

Goemans, M., P. Clarysse, J. Joannes et al. 2004. Catalytic NOx reduction with simultaneous dioxin and furan oxidation. Chemosphere. 54: 1357–1365.

Goldfarb, T.D. and S.J. Harrad. 1991. Consideration of the environmental impact of the volatilization of PCDDs and PCDFs. Chemosphere. 23: 1669–1674.

Goldman, P.J. 1973. Severest acute chloracne. A mass poisoning by 2,3,7,8-tetrachlorodibenzo-p-dioxin. Hautarzt. 24: 149–152. (German)

Goldstein, J.A. and S. Safe. 1989. Mechanism of action and structure-activity relationships for the chlorinated dibenzo-p-dioxins and related compounds. pp. 239–293. *In*: Kimbrough, R.D. and A.A. Jensen (eds.). Halogenated Biphenyls, Naphthalenes, Dibenzodioxins and Related Compounds, 2nd ed. Amsterdam, Elsevier Science Publishers. B.V.

Goncalves, C., A. Dimou, V. Sakkas et al. 2006. Photolytic degradation of quinalphos in natural waters and on soil matrices under simulated solar irradiation. Chemosphere. 64: 1375–1382.

Goto, M. and K. Higuchi. 1969. The symptomatology of Yusho (chlorobiphenyls poisoning) in dermatology. Fukuoka Acta Medica. 60: 409–31. [in Japanese]

Götz, R., E. Schumacher, L.-O. Kjeller et al. 1990. Polychlorinated dibenzo-p-dioxins (PCDDs) and polychlorinated dibenzofurans (PCDFs) in sediments and fish from Hamburg port. Chemosphere. 20: 51–73.

Gouin, T., D. Mackay, K.C. Jones et al. 2004. Evidence for the ‘grasshopper’ effect and fractionation during long-range transport of organic contaminants. Environ. Poll. 128: 139–148.

Granstrom, M.L., R.C. Ahlert and J. Wiesenfeld. 1984. The relationships between the pollutants in the sediments and in the water of the Delaware and Raritan Canal. Water Sci. Technol. 16: 375–380.

Gray, K.A. and R.J. Hilarides. 1995. Radiolytic treatment of dioxin contaminated soils. Radiat. Phys. Chem. 46: 1081–1084.

Green, N.J.L., J. Wood, R.E. Alcock et al. 1999. PCDD/Fs and PCBs in sediment samples from the Venice Lagoon. Organohalogen Compounds. 43: 339–342.

Green, N.J.L., J.J. Jones and K.C. Jones. 2001. PCDD/F deposition time trend to Esthwaite water, U.K., and its relevance to sources. Environ. Sci. Technol. 35: 2882–2888.

Greer, J.S. and G.H. Griwatz. 1980. Ultimate disposal of hazardous materials by reaction with liquid sodium. Control of Hazardous Material Spills. Proceedings of the 1980 National Conference on Control of Hazardous Material Spills. 1: 416–20.

Gregor, D.J. and W.D. Gummer. 1989. Evidence of atmospheric transport and deposition of organochlorine pesticides and polychlorinated biphenyls in Canadian arctic snow. Environ. Sci. Technol. 23: 561–565.

Griffin, R.D. 1986. A new theory of dioxin formation in municipal solid waste combustion. Chemosphere. 15: 1987–1990.

Griffith, J. and R.C. Duncan. 1985. Serum organochlorine residues in Florida citrus workers compared to the National health and nutrition examination survey sample. Bull. Environ. Contam. Toxicol. 35: 411–417.

Gross, M.L., J.O. Lay, P.A. Jr, Lyon et al. 1984. 2,3,7,8-Tetrachlorodibenzo-p-dioxin levels in adipose tissue of Vietnam veterans. Environ. Res. 33: 261–268.

Gullett, B.K., K.R. Bruce and L.O. Beach. 1990. Formation of chlorinated organics during solid waste combustion. Waste Manage. Res. 8: 203–214.

Gullett, B.K., P.M. Lemieux and J.E. Dunn. 1994. Role of combustion and sorbent parameters in prevention of polychlorinated dibenzo-*p*-dioxin and polychlorinated dibenzofuran formation during waste combustion. Environ. Sci. Technol. 28: 107–118.

Hagenmaier, H. and A. Berchtold. 1986. Analysis of waste from production of sodium pentachlorophenolate for polychlorinated dibenzodioxins (PCDD) and dibenzofurans (PCDF). Chemosphere. 15: 1991–1994.

Hagenmaier, H., M. Kraft, H. Brunner et al. 1987. Catalytic effects of fly ash from waste incineration facilities on the formation and decomposition of PCDDs and PCDFs. Env. Sci. Tech. 21: 1080–1084.

Halden, R.U. and D.F. Dwyer. 1997. Biodegradation of dioxin-related compounds: A review. Bioremediation Journal. 1(1): 11–25.

Hall, L.W., M.C. Ziegenfuss, S.A. Fischer et al. 1993. The influence of contaminant and water quality conditions on larval striped bass in the Potomac River and Upper Chesapeake Bay in 1990: An *in situ* study. Arch. Environ. Contam. Toxicol. 24: 1–10.

Hall, L.W., M.C. Ziegenfuss, R.D. Anderson et al. 1995. Use of estuarine water column test for detecting toxic conditions in ambient areas of the Chesapeake Bay watershed. Environ. Toxicol. Chem. 14(2): 267–278.

Hallberg, G.R. 1989. Pesticide pollution of groundwater in the humid United States. Agriculture, Ecosystems and Environment. 26(3-4): 299–367.

Hallett, D.J. and M.G. Brooksbank. 1986. Trends of TCDD and related compounds in the Great Lakes: The Lake Ontario ecosystem. Chemosphere. 15: 1405–1416.

Halsall, C.J., B. Gevao, M. Howsam et al. 1999. Temperature dependence of PCBs in the UK atmosphere. Atmos. Environ. 33: 541–552.

Hanify, J.A., P. Metcalf, C.L. Nobbs et al. 1981. Aerial spraying of 2,4,5-T and human birth malformations: An epidemiological investigation. Science. 212: 349–351.

Hansen, E. and C.L. Hansen. 2003. Substance flow analysis for dioxin 2002. Environmental project no. 811. Danish Environmental Protection Agency, Danish Ministry of the Environment.

Hansen, J.C. 1998. Pollution and Human Health. pp. 775–837. *In*: AMAP Assessment report: Arctic pollution issues. Oslo, Norway.

Harjanto, S., E. Kasai, T. Terui et al. 2002. Behavior of dioxin during thermal remediation in the zone combustion process. Chemosphere. 47: 687–693.

Harless, R.L. and R.G. Lewis. 1982. Quantitative determination of 2,3,7,8-tetrachlorodibenzo-p-dioxin residues by gas chromatography/mass spectrometry. pp. 25–36. *In*: Hutzinger, O. (ed.). Chlorinated Dioxins and Related Compounds, Oxford, London, Pergamon Press.

Harmut, S. Fuhr and J. Paul E. des Rosiers. 1988. Methods of degra&tioq destruction detoxillcalioq and disposal of dioxins and related compounds. Pilot Study on International Information Exchange and Related Compounds (North Atlantic Treaty Organization Committee on the Challenges of Modem Society, Report No. 174, August 1988). pp. 23, 24–25.

Harner, T., J.L. Wideman, L.M. Jantunen et al. 1999. Residues of organochlorine pesticides in Alabama soils. Environ. Pollut. 106(3): 323–332.

Harris, C.A., M.W. Woolridge and A.W.M. Hay. 2001. Factors affecting the transfer of organochlorine pesticide residues to breast milk. Chemosphere. 43: 243–256.

Harrison, N., P. Gangaiya and R.J. Morrison. 1996. Organochlorines in the coastal marine environment of Vanuatu and Tonga. Marine Pollut. Bull. 35: 575–579.

Hashimoto, S., T. Wakimoto and R. Tatsukawa. 1995. Possible natural formation of polychlorinated dibenzo-*p*dioxins as evidenced by sediment analysis from the Yellow Sea, the East China Sea and the Pacific Ocean. Marine Poll. Bull. 5: 341–346.

Hashimoto, S., K. Watanabe, K. Nose et al. 2004. Remediation of soil contaminated with dioxins by subcritical water extraction. Chemosphere. 54: 89–96.

Hatterner-Frey, H.A. and C.C. Travis. 1989. Pentachlorophenol: environmental partioning and human exposure. Archieves of Environmental Contamination and Toxicology. 18: 482–489.

Hay, A. 1982. Toxicology of dioxins. pp. 41–47. *In*: The Chemical Scythe: Lessons of 2,4,5-T and Dioxin. New York, London: Plenum Press.

Hayes, W.J. and A. Curley 1968. Storage and excretion of dieldrin and related compounds. Arch. Environ. Health. 16: 155–162.

Haynes, F. and I. Marnane. 2000. Inventory of dioxin & furan emissions to air, land and water in Ireland for 2000 and 2010. Final report prepared for the Environmental Protection Agency by URS Dames & Moore, Dublin. Environmental Protection Agency, Co. Wexford, Ireland.

Hayward, D.G., J.M. Charles, C. Voss De Bettancourt et al. 1989. PCDD and PCDF in breast milk as correlated with fish consumption in southern California. Chemosphere. 18: 455–468.

Hayward, D.G., M.X. Petreas and L.R. Goldman. 1991. Assessing the risk from 2,3,7,8-TCDD and TCDF in milk packaged in paper. Chemosphere. 23: 1551–1559.

HAZDAT. 1992. Agency for Toxic Substances and Disease Registry (ATSDR), Atlanta, GA. February 13, 1992.

HAZDAT. 1996. Database. Agency for Toxic Substances and Disease Registry (ATSDR), Atlanta, GA.

HazDat. 1998. Database. Agency for Toxic Substances and Disease Registry (ATSDR), Atlanta, GA.

HazDat. 2000. Agency for Toxic Substances and Disease Registry (ATSDR), Atlanta, GA.

Hedley, A.J., L.L. Hui, K. Kypke et al. 2010. Residues of persistent organic pollutants (POPs) in human milk in Hong Kong. Chemosphere. 79(3): 259–265.

Henderson, L.O. and D.G. Patterson Jr. 1988. Distribution of 2,3,7,8-tetrachlorodibenzo-p-dioxin in human whole blood and its association with, and extractability from lipoproteins. Bull. Environ. Contam. Toxicol. 40: 604–611.

Hilarides, R., K. Gray, J. Guzzetta et al. 1994. Radiolytic degradation of 2,3,7,8-TCDD in artificially contaminated soils. Environ. Sci. Technol. 28: 2249–2258.

Hirota, K. and T. Kojima. 2005. Decomposition behavior of PCDD/F isomers in incinerator gases under electron-beam irradiation. Bull. Chem. Soc. Jpn. 78: 1685–1690.

Hites, R. and R. Harless. 1991. Atmospheric transport and deposition of polychlorinated dibenzo-p-dioxins and dibenzofurans. Office of Research and Development, US Environmental Protection Agency, Research Triangle Park, NC. EPA/600/3-91/002.

Hites, R.A. 1990. Environmental behavior of chlorinated dioxins and furans. Act. Chem. Res. 23: 194–201.

Hodson, P.V., M. McWhirter, K. Ralph et al. 1992. Effects of bleached kraft mill effluent in fish in the St. Maurice River, Quebec. Environ. Toxicol. and Chem. 11: 1635–1651.

Hoff, R.M., W.M.J. Strachan, C.W. Sweet et al. 1996. Atmospheric deposition of toxic chemicals to the great lakes: A review of data through 1994. Atmos. Environ. 30(20): 3505–3527.

Hoffman, R., P.A. Stehr-Green, K.B. Webb et al. 1986. Health effects of long-term exposure to 2,3,7,8-tetrachlorodibenzo-p-dioxin. JAMA. 255(15): 2031–2038.

Hollernan, J.W. and A.S. Hammons 1980. Levels of chemical contaminants in nonoccupationally exposed U.S. residents. Oak Ridge, TN: Oak Ridge National Laboratory. Document No. ORNL/EIS 142.

Holmstedt, B. 1980. Prolegomena to Seveso, Ecclesiastes 1:18. Arch. Toxicol. 44: 211–230.

Holoubek, I., A. Kokan, I. Holoubkova et al. 2000. Persistent, Bioaccumulative and Toxic Chemicals in Central and Eastern European Countries -State-of-the-art Report. TOCOEN Report No. 150a. Brno, Czech Republic.

Hooiveld, M., D.J.J. Heederil, M. Kogevinas et al. 1998. Second follow-up of a Dutch cohort occupationally exposed to phenoxy herbicides, chlorophenols, and contaminants. Am. J. Epidemiol. 147: 891–901.

Hosoya, K., K. Kimata, K. Fukunishi et al. 1995. Photodecomposition of 1,2,3,4 and 2,3,7,8-tetrachlorodibenzo-*p*-dioxin (TCDD) in water-alcohol media on a solid support. Chemosphere. 31: 3687–3698.

Hryhorczuk, D.O., P. Orris, J.R. Kominsky et al. 1986. PCB, PCDF, and PCDD exposure following a transformer fire: Chicago. Chemosphere. 15: 1297–1303.

HSDB. 1995. Hazardous Substance Databank. National Library of Medicine, Bethesda, MD. March 13, 1993.

Hsu, S.-T., C.-I. Ma, S.K.S. Hsu et al. 1985. Discovery and epidemiology of PCB poisoning in Taiwan: A four-year followup. Environ. Health Perspect. 59: 5–10.

http://www.mst.dk/

http://www.niehs.nih.gov/news/newsletter/2012/11/science-dioxin/index.htm

Hung, L.-S. and L.L. Jr. Ingram. 1990. Effects of solvents on the photodegradation rates of octachlorodibenzo-*p*-dioxin. Bull. Environ. Contam. Toxicol. 44: 380–386.

Hunt, G., B. Maisel and M. Hoyt. 1990. Ambient concentrations of PCDDs/PCDFs (polychlorinated dibenzodioxins/dibenzofurans) in the South Coast air basin. NTIS PB90-169970.

Hunt, G.T. and B.E. Maisel. 1992. Atmospheric concentrations of PCDDs/PCDFs in southern California. J. Air Wast. Manage. Assoc. 42(5): 672–680.

Hutzinger, O., S. Safe, B.R. Wentzell et al. 1973. Photochemical degradation of di-and octachlorodibenzofuran. Environ. Health Perspect. 5: 267–271.

Hutzinger, O., M.J. Blumich, M. Berg et al. 1985. Sources and fate of PCDDs and PCDFs: An overview. Chemosphere. 14: 581–600.

Hutzinger, O. and H. Fiedler. 1991. Formation of dioxins and related compounds from combustion and incineration processes. pp. 263–434. *In*: Bretthauer, E.W., H.W. Kraus and A. di Domenico (eds.). Dioxin Perspectives—A Pilot Study on International Information Exchange on Dioxins and Related Compounds. Chapter 3, NATO—Challenges of Modern Society, Volume 16, Plenum Press, New York.

Hutzinger, O. and H. Fiedler. 1993. From source to exposure: Some open questions. Chemosphere. 27: 121–129.

Iannuzzi, T.J., S.L. Huntley and N.L. Bonnevie et al. 1995. Distribution and possible sources of polychlorinated biphenyls in dated sediments from the Newark Bay Estuary, New Jersey. Arch. Environ. Contam. Toxicol. 28: 108–117.

IARC. 1977. IARC monographs on the evaluation of the carcinogenic risk of chemicals to man: Some fumigants, the herbicides 2,4-D and 2,4,5-T, chlorinated dibenzodioxins and miscellaneous industrial chemicals. Lyon, France: World Health Organization, International Agency for Research on Cancer. IARC Monogr. 15: 41–102.

IARC. 1997. Polychlorinated dibenzo-*para*-dioxins and polychlorinated dibenzofurans. IARC monographs on the evaluation of carcinogenic risks to humans.Volume 69. WHO, IARC, Lyon, France.

IARC, International Agency for Research on Cancer. 2000. Dioxins/Furans Toxicology Summary, agency for toxic substances and diseases Registry. Oroville, Butte country, California.

Ibrahim, M.S. 2007. Persistent organic pollutants in Malaysia. pp. 629–655. *In*: An Li STGJJPG and K.S.L. Paul (eds.). Developments in Environmental Science. Vol. 7. Elsevier, The Netherlands.

Iida, T., H. Hirakawa, T. Matsueda et al. 1992. Levels of polychlorinated-biphenyls and polychlorinated dibenzofurans in the blood, subcutaneous adipose-tissue and stool of Yusho patients and normal subjects. Toxicological and Environmental Chemistry. 35: 17–24.

Ishii, K., T. Furuichi and Y. Matsuda. 2003. Degradation of dioxins using enzymes and sterilization of *Pseudallescheria boydii.* Organohalogen Compounds. 63: 260–263.

Isosaari, P., T. Tuhkanen and T. Vartiainen. 1997. Dioxin degradation by Fenton's reagent. pp. 53–56. *In*: Vartiainen, T. and H. Komulainen (eds.). 7th Nordic Symposium on Organic Pollutants. Kuopio University Publications C. Natural and Environmental Sciences 68, Kuopio, Finland.

Isosaari, P., O. Laine, T. Tuhkanen et al. 2005. Photolysis of polychlorinated dibenzo-p-dioxins and dibenzofurans dissolved in vegetable oils: Influence of oil quality. Sci. Tot. Environ. 340(1-3): 1–11.

Iwata, H., S. Tanabe, N. Sakai et al. 1993. Distribution of persistent organochlorines in the oceanic air and surface seawater and the role of ocean on their global transport and fate. Environ. Sci. Technol. 27: 1080–1098.
Iwata, H., S. Tanabe, M. Aramoto et al. 1994a. Persistent organochlorine residues in sediments from the Chukchi Sea, Bering Sea and Gulf of Alaska. Mar. Pollut. Bull. 28: 746–753.
Jackson, D.R. and D.L. Bisson. 1990. Mobility of polychlorinated aromatic compounds in soils contaminated with wood-preserving oil. J. Air Waste Manage. Assoc. 40: 1129–1133.
Jacobs, L.W., G.A. O'Connor, M.A. Overcash et al. 1987. Effects of trace organics in sewage sludges on soil-plant systems and assessing their risk to humans. pp. 103–143. *In*: Land application of sludge. Food chain implications. Lewis Publishers, Inc.
Jacobs, M.N., D. Santillo, P.A. Johnston et al. 1998. Organochlorine residues in fish oil dietary supplements: Comparison with industrial grade oils. Chemosphere. 37(9-12): 1709–1721.
James, W.H. 1997. The sex ratio of offspring sired by men exposed to wood preservatives contaminated by dioxin. Scand. J. Work Environ. Health. 23: 69.
Jasinski, J.S. 1989. Multiresidue procedures for the determination of chlorinated dibenzodioxins and dibenzofurans in a variety of foods using capillary gas chromatography-electron-capture detection. J. Chromatogr. 478: 349–367.
Jaward, F.M., J.L. Barber, K. Booij et al. 2004a. Evidence for dynamic air-water coupling and cycling of persistent organic pollutants over the open Atlantic Ocean. Environ. Sci. Technol. 38: 2617–2625.
JECFA. 2001. Joint FAO/WHO Expert Committee on Food Additives, 57th Meeting, Rome, 5–14 June 2001 http://www.fao.org/es/esn/jecfa/jecfa57c.pdf.
Jensen, A.A. 1987. Polychlorobiphenyls (PCBs), polychlorodibenzo-p-dioxins (PCDDs) and polychlorodibenzofurans (PCDFs) in human milk, blood and adipose tissue. Science of the Total Environment. 64: 259–293.
Jian-Hua, Y., C. Tong, Li. Xiao-Dong et al. 2005. Ultraviolet photolysis of dioxins: Octachlorinated dibenzo-p-dioxin and Octachlorodibenzofuran. Journal of Zhejiang University (Engineering Science). 39(7): 1064–1067.
Jirasek, L., J. Kalensky, K. Kubec et al. 1976. Chloracne, porphyria cutanea tarda and other intoxication by herbicides. Hautarzt. 27: 328–333. (German)
Jobb, B., M. Uza, R. Hunsinger et al. 1990. A survey of drinking water supplies in the province of Ontario for dioxins and furans. Chemosphere. 20(10-12): 1553–1558.
Jödicke, B., M. Ende, H. Helge et al. 1992. Fecal excretion of PCDDs/PCDFs in a 3-month-old breast-fed infant. Chemosphere. 25: 1061–1065.
Johnson, J. 1995. Dioxin risk: are we sure yet? Environ. Sci. Technol. 29: 24–29A.
Jones, K. 1995. Diesel engine emissions and the link to human dioxin exposure. Dioxin 95 Secretariat, Edmonton AB Canada. Organohalogen Compounds. 24: 69–74.
Jones, E.L. and H. Krizek. 1962. A technique for testing acnegenic potency in rabbits, applied to the potent acnegen, 2,3,7,8-tetrachlorodibenzo-p-dioxin. J. Invest. Dermatol. 39: 511–517.
Kang, H.K., K.K. Watanabe, J. Breen et al. 1991. Dioxins and dibenzofurans in adipose tissue of US Vietnam veterans and controls. Am. J. Public Health. 81: 344–349.
Kannan, K., S. Tanabe, J.P. Giesy et al. 1997. Organochlorine pesticides and polychlorinated biphenyls in foodstuffs from Asian and oceanic countries. Rev. Environ. Contam. Toxicol. 152: 1–55.
Kannan, K., S. Battula, B.G. Loganathan et al. 2003. Trace organic contaminants, including toxaphene and trifluralin, in cotton field soils from Georgia and South Carolina, USA. Arch. Environ. Contam. Toxicol. 45(1): 30–36.
Kanters, J. and R. Louw. 1996. Thermal and catalysed halogenation in combustion reactions. Chemosphere. 32: 89–97.
Kao, C.M. and M.J. Wu. 2000. Enhanced TCDD degradation by Fenton's reagent preoxidation. J. Hazard. Mater. 74: 197–211.
Kapila, S., A.F. Yanders, C.E. Orazio et al. 1989. Field and laboratory studies on the movement and fate of tetrachlorodibenzo-*p*-dioxin in soil. Chemosphere. 18: 1297–1304.
Karademir, A., M. Bakoglu and S. Ayberk. 2003. PCDD/F removal efficiencies of electrostatic precipitator and wet scrubbers in izaydas hazardous waste incinerator. Fresenius. Environ. Bull. 12: 1228–1232.

Karademir, A., M. Bakoglu, F. Taspinar et al. 2005. Removal of PCDD/Fs from flue gas by a fixed-bed activated carbon filter in a hazardous waste incinerator. Environ. Sci. Technol. 38(4): 1201–1207.

Karasek, F.W. and F.I. Anuska. 1982. Trace analysis of the dioxins. Anal. Chem. 54: 309A–324A.

Karasek, F.W. and L.C. Dickson. 1987. Model studies of polychlorinated dibenzo-p-dioxin formation during municipal refuse incineration. Science. 237: 754–756.

Karlsson, H., D.C.G. Muir, C.F. Teixiera et al. 2000. Persistent chlorinated pesticides in air, water and precipitation from Lake Malawi area, Southern Africa. Environ. Sci. Technol. 34: 4490–4495.

Kasai, E., S. Harjanto, T. Terui et al. 2000. Thermal remediation of PCDD/Fs contaminated soil by zone combustion process. Chemosphere. 41: 857–864.

Kashimoto, T. and H. Miyata. 1986. Differences between Yusho and other kinds of poisoning involving only PCBs. pp. 2–26. *In*: Waid, J.S. (ed.). PCBs and the Environment, Vol. 3. Boca Raton, FL. CRC Press.

Katsuki, S. 1969. Forward, Fukuoka Acta Medica. 60: 403–407.

Kelly, T.J., J.M. Czuczwa, P.R. Sticksel et al. 1991. Atmospheric and tributary inputs of toxic substances to Lake Erie. J. Gt. Lakes Res. 17(4): 504–516.

Kende, A.S., J.J. Wade, D. Ridge et al. 1974. Synthesis and fourier transform carbon-13 nuclear magnetic resonance spectroscopy of new toxic polyhalodibenzo-p-dioxins. J. Org. Chem. 39: 931–937.

Kerger, B., G. Corbett, S. El-Sururi et al. 1995. Validating dermal exposure assessment techniques for dioxin using body burden data and pharmacokinetic modeling. Organohalogen Compounds. 25: 137–141.

Kieatiwong, S., L.W. Nguyen, V.R. Hebert et al. 1990. Photolysis of chlorinated dioxins in organic solvents and on soils. Environ. Sci. Technol. 24: 1575–1580.

Kim, M. and P.W. O'Keefe. 2000. Photodegradation of polychlorinated dibenzo-*p*-dioxins and dibenzofurans in aqueous solutions and in organic solvents. Chemosphere. 41: 793–800.

Kim, S.H., S.Y. Kwak and T. Suzuki. 2006. Photocatalytic degradation of flexible PVC/TiO_2 nanohybrid as an eco-friendly alternative to the current waste landfill and dioxin-emitting incineration of post-use PVC. Polymer. 47: 3005–3016.

Kim, H.H., I. Yamamoto, K. Takashima et al. 2000. Incinerator flue gas cleaning using wet-type electrostatic precipitator. J. Chem. Eng. Jpn. 33: 669–674.

Kimbrough, R.D., C.D. Carter, J.A. Liddle et al. 1977. Epidemiology and pathology of a tetrachlorodibenzodioxin poisoning episode. Arch. Environ. Health. 32(2): 77–85.

Kluyev, N., A. Cheleptchikov, E. Brodsky et al. 2002. Reductive dechlorination of polychlorinated dibenzo-p-dioxins by zerovalent iron in subcritical water. Chemosphere. 46(9-10): 1293–1296.

Knutzen, J. and M. Oehme. 1989. Polychlorinated dibenzofurans (PCDF) and dibenzo-*p*-dioxin (PCDD) levels in organisms and sediments from the Frierfjord, Southern Norway. Chemosphere. 19: 1897–1909.

Koblizkova, M., S. Genualdi, S.C. Lee et al. 2012. Application of sorbent impregnated polyurethane foam (SIP) disk passive air samplers for investigating organochlorine pesticides and polybrominated diphenyl ethers at the global scale. Environ. Sci. Technol. 46(1): 391–396.

Koester, C.J. and R.A. Hites. 1992. Photodegradation of polychlorinated dioxins and dibenzofurans adsorbed to fly ash. Environ. Sci. Technol. 26: 502–506.

Kogevinas, M., R. Saracci, R. Winkelmann et al. 1993. Cancer incidence and mortality in women occupationally exposed to chlorophenoxy herbicides, chlorophenols, and dioxins. Cancer Causes and Control. 4: 547–553.

Kogevinas, M., H. Becher, T. Benn et al. 1997. Cancer mortality in workers exposed to phenoxy herbicides, chlorophenols, and dioxins. Am. J. Epidemiol. 145: 1061–1075.

Kominsky, J.R. and C.D. Kwoka. 1989. Background concentrations of polychlorinated dibenzofurans (PCDFs) and polychlorinated dibenzo-p-dioxins (PCDDS) in office buildings in Boston, Massachusetts. Chemosphere. 18599–18608.

Konstantinov, A. and N.J. Bunce. 1996. Photodechlorination of octachlorodibenzo-*p*-dioxin and octachlorodibenzofuran in alkane solvents in the absence and presence of triethylamine. J. Photochem. Photobiol., A Chem. 94: 27–35.

Konstantinov, A.D., A.M. Johnsto, B.J. Cox et al. 2000. Photolytic method for destruction of dioxins in liquid laboratory waste and identification of the photoproducts from 2,3,7,8-TCDD. Environ. Sci. Technol. 34(1): 143–148.

Koopman-Esseboom, C., D.C. Morse, N. Weisglas-Kuperus et al. 1994a. Effects of dioxins and polychlorinated biphenyls on thyroid hormone status of pregnant women and their infants. Pediatr. Res. 36(4): 468–473.

Koopman-Esseboom, C., M. Huisman, N. Weisglas-Kuperus et al. 1994b. PCB and dioxin levels in plasma and human milk of 418 Dutch women and their infants. Predictive value of PCB congener levels in maternal plasma for fetal and infant's exposure to PCBs and dioxins. Chemosphere. 28(9): 1721–1732.

Krishnamurthy, S. and H.C. Brown. 1980. Selective reductions. 27. Reaction of alkyl halides with representative complex metal hydrides and metal hydrides. Comparison of various hydride reducing agents. J. Org. Chem. 45: 849–856.

Kulkarni, P.S. and J. Crespo. 2008. Dioxins sources and current remediation technologies—A review. Environ. Inter. 34: 139–153.

Kuratsune, M. 1989. Yusho, with reference to Yu-Cheng. pp. 381–400. *In*: Kimbrough, R.D. and A.A. Jensen (eds.). Halogenated Biphenyls, Terphenyls, Naphthalene, Dibenzodioxins and Related Products, 2nd ed. Amsterdam: Elsevier Science Publishers.

Kuratsune, M., H. Yoshimura, Y. Hori et al. (eds.). 1996. YUSHO—A Human Disaster caused by PCBs and Related Compounds, Kyusyu University Press, Fukuoka.

Kusuda, M. 1971. A study on the sexual functions of women suffering from rice-bran oil poisoning. Sanka to Fujinka. 38: 1063–1072. (Japanese)

Kutz, F., S. Strassman and A. Yobs. 1979. Survey of pesticide residues and their metabolites in the general population of the United States. Commission of the European Communities EUR, ISS EUR 5824, Use Biol Specimens Assess Hum Exposure. Environ. Pollut. 267–274.

Kutz, F.W., A.R. Yobs and H.S.C. Yang. 1976. National pesticide monitoring programs. pp. 95–136. *In*: Lee, Re. (ed.). Air Pollution from Pesticides and Agricultural Processes, Cleveland, OH: CRC Press.

LaFleur, L., T. Bousquet, K. Ramage et al. 1990. Analysis of TCD and TCDF on the ppq-level in milk and food sources. Chemosphere. 20(10-12): 1657–1662.

Lakshmanan, M.R., B.S. Campbell, S.J. Chirtel et al. 1986. Studies on the mechanism of absorption and distribution of 2,3,7,8-tetrachlorodibenzo-p-dioxin in the rat. J. Pharmacol. Exp. Ther. 239: 673–677.

Le Bel, G.L., D.T. Williams, F.M. Benoit et al. 1990. Polychlorinated dibenzodioxins and dibenzofurans in human adipose tissue samples from five Ontario municipalities. Chemosphere. 21: 1465–1475.

Levy, C.J. 1988. Agent Orange exposure and post-traumatic stress disorder. J. Nerv. Ment. Dis. 176: 242–245.

Liberti, A., D. Brocco, I. Allegrini et al. 1978. Solar and UV photodecomposition of 2,3,7,8-tetrachloro dibenzo-p-dioxin in the environment. Sci. Tot. Environ. 10(2): 97–104.

Liljelind, P., J. Unsworth, O. Maaskant and S. Marklund. 2001. Removal of dioxins and related aromatic hydrocarbons from flue gas streams by adsorption and catalytic destruction. Chemosphere. 42: 615–623.

Lofroth, G. and Y. Zebuhr. 1992. Polychlorinated dibenzo-p-dioxins (PCDDs) and dibenzofurans (PCDFs) in mainstream and sidestream cigarette smoke. Bull. Environ. Contam. Toxicol. 48: 789–794.

Lohmann, R. and K.C. Jones. 1998. Dioxins and furans in air and deposition: a review of levels, behaviour and processes. Sci. Total Environ. 219: 53–81.

Lohmann, K. and C. Seigneur. 2001. Atmospheric fate and transport of dioxins: local impacts. Chemosphere. 45(2): 161–171.

Lorber, M., P. Saunders, J. Ferrario et al. 1997. A statistical survey of dioxin-like compounds in United States pork fat. Organohalogen Compounds. 32: 80–86.

Lorber, M.N., D.L. Winters, J. Griggs et al. 1998. A national survey of dioxin-like compounds in the United States milk supply. Organohalogen Compounds. 38: 125–129.

Lu, Y.C. and Y.C. Wu. 1985. Clinical findings and immunological abnormalities in Yu-Cheng patients. Environ. Health Perspect. 59: 17–29.
Lundin, L. and S. Marklund. 2007. Thermal degradation of PCDD/F, PCB and HCB in municipal solid waste ash. Chemosphere. 67: 474–481.
Lyman, W.J., W.F. Reehl and D.H. Rosenblatt. 1982. Handbook of chemical property estimation methods. New York, NY: McGraw-Hill Book Company. 4-9: 15–16.
Ma, W.P. and P.W. Brown 1997. Hydrothermal reactions of fly ash with Ca(OH)2 and CaSO4.2H2O. Cem. Concr. Res. 27: 1237–1248.
MacDonald, R.W., M.G. Ikonomou and D.W. Paton. 1998. Historical inputs of PCDDs, PCDFs, and PCBs to a British Columbia interior lake: The effect of environmental controls on pulp mill emissions. Environ. Sci. Technol. 32: 331–337.
Mackay, D., W. Shiu and K.N. Ma. 1992. Illustrated handbook of physical-chemical properties and environmental fate for organic chemicals, Vol. 2. Aromatic Hydrocarbons and Polychlorinated Dibenzofurans and Dibenzodioxins, Lewis Publisher, Chelsea, Mi., USA.
Mackay, D. and F. Wania. 1995. Transport of contaminants to the Arctic: partitioning, processes and models. Sci. Tot. Environ. 160/161: 25–38.
MacPherson, K.A., T.M. Kolic, E.J. Reiner et al. 2003. Practical applications of dual column chromatography of the analysis of dioxin-like and persistent organic pollutants. Organohalogen Compounds. 60: 367–370.
Maier-Schwinning, G. and H. Herden. 1996. Minderungstechniken zur Abgasreinigung von PCDD/PCDF. VDI-Berichte Nr. 1298: 191–229.
Mamontov, A.A., E.A. Mamontova, E.N. Tarasova et al. 2000. Tracing the sources of PCDD/Fs and PCBs to Lake Baikal. Environ. Sci. Technol. 34: 741–747.
Manchester-Neesvig, J.B., A.W. Andren and D.N. Edgington. 1996. Patterns of mass sedimentation and of deposition of sediment contaminated by PCBs in Green Bay. J. Great Lakes Res. 22(2): 444–462.
Manikkam, M., R. Tracey, C. Guerrero-Bosagna et al. 2012. Dioxin (TCDD) induces epigenetic transgenerational inheritance of adult onset disease and sperm epimutations. PLoS One. 7(9): 46249.
Manz, A., J. Berger, J.H. Dwyer et al. 1991. Cancer mortality among workers in chemical plant contaminated with dioxin. Lancet. 338: 959–964.
Marinovich, M., C.R. Sirtori, C.L. Galli et al. 1983. The binding of 2,3,7,8-tetrachlorodibenzo-p-dioxin to plasma lipoproteins may delay toxicity in experimental hyperlipidemia. Chem. Biol. Interact. 45: 393–399.
Marklund, S., C. Rappe, M. Tyslind et al. 1987. Identification of polychlorinated dibenzofurans and dioxins in exhaust from cars run on leaded gasoline. Chemosphere. 16: 29–36.
Marklund, S., R. Andersson, M. Tysklind et al. 1990. Emissions of PCDDs and PCDFs in gasoline and diesel fueled cars. Chemosphere. 20: 553–561.
Marple, L., R. Brunck and L. Throop. 1986a. Water solubility of 2,3,7,8-tetrachlorodibenzo-pdioxin. Environ. Sci. Technol. 20(2): 180–182.
Marple, L., B. Berridge and L. Throop. 1986b. Measurement of the water-octanol partition coefficient of 2,3,7,8-tetra-chlorodibenzo-p-dioxin. Environ. Sci. Technol. 20: 397–399.
Marsalek, J. and H. Schroeter. 1988. Annual loadings of toxic contaminants in urban runoff from the Canadian Great Lakes basin. Water Pollution Research Journal of Canada. 23(3): 360–378.
Marsh, J.M. 1993. Assessment of nonpoint source pollution in stormwater runoff in Louisville, (Jefferson County) Kentucky, USA. Arch. Environ. Contam. Toxicol. 25: 446–455.
Martin, D.B. and W.A. Hartman. 1985. Organochlorine pesticides and polychlorinated biphenyls in sediment and fish from wetlands in the north central United States. J. Assoc. Off. Anal. Chem. 68: 712–717.
Marvin, C.H., S. Painter, M.N. Charlton et al. 2004a. Trends in spatial and temporal levels of persistent organic pollutants in Lake Erie sediments. Chemosphere. 54: 33–40.
Marvin, C.H., S. Painter, D. Williams et al. 2004b. Spatial and temporal trends in surface water and sediment contamination in the Laurentian Great Lakes. Environ. Pollut. 129: 131–144.

Mason, G., K. Farrell, B. Keys et al. 1986. Polychlorinated dibenzo-p-dioxins: Quantitative *in vitro* and *in vivo* structure-activity relationships. Toxicology. 41: 21–31.

Massé, R. and B. Pelletier. 1987. Photochemistry of dibenzo-p-dioxin in organic solvents at 253.7 nm: GC/MS characterization of the photo-transformation products. Chemosphere. 16: 7–17.

Masuda, Y., H. Kuroki and K. Haraguchi. 1985. PCB and PCDF congeners in the blood and tissues of Yusho and Yu-cheng patients. Environ. Health Perspect. 59: 53–58.

Masunaga, S. and J. Nakanishi. 1999. Dioxin impurities in old Japanese agrochemical formulations. Organohalogen Compounds. 41: 41–44.

Matsumura, F., J. Quensen and G. Tsushimoto. 1983. Microbial degradation of TCDD in a model ecosystem. pp. 191–220. *In*: Tucker, R.E., A.L. Young and A.P. Gray (eds.). Human and Environmental Risks of Chlorinated Dioxins and Related Compounds, New York, London, Plenum Press.

May, G. 1973. Chloracne from the accidental production of tetrachlorodibenzodioxin. Br. J. Med. 30: 276–283.

McKee, P., A. Burt, D. McCurvin et al. 1990. Levels of dioxins, furans and other organic contaminants in harbour sediments near a wood preserving plant using pentachlorophenol and creosote. Chemosphere. 20: 1679–1685.

Mckinney, J.D. 1982. Analysis of 2,3,7,8-tetrachlorodibenzo-para-dioxin in environmental samples. Ecol. Bull. (Stockholm). 27: 53–66.

McKinney, J.D., K. Chae, S.J. Oatley et al. 1985a. Molecular interactions of toxic chlorinated dibenzo-p-dioxins and dibenzofurans with thyroxine binding prealbumin. J. Med. Chem. 28: 375–381.

Mckinney, J., P. Albro, M. Luster et al. 1982. Development and reliability of a radioimmunoassay for 2,3,7,8-tetrachlorodibenzo-p-dioxin. pp. 67–77. *In*: Hutzinger, O. (ed.). Chlorinated Dioxins and Related Compounds: Impact on the Environment, Oxford, New York, Pergammon Press.

McLaughlin, D.L., R.G. Pearson and R.E. Clement. 1989. Concentrations of chlorinated dibenzo-p-dioxins (CDD) and dibenzofurans (CDF) in soil from the vicinity of a large refuse incinerator in Hamilton Ontario Canada. Chemosphere. 18: 851–854.

McPeters, A.L. and M.R. Overcash. 1993. Demonstrations of photodegradation by sunlight of 2,3,7,8-tetrachlorodibenzo-p-dioxins in 6 cm soil columns. Chemosphere. 27: 1221–34.

Meijer, S.N., W.A. Ockenden, E. Steinnes et al. 2003a. Spatial and temporal trends of POPs in Norwegian and UK background air: Implications for global cycling. Environ. Sci. Technol. 37(3): 454–461.

Menn, J.J., G.G. Patchet and G.H. Batchelder. 1960. The persistence of trithion, an organophosphorus insecticide in soil. J. Econ. Entomol. 53: 1080.

Meyer, C., D. O'Keefe, D. Hilker et al. 1989. A survey of twenty community water systems in New York State for PCDDs and PCDFs. Chemosphere. 19(1-6): 21–26.

Michalek, J.E., A.J. Rahe and C.A. Boyle. 1998. Paternal dioxin, preterm birth, intrauterine growth retardation, and infant death. Epidemiology. 9: 161–167.

Mill, T. 1985. Dioxins in the environment. pp. 173. *In*: Kamrin, M.A. and P.W. Rodgers (eds.). Hemisphere: Bristol, PA.

Miller, G.C., V.R. Herbert and R.G. Zepp. 1987. Chemistry and photochemistry of low-volatility organic chemicals on environmental surfaces. Environ. Sci. Technol. 21: 1164–1167.

Miller, G.C., V.R. Herbert, M.J. Miille et al. 1989. Photolysis of octachlorodibenzo-p-dioxin on soils: Production of 2,3,7,8-TCDD. Chemosphere. 18: 1265–1274.

Milligan, M.S. and E.R. Altwicker. 1996. Chlorophenol reactions on fly ash. 1. Adsorption/desorption equilibria and conversion to polychlorinated dibenzo-p-dioxins. Environ. Sci. Technol. 30: 225–229.

Milligan, M.S. and E.R. Altwicker. 1996. Chlorophenol reactions on fly ash. 2. Equilibrium surface coverage and global kinetics. Environ. Sci. Technol. 30: 230–236.

Ministry of the Environment, Japan. Results of the Survey on the Body Burden of Dioxins in Humans, Tokyo, 2002.

Mino, Y. and Y. Moriyama. 2001. Possible remediation of dioxin-polluted soil by steam distillation. Chem. Pharm. Bull. 49: 1050–1051.

Mio, H., S. Saeki, J. Kano et al. 2002. Estimation of mechanochemical dechlorination rate of poly(vinyl chloride). Environ. Sci. Technol. 36: 1344–1348.

Mitoma, Y., T. Uda, N. Egashira et al. 2004. Approach to highly efficient dechlorination of PCDDs, PCDFs, and coplanar PCBs using metallic calcium in ethanol under atmospheric pressure at room temperature. Environ. Sci. Technol. 38: 1216–1220.

Miyata, H., K. Takayama, J. Ouaki et al. 1989. Levels of PCDDs, coplanar PCBs and PCDFs in patients with Yusho disease and in the Yusho oil. Chemosphere. 18: 407–41.

Mocarelli, P., A. Marocchi, P. Brambilla et al. 1991. Effects of dioxin exposure in humans at Seveso, Italy. *In*: Banbury Report 35: Biological Basis for Risk Assessment of Dioxin and Related Compounds. Cold Spring Harbor, New York: Cold Spring Harbor Laboratory Press.

Mocarelli, P., L.L. Needham, A. Marocchi et al. 1991. Serum concentrations of 2,3,7,8-tetrachlorodibenzop-dioxin and test results from selected residents of Seveso, Italy. J. Toxicol. Environ. Health. 32: 357–366.

Mocarelli, P., P. Brambilla, P.M. Gerthoux et al. 1996. Change in sex ratio with exposure to dioxin. Lancet. 348: 409.

Monagheddu, M., G. Mulas, S. Doppiu et al. 1999. Reduction of polychlorinated dibenzodioxins and dibenzofurans in contaminated muds by mechanically induced combustion reactions. Environ. Sc. Technol. 33: 2485–2488.

Monirith, I., D. Ueno, S. Takahashi et al. 2003. Asia-Pacific mussel watch: monitoring contamination of persistent organochlorine compounds in coastal waters of Asian countries. Mar. Pollut. Bull. 46: 281–300.

Mori, T. and R. Kondo. 2002. Oxidation of chlorinated dibenzo-p-dioxin and dibenzofuran by white-rot fungus, Phlebia lindtneri. FEMS Microbiol. Lett. 216: 223–227.

Mori, K., H. Matsui, N. Yamaguchi et al. 2005. Multi-component behavior of fixed-bed adsorption of dioxins by activated carbon fiber. Chemosphere. 61(7): 941–6.

Morrison, R.J., N. Harrison and P. Gangaiya. 1996. Organochlorine contaminants in the estuarine and coastal marine environment of the Fiji Islands. Environ. Pollut. 93: 159–167.

Morselli, L., D. Brocco and A. Pirni. 1985. The presence of polychlorodibenzo-p-dioxins (PCDDs), polychlorodibenzofurans (PCDFs), and polychlorobiphenyls (PCBs) in fly ashes from various municipal incinerators under different technological and working conditions. Ann. Chim. 75: 59–64.

Moses, M. and P.G. Prioleau. 1985. Cutaneous histologic findings in chemical workers with and without chloracne with past exposure to 2,3,7,8-tetrachlorodibenzo-p-dioxin. J. Am. Acad. Dermatol. 12: 497–506.

Mucka, V., R. Silber, M. Pospisil et al. 2000. Radiolytic dechlorination of PCBs in presence of active carbon, solid oxides, bentonite and zeolite. Radiat. Phys. Chem. 59: 399–404.

Muir, D.C.G., S. Lawrence, M. Holoka et al. 1992. Partitioning of polychlorinated dioxins and furans between water, sediments and biota in lake mesocosms. Chemosphere. 25(1-2): 119–124.

Muir, D.C.G., D. Tretiak, K. Koczanski et al. 1995b. Spatial and temporal trends of organochlorines in arctic marine mammals. Department of Indian Affairs and Northern Development. Environ. Study. 73.

Muir, D.C.G., E.W. Born, K. Koczansky et al. 2000. Temporal and spatial trends of persistent organochlorines in Greenland walrus (*Odobenus rosmarus rosmarus*). Sci. Total Environ. 245: 73–86.

Mukerjee, D. 1998. Health impact of polychlorinated dibenzo-p-dioxins: A critical review. J. Air & Waste Manage. Assoc. 48: 157–165.

Müller, J.F., C. Gaus, J.A. Prange et al. 2002. Polychlorinated dibenzo-*p*-dioxins and polychlorinated dibenzofurans in sediments from Hong Kong. Mar. Pollut. Bull. 45: 372–378.

Mundy, K.J., R.S. Brown, K. Pettit et al. 1989. Environmental assessment at and around a chemical waste treatment facility I. Measurements on PCDFs and PCDDs. Chemosphere. 19: 381–386.

Murayama, H., Y. Takase, H. Mitobe et al. 2003. Seasonal changes of persistent organic pollutant concentrations in air at Niigata area, Japan. Chemosphere. 52: 683–694.

Murphy, R. and C. Harvey. 1985. Residues and metabolites of selected persistent halogenated hydrocarbons in blood specimens from a general population survey. Environ. Health Perspect. 60: 115–120.

Muto, H. and Y. Takizawa. 1989. Dioxins in cigarette smoke. Arch. Environ. Health. 44: 171–174.

Nagayama, J., Y. Masuda and M. Kuratsune. 1977. Determination of polychlorinated dibenzofurans in tissues of patients with Yusho. Food Cosmet. Toxicol. 15: 195–198.

Nakanishi, Y., N. Shigematsu and Y. Kurita. 1985. Respiratory involvement and immune status in Yusho patients. Environ. Health Perspect. 59: 31–36.

Nakano, T., M. Tsuji and T. Okuno. 1990. Distribution of PCDDs and PCBs in the atmosphere. Atmos. Environ. 24A: 1361–1368.

Napola, A., M.D.R. Pizzigallo, P. Di Leo, M. Spagnuolo et al. 2006. Mechanochemical approach to remove phenanthrene from a contaminated soil. Chemosphere 65: 1583–1590.

Nau, H. 2006. Impacts and impact mechanisms of "dioxins" in humans and animals. Dtsch Tierarztl Wochensch. 113(8): 292–297.

Needham, L.L., D.G. Patterson, C.C. Alley et al. 1987. Polychlorinated dibenzo-p-dioxins and dibenzofurans levels in persons with high and normal levels of 2,3,7,8-tetrachlorodibenzo-p-dioxins. Chemosphere. 16: 2027–2031.

Needham, L.L., P.M. Gerthoux, D.G. Patterson et al. 1994. Half-life of 2,3,7,8-tetra-chlorodibenzo-p-dioxin in serum of Seveso adults: Interim report. Organohalogen Compounds. 21: 81–85.

Neeta, P.T., N. Vaishali, D. Swapnesh et al. 2007. Dioxin formation in pulp and paper mills of India. Environ. Sci. and Poll Res. 14(4): 225–226.

Nestrick, T.J., L.L. Lamparski and D.I. Townsend. 1980. Identification of tetrachlorodibenzo-*p*-dioxin isomers at the 1-ng level by photolytic degradation and pattern recognition techniques. Anal. Chem. 52: 1865–1874.

Nestrick, T.J., L.L. Lamparski, N.N. Frawley et al. 1986. Perspectives of a large scale environmental survey for chlorinated dioxins: Overview and soil data. Chemosphere. 15: 1453–1460.

Nguyen Minh Tue, S. Takahashi, A. Subramanian et al. 2013. Environmental contamination and human exposure to dioxin-related compounds in e-waste recycling sites of developing countries. Environ. Sci. Process Impacts. 15(7): 1326–31.

Nguyen, H.M., B.M. Tu, M. Watanabe et al. 2003. Open dumping site in Asian developing countries: A potential source of polychlorinated dibenz-p-dioxins and polychlorinated dibenzofurans. Environ. Sci. Technol. 37(8): 1493–502.

Nicholsan, K.W., C.L. Rose, D.S. Lee et al. 1993. Behaviour of Polychlorinated Dibenzo-p-dioxins (PCDDs) and Dibenzofurans (PCDFs) in the Terrestrial Environment. A Review Report No. AEA-EE-0519, AEA Environment and Energy.

Nickelsen, M.G., W.J. Cooper C.N. Kurucz et al.1992. Removal of benzene and selected alkyl-substituted benzenes from aqueous solution utilizing continuous high-energy electron irradiation. Environ. Sci. Technol. 26: 144–152.

Niemann, R.A., W.C. Brumley, D. Firestone et al. 1983. Analysis of fish for 2,3,7,8-tetrachlorodibenzo-p-dioxin by electron capture capillary gas chromatography. Anal. Chem. 55: 1497–1504.

Nifuku, M., M. Horvath, J. Bodnar et al. 1997. A study on the decomposition of volatile organic compounds by pulse corona. J. Electrost. 40: 687–692.

Nitnaware, V.C., D. Swapnesh, T. Neeta et al. 2006. Dioxins and furans: Unintentional byproducts of chlorine base industries, Water 2006 organized by NEERI, Nagpur on R&D Frontiers in Water and Wastewater Management, January 20–21, 2006. Nagpur, India.

Niu, J., J. Chen, B. Henkelmann et al. 2003. Photodegradation of PCDD/Fs adsorbed on spruce (*Picea abies* (L.) Karst.) needles under sunlight irradiation. Chemosphere. 50: 1217–1225.

Norwood, C.B., M. Hackett, R.J. Pruell et al. 1989. Polychlorinated dibenzo-p-dioxins and dibenzofurans in selected estuarine sediments. Chemosphere. 18: 553–560.

NTP. 1989. Fifth annual report on carcinogens. Summary. U.S. Department of Health and Human Services, Public Health Services, Research Triangle Park, NC, by Technical Resources, Inc., Rockville, MD.

Nygren, M., M. Hansson, M. Sjoestroem et al. 1988. Development and validation of a method for determination of PCDDs and PCDFs in human blood plasma. A multivariate comparison of blood and adipose tissue levels between Vietnam veterans and matched controls. Chemosphere. 17: 1663–1692.

Obata, S. and H. Fujihira. 1998. Dioxin and NOx control using pilot-scale pulsed corona plasma technology. Combust. Sci. Technol. 133: 3–11.

O'Keefe, P., D. Hilker and C. Meyer. 1984. Tetrachlorodibenzo-p-dioxins and tetrachlorodibenzofurans in Atlantic coast striped bass and in selected Hudson River fish, waterfowl and sediments. Chemosphere. 13: 849–860.

Öberg, L. and C. Rappe. 1992. Biochemical formation of PCDD/F from chlorophenols. Chemosphere. 25: 49–52.

Öberg, L.G., B. Glas, S.E. Swanson et al. 1990. Peroxidase-catalyzed oxidation of chlorophenols to polychlorinated dibenzo-*p*-dioxins and dibenzofurans. Arch. Environ. Contam. Toxicol. 19: 930–938.

Öberg, L.G., R. Andersson and C. Rappe. 1992. *De novo* Formation of Hepta- and Octachlorodibenzo-p-dioxins from Pentachlorophenol in Municipal Sewage Sludge. Organohalogen Compd. 9: 351–354.

Oehme, M., S. Mano, A. Mikalsen et al. 1987. Formation and presence of polyhalogenated and polycyclic compounds in the emissions of small and large scale municipal waste incinerators. Chemosphere. 16: 143–153.

Oehme, M., S. Mano and B. Bjerke. 1989. Formation of polychlorinated dibenzofurans and dibenzo-p-dioxins by production processes for magnesium and refined nickel. Chemosphere. 18: 1379–1389.

Ohsako, S., Y. Miyabara, N. Nishimura et al. 2001. Maternal Exposure to a low dose of 2,3,7,8-tetrachlorodibenzo-*p*-dioxin (TCDD) suppressed the development of reproductive organs of male rats: Dose-dependent increase of mRNA levels of 5-alpha-reductase type 2 in contrast to decrease of androgen receptor in the pubertal ventral prostate. Toxicol. Sci. 60: 132–143.

Okey, A.B., D.S. Riddick and P.A. Harper. 1994. The Ah receptor: Mediator of the toxicity of 2,3,7,8-tetrachlorodibenzo-p-dioxin (TCDD) and related compounds. Toxicol. Lett. 70(1): 1–22.

Oku, A., K. Tomari, T. Kamada et al. 1995. Destruction of PCDDs and PCDFs. A convenient method using alkali-metal hydroxide in 1,3-dimethyl-2-imidazolidinone (DMI). Chemosphere. 31: 3873–3878.

Okumura, M. 1984. Past and current medical states of Yusho patients. Am. J. Ind. Med. 5: 13–18.

Olie, K., P.L. Vermeulen and O. Hutzinger. 1977. Chlorodibenzo-*p*-dioxins and Chlorodibenzofurans are trace components of fly ash and flue gas of some municipal waste incinerators in the netherlands. Chemosphere. 6: 445–459.

Oliver, B.G. and K.D. Nicol. 1982. Chlorobenzenes in sediments, water, and selected fish from Lakes Superior, Huron, Erie, and Ontario. Environ. Sci. Technol. 16: 532–536.

Oliver, B.G. and K.D. Nicol. 1984. Chlorinated contaminants in the Niagara River, 1981–1983. Sci. Total Environ. 3957–70.

Oliver, B.G. and M.N. Charlton. 1984. Chlorinated organic contaminants on settling particulates in the Niagara River vicinity of Lake Ontario. Environmental Science and Technology. 18: 903–908.

Oliver, B.G. and R.A. Bourbonniere. 1985. Chlorinated contaminants in surficial sediments of lakes Huron, St. Clair, and Erie: Implications regarding sources along the St. Clair and Detroit rivers. J. Great Lakes Res. 11: 366–372.

Oliver, B.G. and K.L.E. Kaiser. 1986. Chlorinated organics in nearshore waters and tributaries of the St. Clair River. Water Pollut. Res. J. Can. 21: 344–350.

Oliver, R.M. 1975. Toxic effects of 2,3,7,8-tetrachlorodibenzo-1,4-dioxin in laboratory workers. Br. J. Ind. Med. 32: 49–53.

Orazio, C.E., S. Kapila, R.K. Puri et al. 1992. Persistence of chlorinated dioxins and furans in the soil environment. Chemosphere. 25(7-10): 1469–1474.

Ott, M.G., A. Zober and C. Germann. 1994. Laboratory results for selected target organs in 138 individuals occupationally exposed to TCDD. Chemosphere. 29: 9–11.

Overcash, M.R., A.L. McPeters, E.J. Dougherty et al. 1991. Diffusion of 2,3,7,8-tetrachlorodibenzo-*p*-dioxin in soil containing organic solvents. Environ. Sci. Technol. 25: 1479–1485.

Paasivirta, J., J. Enqvist, S. Raisanen et al. 1977. On the limit of detection of TCDD in gas chromatography. Chemosphere. 6: 355.

Paepke, O., H.T. Quynh and A. Schecter. 2004. Dioxin in Vietnam: Characterisation, monitoring, remediation and effects. Organohalogen Compounds. 66.

Pai-Sheng Cheng, Ming-Sheng Hsu, Edward Ma et al. 2003. Levels of PCDD/FS in ambient air and soil in the vicinity of a municipal solid waste incinerator in Hsinchu. Chemosphere. 52(9): 1389–96.

Palauchek, N. and B. Scholz. 1987. Destruction of polychlorinated dibenzo-p-dioxins and dibenzofurans in contaminated water samples using ozone. Chemosphere. 16: 1857–1863.

Pani, O. and T. Górecki. 2006. Comprehensive two-dimensional gas chromatography (GC × GC) in environmental analysis and monitoring. Anal. Bioanal. Chem. 386 (4).

Päpke, O., M. Ball, Z.A. Lis et al. 1989a. PCDD and PCDF in indoor air of kindergartens in northern W. Germany. Chemosphere. 18: 617–626.

Papke, O. 1998. PCDD/PCDF: Human background data for Germany, a 10-year experience. Environ. Health Perspect. 106(Suppl 2): 723–731.

Park, J.-S., T.L. Wade and S. Sweet. 2001. Atmospheric deposition of organochlorine contaminants to Galveston Bay, Texas. Atmos. Environ. 35: 3315–3324.

Patandin, S., C.I. Lanting, P.G.H. Mulder et al. 1999. Effects of environmental exposure to polychlorinated biphenyls and dioxins on cognitive abilities in Dutch children at 42 months of age. J. Pediatr. 134: 33–41.

Patterson, D.G. Jr., R.E. Hoffman, L.L. Needham et al. 1986a. 2,3,7,8-Tetrachlorodibenzo-p-dioxin levels in adipose tissue of exposed and control persons in Missouri. JAMA. 256: 2683–2686.

Patterson, D.G. Jr., P. Ftirst, L.O. Henderson et al. 1989a. Partitioning of in-viva bound PCDDS-PCDFs among various compartments in whole blood. Chemosphere. 19: 135–142.

Patterson, D.G. Jr., P. Fürst, L.R. Alexander et al. 1989b. Analysis of human serum for PCDDs/PCDFs: A comparison of three extraction procedures. Chemosphere. 19: 89–96.

Patterson, D.G. Jr., G.D. Todd, W.E. Turner et al. 1994. Levels of non-ortho-substituted (coplanar), monoand di-ortho-substituted polychlorinated biphenyls, dibenzo-p-dioxins, and dibenzofurans in human serum and adipose tissue. Environ. Health Perspect. Suppl. 102(1): 195–204.

Pattle Delamore Partners Ltd., 2004.

Paul E. des Rosiers, Chairman. Dioxin Disposal Advisory Group, U.S. EPA, personal communication June 10, 1991.

Paustenbach, D., B. Finley, V. Lau et al. 1991. An evaluation of the inhalation hazard posed by diozin-contaminated soils. J. Air Waste Manage. Assn. 41: 1334–1340.

Paustenbach, D.J., R.J. Wenning, V. Lau et al. 1992. Recent developments on the hazards posed by 2,3,7,8-tetrachlorobenzo-pdioxin in soil: Implications for setting risk-based cleanup levels at residential and industrial sites. J. Toxicol. and Environ. Health. 36: 103–149.

Paustenbach, D.J., R.J. Wenning, D. Mathur et al. 1996. PCDD/PCDFs in urban stormwater discharged to San Francisco Bay, California, USA. Organohalogen Compounds. 28: 111–116.

PCDD/PCDF on Fly Ash of Municipal Waste Incinerators. Chemosphere. 18: 1219–1226.

Pearson, R.F., K.C. Hornbuckle, S.J. Eisenreich et al. 1996. PCBs in Lake Michigan water revisited. Environ. Sci. Technol. 30(5): 1429–1436.

Pearson, R.F., D.L. Swackhamer, S.J. Eisenreich et al. 1997. Concentrations, accumulations, and inventories of toxaphene in sediments of the Great Lakes. Environ. Sci. Technol. 31(12): 3523–3529.

Pearson, R.G., D.L. McLaughlin and W.D. McIlveen. 1990. Concentrations of PCDD and PCDF in Ontario soils from the vicinity of refuse and sewage sludge incinerators and remote rural and urban locations. Chemosphere. 20(10-12): 1543–1548.

Pedersen, L.G., T.A. Darden, S.J. Oatley et al. 1986. A theoretical study of the binding of polychlorinated biphenyls (PCBs), dibenzodioxins, and dibenzofuran to human plasma prealbumin. Journal of Medicinal Chemistry. 29: 2451–2457.

Pelclova, D., P. Urban, J. Preiss et al. 2006. Adverse health effects in humans exposed to 2,3,7,8-tetrachlorodibenzo-p-dioxin (TCDD). Rev. Environ. Hlth. 21(2): 119–138.

Pelizzetti, E., M. Borgarello C. Minero et al. 1988. Photocatalytic degradation of polychlorinated dioxins and polychlorinated biphenyls in aqueous suspensions of semiconductors irradiated with simulated solar light. Chemosphere. 17: 499–510.

Pereira, W.E., C.E. Rostad and M.E. Sisak. 1985. Geochemical investigation of polychlorinated dibenzo-p-dioxins in the subsurface environment at an abandoned wood-treatment facility. Environ. Toxicol. Chem. 4: 629–639.

Pereira, W.E., F.D. Hostettler, J.R. Cashman et al. 1994. Occurrence and distribution of organochlorine compounds in sediment and livers of striped bass (*morone saxatilis*) from the San Francisco Bay-Delta Estuary. Mar. Pollut. Bull. 28(7): 434–441.

Peterson, R. and E. Milicic. 1992. Chemical treatment of dioxin residues from wastewater processing. Chemosphere. 25: 1565–1568.

Petty, J.D., J.N. Huckins and D.B. Martin. 1995. Use of semipermeable membrane devices (SPMDS) to determine bioavailable organochlorine pesticide residues in streams receiving irrigation drainwater. Chemosphere. 30(10): 1891–1903.

Petty, J.D., J.N. Huckins, D.A. Alvarez et al. 2004. A holistic passive integrative sampling approach for assessing the presence and potential impacts of waterborne environmental contaminants. Chemosphere. 54: 695–705.

Pignatello, J.J. and L. Huang. 1993. Degradation of polychlorinated dibenzo-*p*-dioxin and dibenzofuran contaminants in 2,4,5-T by photoassisted iron-catalyzed hydrogen peroxide. Wat. Res. 27: 1731–1736.

Pirke, J., W. Wolfe, D. Patterson et al. 1989. Estimates of the half-life of 2,3,7,8-tetrachlorodibenzop-dioxin in Vietnam veterans of Operation Ranch Hand. J. Toxicol. Environ. Health. 27: 165–171.

Pittman, C.U. and J.B. He. 2002. Dechlorination of PCBs, CAHs, herbicides and pesticides neat and in soils at 25°C using Na/NH3. J. Hazard. Mater. 92: 51–62.

Plimmer, J.R., U.I. Klingebiel, D.G. Crosby et al. 1973. Photochemistry of dibenzo-p-dioxins. pp. 44–54. *In*: Blair, E.H. (ed.). Chlorodioxins-origin and Fate. Washington, DC: Advances in Chemistry Series.

Plumb, R.H. 1987. A comparison of ground water monitoring data from CERCLA and RCRA sites. Ground Water Monit. Rev. 7: 94–100.

Plumb, Jr R.H. 1991. The occurrence of Appendix IX organic constituents in disposal site ground water. Ground Water Monit. Rev. 11: 157–164.

Pocchiari, F., V. Silano and G. Zapponi. 1987. The Seveso accident and its aftermath. pp. 60–78. *In*: Kleindorfer, P. and H. Kunreuther (eds.). Insuring and Managing Hazardous Risks: From Seveso to Bhopal and Beyond. Berlin: Springer-Verlag.

Podoll, T.R., H.M. Jaber and T. Mill. 1986. Tetrachlorodibenzodioxin: rates of volatilization and photolysis in the environment. Environ. Sci. Technol. 20: 490–492.

Pohl, H.R., J. Holler, M. Fay et al. 1998. Focus on dioxins. Organohalogen Compounds. 38: 347–348.

Pohland, A.E. and G.C. Yang. 1972. Preparation and characterization of chlorinated dibenzo-p-dioxins. J. Agric. Food Chem. 20(6): 1093–1099.

Poiger, H. and C. Schlatter. 1980. Influence of solvents and adsorbents on dermal and intestinal absorption of TCDD. Food Cosmet. Toxicol. 18: 477–481.

Poiger, H. and C. Schlatter. 1986. Pharmacokinetics of 2,3,7,8-TCDD in man. Chemosphere. x15: 1489–1494.

Poissant, L., J.-F. Koprivnjak and R. Matthieu. 1997. Some persistent organic pollutants and heavy metals in the atmosphere over a St. Lawrence River valley site (Villeroy) in 1992. Chemosphere. 34(3): 567–585.

Poland, A. and E. Glover. 1973b. Mechanism of toxicity of the chlorinated dibenzo-p-dioxins. Environ. Health Perspect. 5: 245–251.

Poland, A., E. Glover and A.S. Kende. 1976. Stereospecific, high affinity binding of 2,3,7,8-tetrachlorodibenzo-p-dioxin by hepatic cytosol: Evidence that the binding species is a receptor for induction of aryl hydrocarbon hydroxylase. J. Biol. Chem. 251: 4936–4946.

Poland, A., W.F. Greenlee and A.S. Kende. 1979. Studies on the mechanisms of action of the chlorinated dibenzo-p-dioxins and related compounds. Ann. NY Acad. Sci. 320: 214–230.

Poland, A.P., D. Smith, G. Metter et al. 1971. A health survey of workers in a 2,4-D and 2,4,5-T plant with special attention to chloracne, porphyria cutanea tarda, and psychologic parameters. Arch. Environ. Health. 22: 316–327.

Pozo, K., R. Urrutia, M. Mariottini et al. 2014. Levels of persistent organic pollutants (POPs) in sediments from Lenga Estuary, central Chile. Mar. Pollut. Bull. 79(1-2): 338–341.

Prange, J.A., C. Gaus, R. Weber et al. 2003. Assessing forest fire as a potential PCDD/F source in Queensland, Australia. Environ. Sci. Technol. 37: 4325–4329.

Pujadas, E., J. Diaz-Ferrero, R. Marti et al. 2001. Application of the new C18 speedisks to the analysis of polychlorinated dibenzo-p-dioxins and dibenzofurans in water and effluent samples. Chemosphere. 43: 449–454.

Puri, R.K., T.E. Clevenger, S. Kapila et al. 1989. Studies of parameters affecting translocation of tetrachlorodibenzo-p-dioxin in soil. Chemosphere. 18: 1291–1296.

Quaβ, U., M. Fermann and G. Bröker. 2000. The European dioxin emission inventory. Stage II. Volume 1. Executive Summary prepared by the North Rhine Westphalia State Environment Agency on behalf of the European Commission, Directorate General for Environment (DG ENV).

Quass, U., M. Fermann and G. Broker. 2004. The European dioxin air emission inventory project-final results. Chemosphere. 54(9): 1319.

Rainer, M. 1998. Update of PCDD/PCDF-Intake from food in Germany. Chemosphere. 37: 1687–1698.

Ralp Morgan, MODAR, Inc. Personal Communication, Mar. 28, 1991.

Rappe, C. and H.R. Buser. 1980. Chemical properties and analytical methods. pp. 41–80. *In*: Kimbrough, R.D. (ed.). Halogenated Biphenyls, Terphenyls, Naphthalenes, Dibenzodioxins and Related Products, Amsterdam, Oxford, New York, Elsevier Science Publishers.

Rappe, C., P.-A. Bergovist and S. Marklind. 1985. Analysis of polychlorinated dibenzofurans and dioxins in ecological samples. pp. 135–138. *In*: Keith, L.H., C. Rappe and G. Choudhary (eds.). Chlorinated Dioxins and Dibenzofurans in the total Environment II, Boston, Butterworth Publishers.

Rappe, C., P.-A. Bergovist and L.-O. Kjeller. 1989. Levels, trends and patterns of PCDDs and PCDFs in Scandinavian environmental samples. Chemosphere. 18: 651–658.

Rappe, C., S. Marklund, L.O. Kjeller et al. 1989b. Long-range transport of PCDDs and PCDFs on airborne particles. Chemosphere. 18: 1283–1290.

Rappe, C., L.O. Kjeller, S.E. Kulp et al. 1991. Levels, profile and pattern of PCDDs and PCDFs in samples related to the production and use of chlorine. Chemosphere. 23: 1629–1636.

Rappe, C. 1991. Sources of and human exposure to PCDDs and PCDFs. pp. 121–131. *In*: Banbury Report 35: Biological basis for risk assessment of dioxins and related compounds. Cold Spring Harbor Laboratory.

Rappe, C. 1993. Sources of exposure, environmental concentrations and exposure assessment of PCDDs and PCDFs. Chemosphere. 27: 211–225.

Ray, L.E., H.E. Murray and C.S Giam. 1983. Organic pollutants in marine samples from Portland, Maine. Chemosphere. 12: 1031–1038.

Rayne, S., P. Wan, M.G. Ikonomou et al. 2002. Photochemical mass balance of 2,3,7,8-TeCDD in aqueous solution under UV light shows formation of chlorinated dihydroxybiphenyls, phenoxyphenols, and dechlorination products. Environ. Sci. Technol. 36: 1995–2002.

Reed, L.W., G.T. Hunt, B.E. Maisel et al. 1990. Baseline assessment of PCDDs/PCDFs in the vicinity of the Elk River, Minnesota generating station. Chemosphere. 21: 159–172.

Reggiani, G. 1980. Acute human exposure to TCDD in Seveso, Italy. J. Toxicol. Environ. Health. 6: 27–43.

Reichrtova, E., V. Prachar, L. Palkovicova et al. 2001. Contamination of human placentas with organochlorine compounds in five Slovak regions related to different environmental characteristics. Fres. Environ. Bull. 10: 772–776.

Reid, N.W., D.B. Orr, M.N. Shackleton et al. 1990. Monitoring dioxins and dibenzofurans in precipitation in Ontario. Chemosphere. 20: 1467–1472.

Ribes, A., J.O. Grimalt, C.J.T. Garcia et al. 2002. Temperature and organic matter dependence of the distribution of organochlorine compounds in mountain soils from the subtropical Atlantic (Teide, Tenerife Island). Environ. Sci. Technol. 36: 1879–1885.

Richter, B.E., B.A. Jones, J.L. Ezzell et al. 1996. Accelerated solvent extraction: A technique for sample preparation. Anal. Chem. 68: 1033–1039.

Rippen, G. and H. Wesp. 1993. Kale uptake of of PCDD/PCDF, PCB and PAH under field conditions: Importance of gaseous dry deposition. Presented at Dioxin 93. 13th International Symposium on Chlorinated Dioxins and Related Compounds, Vienna, Austria, September 1993.

Ritterbush, J., W. Lorenz and M. Bahadir. 1994. Determination of polyhalogenated dibenzo-p-dioxins and dibenzofurans in analytical laboratory waste and their decomposition by UV-photolysis. Chemosphere. 29: 1829–1838.

Robert, L. Peterson and Stephen L. New. (undated) Dioxin Destruction with APEG-PLUS Chemical Dechlorination. Galson Remediation Corp.

Rocha, E.M.R., V.J.P. Vilar, A. Fonseca et al. 2011. Landfill leachate treatment by solar-driven AOPs. Sol. Energy. 85: 46–56.

Rodgers, P.W. and W.R. Swain. 1983. Analysis of polychlorinated biphenyl (PCB) loading trends in Lake Michigan. J. Great Lakes Res. 9: 548–558.

Rogan, W.J. 1989. Yu-Cheng. pp. 401–415. *In*: Kimbrough, R.D. and A.A. Jensen (eds.). Halogenated Biphenyls, Terphenyls, Naphthalenes, Dibenzodioxins and Related Products, 2nd ed. Amsterdam: Elsevier Science Publishers.

Rohde, S., G.A. Moser and O. PΣpke et al. 1997. Fecal clearance of PCDD/Fs in occupationally exposed persons. Organohalogen Compounds. 33: 408–413.

Rowlands, J.C. and J.A. Gustafsson. 1997. Aryl hydrocarbon receptor-mediated signal transduction. Crit. Rev. Toxicol. 27(2): 109–34.

Rubey, W., B. Dellinger, D.L. Hall et al. 1985. High-temperature gas-phase formation and destruction of polychlorinated dibenzofurans. Chemosphere. 14: 1483–1494.

Ryan, J.J., A. Schecter, Y. Masuda et al. 1987a. Comparison of PCDDs and PCDFs in the tissue of Yusho patients with those from the general population in Japan and China. Chemosphere. 16: 2017–2025.

Ryan, J.J., R. Lizotte and D. Lewis. 1987b. Human tissue levels of PCDDs and PCDFs from a fatal pentachlorophenol poisoning. Chemosphere. 16: 1989–l996.

Ryan, J.J., R. Lizotte, L.G. Panopio et al. 1989. The effect of strong alkali on the determination of polychlorinated dibenzofurans PCDFs and polychlorinated dibenzo-p-dioxins PCDDs. Chemosphere. 18: 149–154.

Ryan, J.J., D. Levesque, L.G. Panopio et al. 1993a. Elimination of polychlorinated dibenzofurans (PCDFs) and polychlorinated biphenyls (PCBs) from human blood in the Yusho and Yu-Cheng rice oil poisonings. Arch. Environ. Contam. Toxicol. 24: 504–512.

Ryhage, R. 1964. Use of mass spectrometer as a detector and analyzer for effluents emerging from high temperature gas liquid chromatography columns. Anal. Chem. 36: 759–764.

Safe, S. 1990. Polychlorinated dibenzofurans: Environmental impact, toxicology and risk assessment. pp. 283–327. *In*: J. Saxena (ed.). Hazard Assessment of Chemicals—Current Developments. Washington, DC: Hemisphere Publishing Co. Vol. 7.

Safe, S. 1991. Polychlorinated dibenzofurans: Environmental impact, toxicology, and risk assessment. Toxic. Substances Journal. 11: 177–222, 287.

Safe, S. 1994. Polychlorinated biphenyls (PCBs): Environmental impact, biochemical and toxic responses, and implications for risk assessment. Crit. Rev. Toxicol. 24(2): 87–149.

SAI SYSTEMS APPLICATIONS, INC. 1980. Human exposure to atmospheric concentration of selected chemicals, Springfield, Virginia, National Technical Information Service, Vol. 1 (Report prepared for US Environmental Protection Agency, Research Triangle Park, North Carolina) (PB 81-193252).

Sako, T., T. Sugeta, K. Otake et al. 1997. Decomposition of dioxins in fly ash with supercritical water oxidation. J. Chem. Eng. Jpn. 30: 744–747.

Sako, T., T. Sugeta, K. Otake et al. 1999. Dechlorination of PCBs with supercritical water hydrolysis. J. Chem. Eng. Jpn. 32: 830–832.

Sako, T., S. Kawasaki, H. Noguchi et al. 2004. Destruction of dioxins and PCBs in solid wastes by supercritical fluid treatment. Organohalog. Compd. 66: 1187–1193.

Sakurai, T., J.-G. Kim, N. Suzuki et al. 2000. Polychlorinated dibenzo-*p*-dioxins and dibenzofurans in sediment, soil, fish, shellfish and crab samples from Tokyo Bay area, Japan. Chemosphere. 40: 627–640.

Saldana, M.D.A., V. Nagpal and S.E. Guigard. 2005. Remediation of contaminated soils using supercritical fluid extraction: a review (1994–2004). Environ. Technol. 26: 1013–1032.

Sanger, V.L., L. Scoot, A. Hamdy et al. 1958. Alimentary toxemia in chickens. J. Am. Vet. Med. Assoc. 133(3): 172–176.

Saracci, R., M. Kogevinas, P.A. Bertazzi et al. 1991. Cancer mortality in workers exposed to chlorphenoxy herbicides and chlorophenols. Lancet. 338: 1027–1032.

Sauer, T.C. Jr, G.S. Durell, J.S. Brown et al. 1989. Concentrations of chlorinated pesticides and PCBs in microlayer and seawater samples collected in open-ocean waters off the U.S. East Coast and in the Gulf of Mexico. Marine Chemistry. 27: 235–257.

Schecter, A., T. Tiernan, F. Schaffner et al. 1985a. Patient fat biopsies for chemical analysis and liver biopsies for ultrastructural characterization after exposure to polychlorinated dioxins, furans and PCBs. Environ. Health Perspect. 60: 241–254.

Schecter, A., J.J. Ryan and G. Gitlitz. 1986. Chlorinated dioxin and dibenzofuran levels in human adipose tissues from exposed and control populations. pp. 51–56. *In*: Rappe, C., G. Choudhury, L.H. Keith (eds.). Chlorinated Dioxins and Dibenzofurans in Perspective. Chelsea, MI: Lewis Publishers, Inc.

Schecter, A., J.J. Ryan and J.D. Constable. 1986a. Chlorinated dibenzo-p-dioxin and dibenzofuran levels in human adipose tissue and milk samples from the north and south of Vietnam. Chemosphere. 15: 1613–1620.

Schecter, A. 1987. The Binghamton state office building PCB transformer incident 1981–1987. Chemosphere. 16: 2155–2160.

Schecter, A., J.D. Constable, S. Arghestani et al. 1987a. Elevated levels of 2,3,7,8-tetrachlorodibenzo-p-dioxin in adipose tissue of certain U.S. veterans of the Vietnam war. Chemosphere. 16: 1997–2001.

Schecter, A. and T.A. Gasiewicz. 1987a. Health hazard assessment of chlorinated dioxins and dibenzofurans contained in human milk. Chemosphere. 16: 2147–2154.

Schecter, A. and T.A. Gasiewicz. 1987b. Human breast milk levels of dioxins and dibenzofurans: Significance with respect to current risk assessments. ACS Symp Ser. 338: 162–173.

Schecter, A. and J.J. Ryan. 1989. Blood and adipose tissue levels of PCDDs-PCDFs over three years in a patient after exposure to polychlorinated dioxins and dibenzofurans. 1989. Chemosphere. 18: 635–642.

Schecter, A., P. Ftirst, J.J. Ryan et al. 1989d. Polychlorinated dioxin and dibenzofuran levels from human milk from several locations in the United States, Germany and Vietnam. Chemosphere. 19: 979–984.

Schecter, A., P. Fürst, J.J. Ryan et al. 1989e. Polychlorinated dioxins and dibenzofurans levels from human milk from several locations in the United States, Germany, and Vietnam. Chemosphere. 19: 979–984.

Schecter, A., O. Papke and M. Ball. 1990a. Evidence for transplacental transfer of dioxins from mother to fetus: Chlorinated dioxinal dibenzofuran levels in the livers of stillborn infants. Chemosphere. 21: 1017–1022.

Schecter, A., P. Fürst, C. Ftirst et al. 1990c. Levels of dioxins, dibenzofurans and other chlorinated xenobiotics in human milk from the Soviet Union. Chemosphere. 20: 927–934.

Schecter, A. 1991. Dioxins and related chemicals in humans and in the environment. pp. 169–214. *In*: Banbury report 35: Biological Basis for Risk Assessment of Dioxins and Related Compounds. Cold Spring Harbor Laboratory.

Schecter, A., O. Papke, M. Ball et al. 1991a. Partitioning of dioxins and dibenzofurans: Whole blood, blood plasma and adipose tissue. Chemosphere. 23: 1913–1919.

Schecter, A., P. Fürst, C. Ftirst et al. 1991b. Dioxins, dibenzofurans and selected chlorinated organic compounds in human milk and blood from Cambodia, Germany, Thailand, the USA, the USSR, and Vietnam. Chemosphere. 23: 1903–1912.

Schecter, A., J.J. Ryan, O. PΣpke et al. 1993. Elevated dioxin levels in the blood of male and female Russian workers with and without chloracne 25 years after phenoxyherbicide exposure: the UFA "Khimprom" incident. Chemosphere. 27: 253–258.

Schecter, A., J. Startin, C. Wright et al. 1993. Dioxin levels in food from the United States with estimated daily intake. pp. 93–96. *In*: Fiedler, H., H. Frank, H. Otto et al. (eds.). Organohalogen Compounds, Vol. 13. Vienna: Federal Environmental Agency.

Schecter, A. 1994. Elevated dioxin blood levels in British chemical workers. J. Occup. Med. 36(12): 1283–1287.

Schecter, A., J. Startin, C. Wright et al. 1994d. Congener-specific levels of dioxins and dibenzofurans in U.S. food and estimated daily dioxin toxic equivalent intake. Environ. Health Perspectives. 102(11): 962–966.

Schecter, A., P. Cramer, K. Bogges et al. 1996. Dioxin intake from US food: Results from new nationwide survey. Organohalgen Compds. 28: 320–324.

Schecter, A.J. 1983. Contamination of an office building in Binghamton, New York by PCBs, dioxins, furans and biphenylenes after an electrical panel and electrical transformer incident. Chemosphere. 12: 669–680.

Schiestl, R.H., J. Aubrecht, W.Y. Yap et al. 1997. Polychlorinated biphenyls and 2,3,7,8-tetrachlorodibenzo-p-dioxin induce intrachromosomal recombination *in vitro* and *in vivo*. Cancer Research. 57(19): 4378–4383.

Schmittle, S.C., H.M. Edwards and D. Morris. 1958. A disorder of chickens probably due to a toxic feed-preliminary report. J. Am. Vet. Med. Assoc. 132: 216–219.

Schoula, R., J. Hajšlová, V. Bencko et al. 1996. Occurrence of persistent organochlorine contaminants in human milk collected in several regions of Czech Republic. Chemosphere. 33: 1485–1494.

Schuler, F., S.P. Schmid and Ch. Schlatter. 1998. Photodegradation of polychlorinated dibenzo-p-dioxins and dibenzofurans in cuticular waxes of laurel cherry (*Prunus laurocerasus*). Chemosphere. 36: 21–34.

SCOOP. 2000. Assessment of dietary intake of dioxins and related PCBs by the population of EU Member States, 7 June 2000. European Commission, Reports on tasks for scientific cooperation, Task 3.2.5. http://europa.eu.int/comm/dgs/health_consumer/library/ pub/pub08_en.pdf.

Sedman, R.M. and J.R. Esparza. 1991. Evaluation of the public health risks associated with semivolatile metal and dioxin emissions from hazardous waste incinerators. Environmental Health Perspectives. 94: 181–187.

Seung-Kyu Kim, Jong Seong Khim, Kyu-Tae Lee et al. 2007. Emission, Contamination and Exposure, Fate and Transport, and National Management Strategy of Persistent Organic Pollutants in South Korea. Chapter 2. *In*: Li, A., S. Tanabe, G. Jiang, J.P. Giesy and P.K.S. Lam (eds.). Developments in Environmental Science, Volume 7.

Shigematsu, N., Y. Norimatsu, T. Ishibashi et al. 1971. Clinical and experimental studies on respiratory involvement in chlorobiphenyls poisoning. Fukuoka Igaku Zasshi. 62: 150–156. (Japanese)

Shigematsu, N., S. Ishimaru, T. Ikeda et al. 1977. Further studies on respiratory disorders in polychlorinated biphenyls (PCB) poisoning. Fukuoka Ishi. 68: 133–138. (Japanese)

Shiomitsu, T., A. Hirayama, T. Iwasaki et al. 2002. Volatilization and decomposition of dioxin from fly ash with agitating fluidized bed heating chamber. NKK Tech. Rev. 86: 25–29.

Shiu, W.Y., W. Doucette, F.A.P.C. Gobas et al. 1988. Physical-chemical properties of chlorinated dibenzo-p-dioxins. Environ. Sci. Technol. 22: 651–658.

Silk, P.J., G.C. Lonergan, T.L. Arsenault et al. 1997. Evidence of natural organochlorine formation in Peat Bog. Chemosphere. 12: 2865–2880.

Silva, A., C. Delerue-Matos and A. Fiuza. 2005. Use of solvent extraction to remediate soils contaminated with hydrocarbons. J. Hazard. Mater. 124: 224–229.

Sinkkonen, S. and J. Paasivirta. 2000. Degradation half-life times for PCDDs, PCDFs and PCBs for environmental fate modeling. Chemosphere. 40: 943–949.

Slonecker, P.J., J.R. Pyle and J.S. Cantrell. 1983. Identification of polychlorinated dibenzo-p-dioxin isomers by powder X-ray diffraction with electron capture gas chromatography. Anal. Chem. 55: 1543–1547.

Smith, L.M., T.R. Schwartz, K. Feltz et al. 1990a. Determination and occurrence of AHH-active polychlorinated-biphenyls, 2,3,7,8-tetrachloro-para-dioxin and 2,3,7,8-tetrachlorodibenzofuran in Lake Michigan sediment and biota—the question of their relative toxicological significance. Chemosphere. 21: 1063–1085.

Smith, R.M., P.W. O'Keefe, D.R. Hilker et al. 1986a. Determination of picogram per cubic meter concentrations of tetra- and pentachlorinated dibenzofurans and dibenzo-p-dioxins in indoor air by high-resolution gas chromatography/high-resolution mass spectrometry. Anal. Chem. 58: 2414–2420.

Smith, R.M., P.W. O'Keefe, K.M. Aldous et al. 1990b. Chlorinated dibenzofurans and dioxins in atmospheric samples from cities in New York. Environ. Sci. Technol. 24: 1502–1506.

Smith, R.M., P. O'Keefe, K. Aldous et al. 1992. Measurement of PCDFs and PCDDs in air samples and lake sediments at several locations in upstate New York. Chemosphere. 25(1-2): 95–98.

Snow, N.H. 2005. Fast gas chromatography with short columns: Are speed and resolution mutually exclusive? Journal of Liquid Chromatography & Related Technologies. 27(7): 1317–1330.

Stanley, J.S., K.M. Bauer, K. Turman et al. 1989. Determination of body burdens for polychlorinated dibenzo-p-dioxins (PCDDs) and polychlorinated dibenzofurans (PCDFs) in California residents. Air Resources Board, State of California, Sacramento, CA. NTIS PB90-148289.

Staples, C.A., A. Werner and T. Hoogheem. 1985. Assessment of priority pollutant concentrations in the United States using STORET database. Environ. Toxicol. Chem. 4: 131–142.

Steer, P.I., C. Tashiro, R. Clement et al. 1990. Ambient air sampling of polychlorinated dibenzop-dioxins and dibenzofurans in Ontario: Preliminary results. Chemosphere. 20: 1431–1437.

Stehr-Green, P.A. 1989. Demographic and seasonal influences on human serum pesticide residue levels. J. Toxicol. Environ. Health. 27: 405–421.

Stephens, R.D. 1986. Transformer fire. Chemosphere. 15: 1281–1289.

Stevens, R.J.J. and M.A. Neilson. 1989. Interlake and intralake distributions of trace organic contaminants in surface waters of the Great Lakes North America. Journal of Great Lakes Research. 15(3): 377–393.

Stieglitz, L., G. Zwick, J. Beck et al. 1989. On the *De novo*-Synthesis of PCDD/PCDF on fly ash of municipal waste incinerators. Chemosphere. 18: 1219–1226.

Stockbauer, J.W., R.E. Hoffman, W.F. Schramm et al. 1988. Reproductive outcomes of mothers with potential exposure to 2,3,7,8-tetrachlorodibenzo-p-dioxin. Am. J. Epidemiol. 128: 410–419.

Strachan, W.M.J. 1990. Atmospheric deposition of selected organochlorine compounds in Canada. pp. 233–240. *In*: Kurtz, D.A. (ed.). Long Range Transport of Pesticides, 195th National Meeting of the American Chemical Society held jointly with the Third Chemical Congress of North America, Toronto, Ontario, Canada, June 1988. Washington, DC: American Chemical Society.

Strandberg, B., B. van Bavel, P.-A. Berqvist et al. 1998a. Occurrence, sedimentation and spatial variations of organochlorine contaminants in settling particulate matter and sediments in the northern part of the Baltic Sea. Environ. Sci. Technol. 32: 1754–1759.

Stringer, G.E. and R.G. McMynn. 1960. Three years use of toxaphene as a fish toxicant in British columbia. Can. Fish Culturist. 28: 37.

Stubin, A.I., T.M. Brosnan, K.D. Porter et al. 1996. Organic priority pollutants in New York City municipal wastewaters: 1989–1993. Water Environ. Res. 68: 1037–1044.

Suskind, R.R. and V.S. Hertzberg. 1984. Human health effects of 2,4,5-T and its toxic contaminants. J. Am. Med. Assoc. 251: 2372–2380.

Suskind, R.R. 1985. Chloracne, "the hallmark of dioxin intoxication." Scand. J. Work Environ. Health. 11: 165–171.

Svensson, B.G., A. Nilsson, M. Hansson et al. 1991. Exposure to dioxins and dibenzofurans through the consumption of fish. N. Engl. J. Med. 324: 8–12.

Swallow, K.C. et al. 1990. Behavior of Metal Compounds in the Supercritical Water Oxidation Process. Paper presented at the 20th Intersociety Conference on Environmental Systems of the Engineering Society for Advancing Mobility, Land, Sea, Air, and Space; Williamsburg, VA, July 9–12.

Sweeney, M.H., M.A. Fingerhut, J. Arezzo et al. 1993. Peripheral neurophathy after occupational exposure to 2,3,7,8-tetrachlorodibenzo-p-dioxin (TCDD). Am. J. Ind. Med. 23: 845–858.

Szeto, S.Y. and P.M. Price. 1991. Persistence of pesticide residues in mineral and organic soils in the Fraser Valley of British Columbia. J. Agric. Food Chem. 39: 1679–1684.

Taioli, E., R. Marabelli, G. Scortichini et al. 2005. Human exposure to dioxins through diet in Italy. Chemosphere. 61(11): 1672–1676.

Takada, S., M. Nakamura, T. Matsueda et al. 1996. Degradation of polychlorinated dibenzo-*p*-dioxins and dibenzofurans by the white rot fungus *Phanerochaete sordida* YK-624. Appl. Environ. Microbiol. 62: 4323–4328.

Tashiro, C., R.E. Clement, S. Davies et al. 1990. Water round robin parts-per-quadrillion determination of PCDDs and PCDFs. Chemosphere. 20: 1313–1318.

Tate, C.M. and J.S. Heiny. 1996. Organochlorine compounds in bed sediment and fish tissue in the South Platte River Basin, USA, 1992–1993. Arch. Environ. Contam. Toxicol. 30: 62–78.

Taylor, K.Z., D.S. Waddell, E.J. Reiner et al. 1995. Direct elution of solid phase extraction disks for the determination of polychlorinated dibenzo-p-dioxins and polychlorinated dibenzofurans in effluent samples. Anal. Chem. 67: 1186–1190.

Taylor, M.L., T.O. Tiernan, B. Ramalingam et al. 1985. Synthesis, isolation, and characterization of the tetrachlorinated dibenzo-p-dioxins and other related compound. pp. 17–35. *In*: Keith, L., C. Rappe and G. Choudhary (eds.). Chlorinated Dioxins and Dibenzofurans in the Total Environment. II, Boston, Butterworth Publishers.

Terry, B. Thomason and Michael Modell. 1984. Supercritical Water Destruction of Aqueous Wastes. Hazardous Waste. 1(4): 465.

Thacker, N.P., V.C. Nitnaware, G.H. Pandya et al. 2007. Characterization of 2,3,7,8-substitutated chlorodibenzo-p-dioxin in soil samples using LRGC-MS/MS. Asian Jour. of Chem. 19(2): 1122–1130.

The UV/Oxidation Handbook. 1994. Solarchem Environmental Systems, Markham, Ontario, Canada.

Thibodaux, L.J. 1983. Offsite transport of 2,3,7,8-tetra-chlorodibenzo-p-dioxin from a production disposal facility. pp. 75–85. *In*: Choudhary, G., L.H. Keith and C. Rappe (eds.). Chlorinated Dioxins and Dibenzofurans in the total Environment, Boston, Butterworth Publishers.

Thiess, A.M., R. Frentzel-Beyme and R. Link 1982. Mortality study of persons exposed to dioxin in a trichlorophenol-process accident that occurred in the BASF/AG on November 17, 1953. Am. J. Ind. Med. 3: 179–189.

Thoma, H. 1988. PCDD/F-concentrations in chimney soot from house heating systems. Chemosphere. 17: 1369–1379.

Thoma, H., W. Muecke and G. Kauert. 1990. Comparison of the polychlorinated dibenzo-p-dioxin and dibenzofuran in human tissue and human liver. Chemosphere. 20: 433–442.

Thomas, R.L. and R. Frank. 1981. PCBs in sediment and fluvial suspended solids in the Great Lakes. pp. 245–267. *In*: Mackay D. et al. (eds.). Physical Behavior of PCBs in the Great Lakes. Ann. Arbor, MI: Ann Arbor Science Press.

Thompson, Jr H.C., D.C. Kendall, W.A. Korfmacher et al. 1986. Assessment of the contamination of a multibuilding facility by polychlorinated biphenyls, polychlorinated dibenzo-p-dioxins, and polychlorinated dibenzofurans. Environ. Sci. and Tech. 20: 597–603.

Thompson, T.S., R.E. Clement, N. Thornton et al. 1990. Formation and emission of PCDDs/PCDFs in the petroleum refining industry. Chemosphere. 20(10-12): 1525–32.

Tiernan, T.O. 1983. Analytical chemistry of polychlorinated dibenzo-p-dioxins and dibenzofurans: A review of the current status. pp. 211–237. *In*: Choudhary, G., L. Keith and C. Rappe (eds.). Chlorinated Dioxins and Dibenzofurans in the total Environment, Boston, Butterworth Publishers.

Tiernan, T.O., M.L. Taylor, J.H. Garret et al. 1983. Chlorodibenzodioxins, chlorodibenzofurans and related compounds in the effluents from combustion processes. Chemosphere. 12: 595–606.

Tiernan, T.O., M.L. Taylor, G.F. VanNess et al. 1984. Analyses of human tissues for chlorinated dibenzo-p-dioxins and chlorinated dibenzofurans: The state of the art. pp. 31–56. *In*: Lowrance, W.W. (ed.). Public Health Risks of the Dioxins. New York, NY: Rockefeller University.

Tiernan, T.O., M.L. Taylor, J.H. Garrett et al. 1985. Sources and fate of polychlorinated dibenzodioxins, dibenzofurans and related compounds in human environments. Environ. Health Perspect. 59: 145–158.

Tiernan, T.O., D.J. Wagel, G.F. Vanness et al. 1989b. PCDD/PCDF in the ambient air of metropolitan area in the U.S. Chemosphere. 19: 541–546.

Todaka, T., H. Hirakawa, T. Hori et al. 2006. Concentrations of polychlorinated dibenzo-p-dioxins, polychlorinated dibenzofurans, and non-ortho and mono-ortho polychlorinated biphenyls in blood of Yusho patients. Chemosphere. 66(10): 1983–1989.

Tolosa, I., J.M. Bayona and J. Albaiges. 1995. Spatial and temporal distribution, fluxes, and budgets of organochlorinated compounds in Northwest Mediterranean sediments. Environ. Sci. Technol. 29: 2519–2572.

Travis, C.C. and H.A. Hattemer-Frey. 1989. A perspective on dioxin emissions from municipal solid waste incinerators. Risk Anal. 9: 91–97.

TRI98. 2000. Toxic Chemical Release Inventory. National Library of Medicine, National Toxicology Information Program, Bethesda, MD.

Troyanskaya, A.F., N.A. Rubtsova, D.P. Moseeva et al. 2003. Contamination of natural matrixes with persistent organic pollutants as a result of wood treatment in the northern regions of Russia. Organohalogen Compounds. 62: 61–64.

Tsutsumi, T., T. Yanagi, M. Nakamura et al. 2001. Update of daily intake of PCDD, PCDFs, and dioxin-like PCBs from food in Japan. Chemosphere. 45: 1129–1137.

Tundo, P., A. Perosa, M. Selva et al. 2001. A mild catalytic detoxification method for PCDDs and PCDFs. Appl. Catal. B-Environ. 32: L1–L7.

Tuppurainen, K., I. Halonen, P. Ruokojärvi et al. 1998. Formation of PCDDs and PCDFs in municipal waste incineration and its inhibition mechanisms: A review. Chemosphere. 36: 1493–1511.

Tysklind, M., K. Lundgren and C. Rappe. 1993. Ultraviolet absorption characteristics of all tetra-to octachlorinated dibenzofurans. Chemosphere. 27: 535–546.

U.S. Congress. 1991. Dioxin treatment technologies background paper. Office of Technology Assessment. OTA-BP-0-93. Washington, DC: U.S. Government Printing Office. November 1991.

Ukisu, Y. and T. Miyadera. 2003. Hydrogen-transfer hydrodechlorination of polychlorinated dibenzo-*p*-dioxins and dibenzofurans catalyzed by supported palladium catalysts. Appl. Catal. B-Environ. 40: 141–149.

UNEP. 1999. Dioxin and furan Inventories–National and Regional emissions of dioxins and furans. UNEP Chemicals, Geneva. p. 100.

UNEP. 2002a. Region III (Europe) Report. United Nations Environment Program.

UNEP. 2002b. Region V (Sub-Saharan Africa) Report. United Nations Environment Program.

USAF. 1991. Air Force health study: An epidemiological investigation of health effects in Air Force personnel following exposure to herbicides. Brooks Air Force Base, TX: U.S. Air Force, Chapters 1–5, 18, 19.

USDA. 1995. U.S. Department of Agriculture, National Agricultural Pesticide Impact Assessment Program (NAPIAP), Reregistration Notification Network (RNN). 3(11): l-l to l-4.

USEPA. 1990. Background document to the integrated risk assessment for dioxins and furans from chlorine bleaching in pulp and paper mills. Washington, D.C.: U.S. Environmental Protection Agency, Office of Toxic Substances. EPA 560/5-90-014.

USEPA. 1994a. Estimating exposure to dioxin-like compounds. Volumes I, II and III. Review Draft. EPA/600/688/OO5C a,b,c.

USEPA. 1994b. Method 1613, Revision B: Tetra-through Octachlorinated Dioxins and Furans by Isotope Dilution HRGC/HRMS. EPA 821-B94-0059. Office of Water, US Environmental Protection Agency, Washington, DC.

USEPA. 1998. The Inventory of Sources of Dioxin in the United States. External Review Draft. EPA/600/P-98/002Aa. Office of Research and Development, Washington D.C.

USEPA. 2000. Exposure and human health reassessment of 2,3,7,8-tetrachlorodibenzop-dioxin (TCDD) and related compounds. Part III: Integrated summary and risk characterization for 2,3,7,8-tetrachlorodibenzo-p-dioxin (TCDD) and related compounds. Science Advisory Board Review Draft (September 2000). EPA report no. 600/P-00/001Bg. United States Environmental Protection Agency, National Center for Environmental Assessment, Office of Research and Development, Washington, DC.

USGS. 1985. Pesticides in the nations rivers, 1975–1980, and implications for future monitoring. U.S. Geological Survey Water Supply Paper 2271.

Van den Berg, M., F.W.M. Van der Wielen, K. Olie et al. 1986. The presence of PCDDs and PCDFs in human breast milk from the Netherlands. Chemosphere. 15: 693–706.

Van den Berg, M., L.S. Birnbaum, M. Denison et al. 2006. The 2005, World Health Organization reevaluation of human and mammalian toxic equivalency factors for dioxins and dioxin-like compounds. Toxicol. Sci. 93(2): 223–241.

Van Metre, P., E. Callender and C. Fuller. 1997. Historical trends in organochlorine compounds in river basins identified using sediment cores from reservoirs. Environ. Sci. Technol. 31: 2339–2344.

Van Metre, P. and E. Callender. 1997. Water quality trends in White Rock Creek Basin from 1912–1994 identified using sediment cores from White Rock Lake Reservoir, Dallas, Texas. Jour. of Paleolim. 17: 239–249.

Van Metre, P.C., J.T. Wilson, E. Callender et al. 1998. Similar rates of decrease of persistent, hydrophobic and particle-reactive contaminants in riverine systems. Environ. Sci. Technol. 32: 3312–3317.

Van Oostdam, J., A. Gilman, E. Dewailly et al. 1999. Human health implications of environmental contaminants in Arctic Canada: A review. Sci. Total Environ. 230: 1–82.

van Ysacker, P.G., H.G. Janssen, H.M.J. Snijders et al. 1995. Electron capture detection in high-speed narrow-bore capillary gas chromatography: Fast and sensitive analysis of PCBs and pesticides. J. High Resol. Chromatogr. 18: 397–402.

Vena, J., P. Boffetta, H. Becher et al. 1998. Exposure to dioxin and nonneoplastic mortality in the expanded IARC international cohort study of phenoxy herbicide and chlorophenol production workers and sprayers. Environ. Health Perspect. 106(Suppl 2): 645–653.

Villeneuve, J.P. and C. Cattini. 1986. Input of chlorinated hydrocarbons through dry and wet deposition to the western Mediterranean. Chemosphere. 15: 115–120.

Vis, P.I.M. and P. Krijger. 2000. Full-scale thermal desorption of soil contaminated with chlorinated compounds. pp. 1–6. *In*: Wickramanayake, G.B. and R.E. Hinchee (eds.). Physical, Chemical, and Thermal Technologies. The Second International Conference on Remediation of Chlorinated and Recalcitrant Compounds. Monterey, California, May 22–25, 2000. Battelle Press, Columbus, Ohio, USA.

Vogg, H. and L. Stieglitz. 1986. Thermal-behavior of PCDD/PCDF in fly-ash from municipal incinerators. Chemosphere. 15: 1373–1378.

Vogg, H., M. Metzger and L. Stieglitz. 1987. Recent Findings on the Decomposition of PCDD/PCDF in Municipal Waste Incineration. Waste Manage. Res. 5: 285–294.

Vollmuth, S. and R. Niessner. 1995. Degradation of PCDD, PCDF, PAH, PCB and chlorinated phenols during the destruction-treatment of landfill seepage water in laboratory model reactor (UV, ozone, and UV/ozone). Chemosphere. 30: 2317–2331.

Waddell, D., B. Chittim, R. Clement et al. 1990. Database of PCDD/PCDF levels in ambient air and in samples related to the pulp and paper industry. Chemosphere. 20: 1463–1466.

Wade, T.L., E.L. Atlas, J.M. Brooks et al. 1988. NOAA Gulf of Mexico status and trends program: Trace organic contaminant distribution in sediments and oysters. Estuaries. 11(3): 171–179.

Wagenaar, W.J., E.J. Boelhouwers, H.A.M. de Kok et al. 1995. A comparative study of the photolytic degradation of octachlorodi-benzofuran (OCDF) and octachlorobidenzo-p-dioxin. Chemosphere. 31: 2983–2992.

Wagner, H.C., K.-W. Schramm and O. Hutzinger. 1990. Biogenes polychloriertes Dioxin aus Trichlorphenol. UWSF—Z Umweltchem Ökotox. 2: 63–65.

Wang, D.K.W., D.H. Chiu, P.K. Leung et al. 1983. Sampling and analytical methodologies for PCDDs and PCDFs in incinerators and wood burning facilities. Environ. Sci. Res. 26: 113–126.

Wang, J. 2004. Treatment of pesticides industry waste. Revised and expanded black and veatch, concord, California, USA. pp. 1005–1049.

Wania, F. and D. Mackay. 1996. Tracking the distribution of persistent organic pollutants. Environ. Sci. Technol. 30: 390A–396A.

Watts, R.J., B.R. Smith and G.C. Miller. 1991. Catalyzed hydrogen peroxide treatment of octachlorodibenzo-*p*-dioxin (OCDD) in surface soils. Chemosphere. 23: 949–955.

Wearne, S.J., N. Harrison, M.G. de M Gem et al. 1996. Time trends in human dietary exposure to PCDDs, PCDFs and PCBs in the UK. Organohalogen Compounds. 30: 1–6.

Weaver, L., C.G. Gunnerson, A.W. Briedenbach et al. 1965. Chlorinated hydrocarbon pesticides in major U.S. river basins. Public Health Rep. 80: 481–493.

Webb, K.B., R.G. Evans, A.P. Knutsen et al. 1989. Medical evaluation of subjects with known body levels of 2,3,7,8-tetrachlorodibenzo-p-dioxin. J. Toxicol. Environ Health. 28: 183–193.

Weber, R., S. Yoshida and K. Miwa. 2002. PCB destruction in subcritical and supercritical water evaluation of PCDF formation and initial steps of degradation mechanisms. Environ. Sci. Technol. 36: 1839–1844.

Weber, R., C. Gaus, M. Tysklind et al. 2008. Dioxin- and POPcontaminated sites—contemporary and future relevance and challenges. Environ. Sci. Pollut. Res. 15: 363–393.

Weerasinghe, N.C.A., A.J. Schecter, J.C. Pan et al. 1986. Levels of 2,3,7,8-tetrachlorodibenzo-p-dioxin (2,3,7,8-TCDD) in adipose tissue of Vietnam veterans seeking medical assistance. Chemosphere. 15: 1787–1794.

Weisiger, R., J. Gollan and R. Ockner. 1981. Receptor for albumin on the liver cell surface may mediate uptake of fatty acids and other albumin-bound substances. Science. 211: 1048–1050.

Welschpausch, K., M.S. McLachlan and G. Umlauf. 1995. Determination of the principal pathways of polychlorinated dibenzo-p-dioxins and dibenzofurans to Lolium multiflorum, Welsh ray grass. Environ. Sci. Technol. 24(4): 1090–1098.

Wenborn, M., K. King, D. Buckley-Golder et al. 1999. Releases of dioxins and furans to land and water in Europe. Final Report Issue 2. Produced for Landesumweltamt Nordrhein-Westfalen, Germany, on behalf of European Commission, DG Environment. AEA Technology Environment, Oxfordshire, UK.

Wendling, J.M., R.G. Orth and H. Poiger. 1990. Determination of [3H]-2,3,7,8-tetrachlorodibenzo-p-dioxin in human feces to ascertain its relative metabolism in man. Anal. Chem. 62: 796–800.

Whitemore, R.C., L.E. LaFleur, W.J. Gillespie et al. 1990. US EPA/Paper industry cooperative dioxin study: The 104 mill study. Chemosphere. 20: 1625–1632.

Whiteside, T. 1970. Defoliation. Ballantine Books, New York. 1–168.

Whitlock, J.P. 1987. The regulation of gene expression by 2,3,7,8-tetrachlorodibenzo-p-dioxin. Pharmacol. Rev. 39: 147–161.

WHO. 1989. International Programme on Chemical Safety. Environmental Health Criteria 91: Aldrin and Dieldrin. Geneva, Switzerland: World Health Organization. http://www.inchem.org/documents/ehc/ehc/ehc91.htm.

WHO. 1997. WHO Toxic Equivalency Factors (TEFs) for Dioxin-like Compounds for Humans and Wildlife. 15–18 June 1997, Stockholm, Sweden.

WHO. 1998. WHO toxic equivalent factors (TEFs) for dioxin-like compounds for humans and wildlife. Environ. Health Perspect. 106(12): 775–792.

WHO. Temporary Adviser Group. 2000. Consultation on assessment of the health risk of dioxins: re-evaluation of the tolerable daily intake (TDI): Executive Summary. Food Addit. Contam. 17: 223–240.

Williams, D.T., G.L. LeBel and E. Junkins. 1984. A comparison of organochlorine residues in human adipose tissue autopsy samples from two Ontario municipalities. J. Toxicol. Environ. Health. 13(1): 19–29.

Williams, D.T., G.L. LeBel and E. Junkins. 1988. Organohalogen residues in human adipose autopsy samples from six Ontario municipalities. J. Asso. Off. Anal. Chem. 71(2): 410–414.

Williams, W.M., P.W. Holden, D.W. Parsons et al. 1988b. Pesticides in groundwater data base: 1988 interim report. NTIS no. PB89-164230 AS.

Wilson, D. and J.D. Yarbrough. 1988. Autoradiographic analysis of hepatocytes in mirex-induced adaptive liver growth. Am. J. Physiol. 255: G132–G139.

Windal, I., S. Hawthorne and E. De Pauw. 1999. Subcritical water degradation of dioxins. Organohalogen Compd. 40: 591–594.

Windgasse, G. and L. Dauerman. 1992. Microwave treatment of hazardous wastes—removal of volatile and semivolatile organic contaminants from soil. J. Microw. Power. Electromagn. Energy. 27: 23–32.

Winters, D., D. Cleverly, K. meier et al. 1996. A statistical survey of dioxin-like compounds in United States beef: a progress report. Chemosphere. 32(3): 469–478.

Wipf, H.K., E. Homberger, N. Neuner et al. 1978. Field trials of photodegradation of TCDD on vegetables after spraying with vegetable oil. pp. 201–217. *In*: Cattabeni, F., A. Cavallaro and G. Galli (eds.). Dioxin—Toxicological and Chemical Aspects. SP Medical & Scientific Books, a Division of Spectrum Publications, New York, USA.

Wittich, R.-M. 1998. Degradation of dioxin-like compounds by microorganisms. Appl. Microbiol. Biotechnol. 49: 489–499.

Wittsiepe, J., P. Schrey, U. Ewers et al. 2000. Decrease of PCD/F levels in human blood from Germany over the past ten years (1989–1998). Chemosphere. 40: 1103–1109.

Wolfe, W.H., G.D. Lathrop, R.A. Albanese et al. 1985. An epidemiologic investigation of health effects in air force personnel following exposure to herbicides and associated dioxins. Chemosphere. 14: 707–716.

Wolfe, W.H., J.E. Michalek, J.C. Miner et al. 1994. Determinants of TCDD half-life in veterans of Operation Ranch Hand. J. Toxicol. Environ. Health. 41: 481–488.

Wolff, M.S., G.S. Berkowitz, S. Brower et al. 2000a. Organochlorine exposures and breast cancer risk in New York City women. Environ. Res. A. 84: 151–161.

Worobec, S.M. and J.P. DiBeneditto. 1984. Perspectives on occupational dermatoses. pp. 253–268. *In*: Drill, V.A. and P. Lazar (eds.). Cutaneous Toxicity. New York, NY: Raven Press.

Wu, C.H., G.P. Chang-Chien and W.S. Lee. 2005. Photodegradation of tetra and hexachlorodibenzo-p-dioxins. J. Hazard. Mater. 120(1-3): 257–63.

Wu, W.Z., K.-W. Schramm, B. Henkelmann et al. 1997. PCDD/Fs, PCBs, HCHs and HCB in sediments and soils of Ya-Er Lake area in China: results on residual levels and correlation to the organic carbon and the particle size. Chemosphere. 34: 191–202.

www.environmental-expert.com/events/r2000/ r2000.htm

www.mst.dk/udgiv/publikationer/2003/87-7972-570-8/html/

Wyne, B. 1987. Implementation of Article 8 of the Directive 82/501/EEC: A Study of Public Information. Commission of the European Communities Contract 86B.

Yanders, A.F., C.E. Orazio, R.K. Puri et al. 1989. On translocation of 2,3,7,8-tetrachlorodibenzo-p-dioxin: Time dependent analysis at the times beach experimental site. Chemosphere. 19: 429–432.

Yao, Y., S. Masunaga, H. Takada et al. 2002. Identification of polychlorinated dibenzo-*p*-dioxin, dibenzofuran and coplanar polychlorinated biphenyl sources in Tokyo Bay, Japan. Environ. Toxicol. Chem. 21: 991–998.

Yasuhara, A., T. Katami, T. Okuda et al. 2001. Formation of dioxins during the combustion of newspapers in the presence of sodium chloride and poly (vinyl chloride). Environ. Sci. Technol. 35(7): 1373–1378.

Ylitalo, G.M., J. Buzitis and M.M. Krahn. 1999. Analyses of tissues of eight marine species from Atlantic and Pacific coasts for dioxin-like chlorobiphenyls (Cbs) and total Cbs. Arch. Environ. Contam. Toxicol. 37: 205–219.

Yoshimura, T. 2003. Yusho in Japan. Industrial Health. 41: 139–148.

Young, A.L. 1984. Analysis of dioxins and furans in human adipose tissue. Proceedings on the symposium on Public Health Risks Dioxins. 63–75.

Yousefi, Z. and R.W. Walters. 1987. Use of soil columns to measure sorption of dioxins to soils. Toxic. Hazard Wastes. 19: 181–193.

Yuan, G.L., J.X. Qin, J. Li et al. 2014. Persistent organic pollutants in soil near the Changwengluozha glacier of the Central Tibetan Plateau, China: Their sorption to clays and implication. Sci. Total Environ. 472: 309–315.

Zacharewski, T., L. Safe, S. Safe et al. 1989. Comparative analysis of polychlorinated dibenzop-dioxin and dibenzofuran congeners in Great Lakes fish extracts by gas chromatography-mass spectrometry and *in vitro* enzyme induction activities. Environ. Sci. and Technol. 23: 730–735.

Zanotto, E., R.E. Alcock, S. Della Sala et al. 1999. PCDD/Fs in Venetian foods and a quantitative assessment of dietary intake. Organohalogen Compds. 44: 13–16.

Zeng, E., C. Yu and K. Tran. 1999. *In situ* measurements of chlorinated hydrocarbons in the water column off the Palos Verdes Peninsula, California. Environ. Sci. Technol. 33: 392–398.

Zhang, L., T. Zhang, L. Dong et al. 2013. Assessment of halogenated POPs and PAHs in three cities in the Yangtze River Delta using high-volume samplers. Sci. Total Environ. 454-455: 619–626.

Zhao, C., K. Hirota, M. Taguchi et al. 2007. Radiolytic degradation of octachlorodibenzo-*p*-dioxin and octachloro-dibenzofuran in organic solvents and treatment of dioxin-containing liquid wastes. Radiation Physics and Chemistry. 76(1): 37–45.

Zhou, Y.X., P. Yan, Z.X. Cheng et al. 2003. Application of non-thermal plasmas on toxic removal of dioxin-contained fly ash. Powder Technol. 135: 345–353.

Zieliński, M., J. Kamińska, M. Czerska, D. Ligocka and M. Urbaniak. 2014. Levels and sources of PCDDs, PCDFs and dl-PCBs in the water ecosystems of central Poland—A mini review. Int. Jour. of Occup. Med. and Environ. Hlth. 27(6): 902–918.

Zitko, V. 1992. Patterns of 2,3,7,8-substituted chlorinated dibenzodioxins and dibenzofurans in aquatic fauna. Sci. of the Total Environment. 111: 95–108.

Zober, A., P. Messerer and P. Huber. 1990. Thirty-four-year mortality follow-up of BASF employees exposed to 2,3,7,8-TCDD after the 1953 accident. Int. Arch. Occup. Environ. Health. 62: 139–157.

Zober, A., M. Ott, I. Fleig et al. 1993. Cytogenic studies in lymphocyte of workers exposed to 2,3,7,8-TCDD. Int. Arch. Occup. Environ. Health. 65: 157–161.

Zook, D.R. and C. Rappe. 1994. Environmental sources, distribution, and fate of polychlorinated dibenzodioxins, dibenzofurans, and related organochlorines. pp. 80–113. *In*: Schecter, A. (ed.). Dioxins and Health. New York: Plenum Press.

Index

S

T

V